Lecture Notes in Mechanical Engineering

Lecture Notes in Mechanical Engineering (LNME) publishes the latest developments in Mechanical Engineering—quickly, informally and with high quality. Original research reported in proceedings and post-proceedings represents the core of LNME. Volumes published in LNME embrace all aspects, subfields and new challenges of mechanical engineering. Topics in the series include:

- Engineering Design
- Machinery and Machine Elements
- Mechanical Structures and Stress Analysis
- Automotive Engineering
- Engine Technology
- Aerospace Technology and Astronautics
- Nanotechnology and Microengineering
- Control, Robotics, Mechatronics
- MEMS
- Theoretical and Applied Mechanics
- Dynamical Systems, Control
- Fluid Mechanics
- Engineering Thermodynamics, Heat and Mass Transfer
- Manufacturing
- Precision Engineering, Instrumentation, Measurement
- Materials Engineering
- Tribology and Surface Technology

To submit a proposal or request further information, please contact the Springer Editor of your location:

China: Ms. Ella Zhang at ella.zhang@springer.com
India: Priya Vyas at priya.vyas@springer.com
Rest of Asia, Australia, New Zealand: Swati Meherishi at swati.meherishi@springer.com
All other countries: Dr. Leontina Di Cecco at Leontina.dicecco@springer.com

To submit a proposal for a monograph, please check our Springer Tracts in Mechanical Engineering at https://link.springer.com/bookseries/11693 or contact Leontina.dicecco@springer.com

Indexed by SCOPUS. All books published in the series are submitted for consideration in Web of Science.

More information about this series at https://link.springer.com/bookseries/11236

José Machado · Filomena Soares ·
Justyna Trojanowska · Erika Ottaviano ·
Petr Valášek · Mallikarjuna Reddy D. ·
Eduardo André Perondi ·
Yevheniia Basova
Editors

Innovations in Mechanical Engineering II

Springer

Editors
José Machado
Mechanical Engineering Department
University of Minho
Guimarães, Portugal

Justyna Trojanowska
Poznan University of Technology
Poznan, Wielkopolskie, Poland

Petr Valášek
Faculty of Engineering
Czech University of Life Sciences Prague
Prague, Czech Republic

Eduardo André Perondi
Mechanical Engineering Department
Federal University of Rio Grande do Sul
Porto Alegre, Brazil

Filomena Soares
Department of Industrial Electronics
University of Minho
Guimarães, Portugal

Erika Ottaviano
Department of Civil and Mechanical Engineering
University of Cassino and Southern Lazio
Cassino, Italy

Mallikarjuna Reddy D.
Faculty of Mechanical Engineering
Vellore Institute of Technology University
Vellore, India

Yevheniia Basova
NTU "Kharkiv Polytechnic Institute"
Kharkiv, Ukraine

ISSN 2195-4356 ISSN 2195-4364 (electronic)
Lecture Notes in Mechanical Engineering
ISBN 978-3-031-09381-4 ISBN 978-3-031-09382-1 (eBook)
https://doi.org/10.1007/978-3-031-09382-1

This Springer imprint is published by the registered company Springer Nature Switzerland AG
The registered company address is: Gewerbestrasse 11, 6330 Cham, Switzerland

Preface

This volume of Lecture Notes in Mechanical Engineering gathers selected papers presented at the Second International Scientific Conference (ICIE'2022), held in Guimarães, Portugal, on 28–30 June 2022. The conference was organized by the School of Engineering of University of Minho, throughout MEtRICs and Algoritmi Research Centres.

The aim of the conference was to present the latest engineering achievements and innovations and to provide a chance for exchanging views and opinions concerning the creation of added value for the industry and for the society. The main conference topics include (but are not limited to):

- Innovation
- Industrial Engineering
- Mechanical Engineering
- Mechatronics Engineering
- Systems and Applications
- Societal Challenges
- Industrial Property

The organizers received 139 contributions from 16 countries around the world.

After a thorough peer-review process, the committee accepted 81 papers written by 335 authors from 15 countries for the conference proceedings (acceptance rate of 58%), which were organized in three volumes of the Springer Lecture Notes in Mechanical Engineering.

This volume, with the title "Innovations in Mechanical Engineering II", is specifically spanning from advanced materials and composites, optimization of manufacturing and production processes and converging issues and technologies in additive manufacturing and Industry 4.0. It covers applications in the transport and automotive, and medical and education sector, among others. Last but not least, it analyses important issues proposing a good balance of theoretical and practical aspects. This book consists of 30 chapters, prepared by 146 authors from ten countries.

Extended versions of selected best papers from the conference will be published in the following journals: Sensors, Applied Sciences, Machines, Management and Production Engineering Review, International Journal of Mechatronics and Applied Mechanics, SN Applied Sciences, Dirección y Organización, and Smart Science, Business Systems Research and International Journal of E-Services and Mobile Applications.

A special thank to the members of the International Scientific Committee for their hard work during the review process.

We acknowledge all that contributed to the staging of ICIE'2022: authors, committees and sponsors. Their involvement and hard work were crucial to the success of ICIE'2022.

June 2022

José Machado
Filomena Soares
Justyna Trojanowska
Erika Ottaviano
Petr Valášek
Mallikarjuna Reddy D.
Eduardo Perondi
Yevheniia Basova

Contents

Development of Guidelines for the Design of Functionally Integrated Molds for a Novel Vacuum Assisted Resin Infusion Process 1
Sven Meißner, Paul Falkenhain, Rafael Cordeiro, and Sebastian Scholz

Challenges and Opportunities of Industry 4.0 at Mold Production Engineering and Management 13
Arminda Pata and Agostinho Silva

Influence of Mechanical Processing on the Durability of Parts in Additive Manufacturing Conditions 24
Viktoriya Pasternak, Oleg Zabolotnyi, Kostiantyn Svirzhevskyi, Irina Zadorozhnikova, and José Machado

ISO 2631 Vibration Analysis in Portuguese Alfa Pendular Trains 36
Patrícia Silva, Eurico Seabra, and Joaquim Mendes

Noise Attenuation Analysis of a Multilayer Wall 47
Nicolae Adrian Nițu, Marcelin Benchea, Gelu Ianuş, and Carmen Bujoreanu

Improving Procedures for Production and Maintenance Control Towards Industry 4.0 Implementation 58
Rúben D. F. S. Costa, Marta L. S. Barbosa, Francisco J. G. Silva, José C. Sá, Luís P. Ferreira, and Bebiana Pinto

Industry 4.0 Technologies for Maintenance Management – An Overview 68
Małgorzata Jasiulewicz-Kaczmarek and Katarzyna Antosz

Design Concept of a Non-invasive Tagging Device for Blue Sharks 80
José Azevedo, Tiago Bartolomeu, Senhorinha Teixeira, and José Teixeira

Aspects Regarding Subtractive and Additive CAD-CAM Processes Applied in Prosthetic Dentistry 91
Edgar Moraru, Grigore Octavian Dontu, Tudor Catalin Apostolescu, Laurentiu Adrian Cartal, Bogdan Grămescu, and Nazem Dawod

Design and Performance Assessment of a Double-Screen Trommel for MSW Separation 103
Luís D. Forte, Ricardo Faria, and Ana C. Ferreira

A Comparative Study of Weld Bead Geometry of AISI 1023 with Varying Plate Width Using SAW and ASAW 116
Tejendra Singh Singhal, Jinesh Kumar Jain, Manoj Kumar, and Rajeev Agrawal

Ecofriendly Technologies of Fish Oils and Collagen Extraction from Fishery Waste – A Review 136
Nádia Valério, Margarida Soares, Maria Cândida Vilarinho, and Joana Carvalho

Study Concerning the Correspondence Between Optical Materials and Optical Coatings Using Matlab 148
Dana Rizescu and Ciprian Ion Rizescu

Bat Algorithm for Discrete Optimization Problems: An Analysis 161
Bruno Sousa, Rita Guerreiro, André S. Santos, João A. Bastos, Leonilde R. Varela, and Marlene F. Brito

Geometrical Metrology in Food Product Measurements 173
M. Wieczorowski, N. Swojak, M. Szelewski, A. Pereira, T. G. Mathia, and B. Gapinski

Analysis and Validation of a CFD Simulation of the Wind Through a Horizontal Axis Wind Turbine as an Actuator Disc with a Porous Jump Condition 187
Diogo Silva, João Silva, Paulo Pinto, Senhorinha Teixeira, and José Teixeira

Eutectic Modification of Al-Si Alloys by Using $AlSr_{10}$ 200
Iryna Hren, Sylvia Kusmierczak, Nataša Naprstková, and Roman Horky

Flow Visualizations in a PDMS Cerebral Aneurysm Biomodel 209
Andrews Souza, Maria Sabrina Souza, Conrado Ferrera, Hélder Puga, Rui Lima, and João Ribeiro

What Times Should Be Used for fNIRS and EEG Protocol? 216
Iara Margolis, Rute Silva, Diego Pinal, Nelson Costa, and Bernardo Providência

Inclusive Platform: GUS for Social Inclusion and Competencies Development in Cerebral Palsy ... 228
Gil Dias, Filomena Soares, Vítor Carvalho, Armanda Pereira, and Tiago Martins

Kinematic Analysis from Solar Panel Tracking System ... 240
Andreea Dana Alionte, Cristian Canacheu, Alexandru Costin Ciobanu, Ion Murzac, Dana Mirela Văleanu, and Liviu Marian Ungureanu

Continuous Improvement of Logistics Processes in a Production Line of Electric Buses Chassis ... 250
Beatriz Especial Gomes, Maria Teresa Pereira, Marisa Oliveira, and Fernanda A. Ferreira

3D Print Technology Used in High Technical Education ... 259
Jan Novotný, Martin Jaskevič, and Tomáš Vysloužil

Is Critical Thinking a Skill or a Way to Develop Skills? An Overview in Engineering Education ... 266
Celina P. Leão and Ana C. Ferreira

Solving Manufacturing Orders Scheduling Problem Using Annealing Simulation ... 279
Kamil Musiał, Artem Balashov, Anna Burduk, and Dominika Rysińska-Wojtasik

Dynamic Sectorization - Conceptualization and Application ... 293
Filipe Soares de Sousa, Maria Margarida Lima, Elif Göksu Öztürk, Pedro Filipe Rocha, Ana Maria Rodrigues, José Soeiro Ferreira, Ana Catarina Nunes, and Cristina Oliveira

Research of Features of Oxide Coatings Deposition by D–gun Spraying ... 305
Oleksiy Romanchenko, Olga Porkuian, Yuriy Kharlamov, Volodymyr Sokolov, and Oleg Krol

Key Performance Indicators: A Perspective on Developing and Implementing Performance Indicators in a Textile Company Warehouse ... 317
Ana Francisca Costa, Ana Pinto, Ana Catarina Mendes, Ana Rita Quinteiro, Carla Coelho, Carla Meira, Isac Pereira, Jorge Marinho, José Barbosa, Natacha Cunha, Rui Miranda, Nuno Drumond, and Rui M. Lima

Composite Materials Based on GPPS/PE-HD Matrix Used in Automotive Industry ... 329
Lucia Knapčíková, Justyna Trojanowska, Matúš Martiček, and Jozef Husár

On the Suitability of Instrumented Orthogonal Cutting Tests Towards Cutting Load Modelling in Turning Operations 339
T. Silva, V. Sousa, F. Silva, and A. M. P. Jesus

Author Index 351

Development of Guidelines for the Design of Functionally Integrated Molds for a Novel Vacuum Assisted Resin Infusion Process

Sven Meißner(✉), Paul Falkenhain, Rafael Cordeiro, and Sebastian Scholz

Fraunhofer Institute for Machine Tools and Forming Technology IWU, 02763 Zittau, Saxony, Germany

sven.meissner@iwu.fraunhofer.de

Abstract. In the present study, a newly developed vacuum assisted resin infusion process (FuVak) is described in which the conventional, disposable multilayer vacuum setup is replaced by reusable silicone vacuum hoods with integrated flow channels. In order to ensure high-quality impregnation of the fiberglass textile structure in a short time, while avoiding the use of disposable layers like flow aids, a suitable design of the flow channel geometry and their positioning on the vacuum hood must be determined. To study the effect of the channel cross-section, an experimental setup with varying geometries in laboratory-scale was designed, constructed, and tested. Optical detection of the flow front during the Infusion process and later examining of porosity distribution on the cured part were performed. The relationship between channel cross-section, flow speed and final part quality (porosity distribution) were determined and discussed. Based on this, an industry-scale, 2 m long part with higher complexity was fabricated with the novel process, validating the process ability to control of the flow front and quality of the composite.

Keywords: Fiber reinforced plastic · Vacuum assisted resin infusion · Functional integration

1 Introduction and State of the Art

The vacuum assisted resin infusion (VARI) is an established process for the production of fiber-reinforced plastic (FRP) components with small and medium batch sizes. This is a manufacturing process in which dry reinforcement textile layers (e.g. Glass fibers, carbon fibers) are placed into a mold and subsequently impregnated with a reactive resin system, resulting in a final structural part after curing [1].

Traditional VARI uses a multi-layer vacuum setup, in which all layers are single-use and must be discarded after the production of a single part (see Fig. 1). Additionally to the high level of waste material, the layers must be manually placed at the start of every cycle of production of a single part, taking substantial time and demanding specialized labor. Furthermore, this manual step is a source of variability in the part quality [2]. For this reason, there is a need for a faster and more reproducible manufacturing process.

J. Machado et al. (Eds.): icieng 2022, LNME, pp. 1–12, 2023.
https://doi.org/10.1007/978-3-031-09382-1_1

The goal of the research presented here is to improve on the current standard VARI processes by developing not only a reusable silicon vacuum hood, but also integrating flow channels in its internal surface, so that the flow of the resin is optimized for the best product quality and minimal infusion time. To achieve this the geometry of these flow channels, both the cross-section and their arrangement on the surface, must be evaluated and tested. Furthermore, the initial results are validated in a mold for a large and complex part. In this context, the newly developed VARI process described here represents an important technological step.

The basic idea of the FuVak-process is based on the use of functionally integrated, reusable silicone vacuum hoods (see Fig. 1) [3]. Compared to the conventional VARI-process, it enables a reduction of expendables (peel ply, flow aid, vacuum bagging film, resin entrance lines, vacuum channels, sealing tape, etc.), which in turn means a significant saving of resources, reduced cycle times, greater reproducibility, and less person-hours required from the workers. In the conventional VARI process, a separate, multilayer vacuum buildup must be laid for each component. In the FuVak-process, only one silicone vacuum hood has to be produced, which can then be used to produce at least 500 parts of the same geometry.

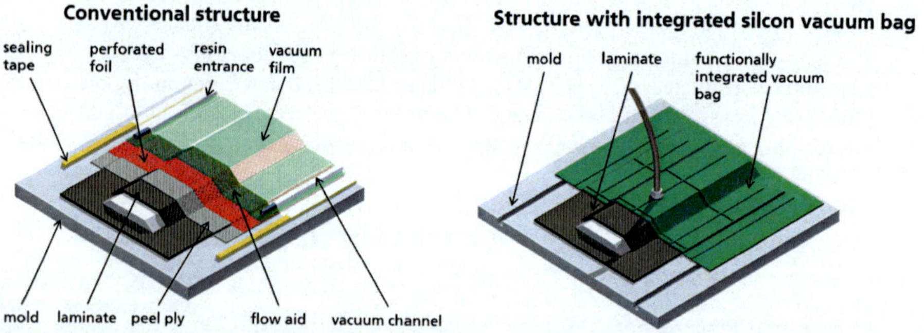

Fig. 1. Multi-layer vacuum build-up in the conventional-VARI-process (left) and simplified vacuum build-up in the FuVak-process (right) [3]

The integration of fluid-dynamically optimized flow channels in the silicone vacuum hood allows fast and reproducible resin impregnation of the textile reinforcement structure. This eliminates the time-consuming, material-intensive and not entirely error-free application of expendables to build up the resin-/vacuum system.

Numerical methods validated according to the current state of science and technology exist and their applicability to conventional VARI methods has been proven [4–6]. Nevertheless, there is recent research that has found that the physics of resin flow may be different for some applications and a conventional approach may not apply to all resin filling operations [7–9]. Due to the novel approach of the FuVak process to integrate flow channels into the silicon vacuum hood, it is necessary to verify to what extent a classical modeling approach is valid for this application. It may be necessary to develop a modified calculation approach that describes the resin spreading in the flow channels as well as the impregnation of the reinforcement textile with sufficient accuracy. These steps are outside the scope of this work but are being currently developed by the authors.

2 Experimental Determination of the Optimum Flow Channel Geometry

2.1 Experimental Setup and Parameters

To investigate how the infusion channel cross-section affects the flow front and the impregnation quality, a silicone vacuum hood with varying resin infusion channels was designed (See Fig. 2 and Table 1). By means of a large volume linear inlet sprue (channel cross section "1"), the low viscosity resin is infiltrated through the multilayer structure of fiberglass weaved fabric, which is kept under vacuum by the outer evacuation ring and outlet linear channel (channel cross section "2"). The flow front can be observed through the partially transparent silicone of the silicone vacuum bag.

Five layers of bidirectional (BD) fiberglass 260 g/m^2 fabric with known permeability were used in the experimental setup. The unsaturated polyester resin was infused through a 12 mm diameter infusion nozzle attached to the silicone vacuum bag at atmospheric pressure. The vacuum was drawn on the other side and at the right channel down to a pressure of 0.15 bar. The test parameters as well as textile and resin properties are shown in Table 2.

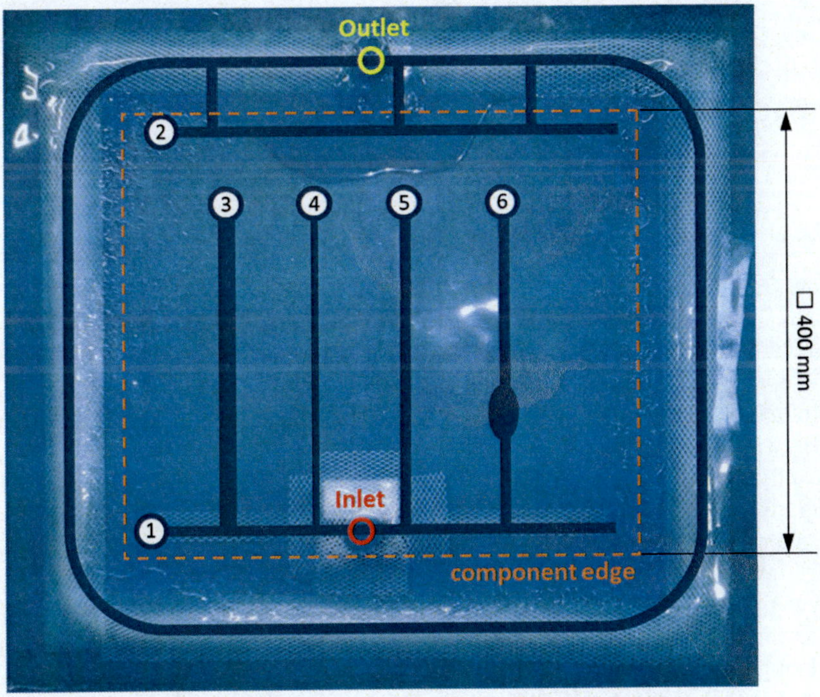

Fig. 2. Schematic representation of the experimental setup, functionally integrated silicone vacuum hood with varying flow channels

Table 1. Channel geometries on silicone vacuum hood with varying flow channels

Number	Width [mm]	Height [mm]	Notes
1	6.0	4.5	Inlet sprue channel
2	6.0	1.5	Outlet vacuum channel
3	9.0	1.5	Infusion channel
4	4.0	1.5	Infusion channel
5	6.0	3.0	Infusion channel
6	6.0	1.5	Infusion channel with node

Table 2. Experimental parameters for VARI permeation test

Infusion pressure p_{inf}	15.3 kPa
Ambient pressure p_{atm}	101.3 kPa
Resin viscosity η (at 20 °C)	400 mPa·s
Textile thickness per layer	1 mm
Total Number of textile layers	5
Total textile wight per unit area	1307 g/m^2
Fiber volume content ϕ_0	54%
Permeability K_0	1.695 10^{-11} m^2

2.2 Execution and Evaluation of Experiments

The flow front progress during the VARI process was recorded optically (see Fig. 3) and clearly shows that the infusion velocity of the resin is dependent on the channel cross-section. While the resin flows rapidly in the central infusion channels "4" and "5", it is considerably slowed down in channels "3" and "6". As a result of the vacuum in the infusion channel, a strong contraction of the silicone material takes place in the case of a wide and flat channel cross-section, as is the case in the channel "3" which significantly reduces the flow cross-section and increases the flow resistance (see Fig. 4). In comparison, channel "4" is considerably narrower, limiting the bending arc of the silicon material above it. The quickest flow in on the channel "5", which is the one with the highest cross-section and an intermediate width between the two just discussed, minimizing this constriction of the channel. The infusion channel "6", with 6 mm widt, shows a quick flow up to the node, and at this point the flow is greatly restricted. The reason being the same channel constriction, as in the channel "6", but this time localized.

High flow velocities in the channel result in a resin flow front parallel to the flow channel and low flow velocities result in a wedge-shaped spread in the reinforcement textile (see Fig. 5).

When two parallel flow fronts meet, a weld line occurs [6]. The effect is characterized by incomplete impregnation due to air trapped between the resin flow fronts, leading

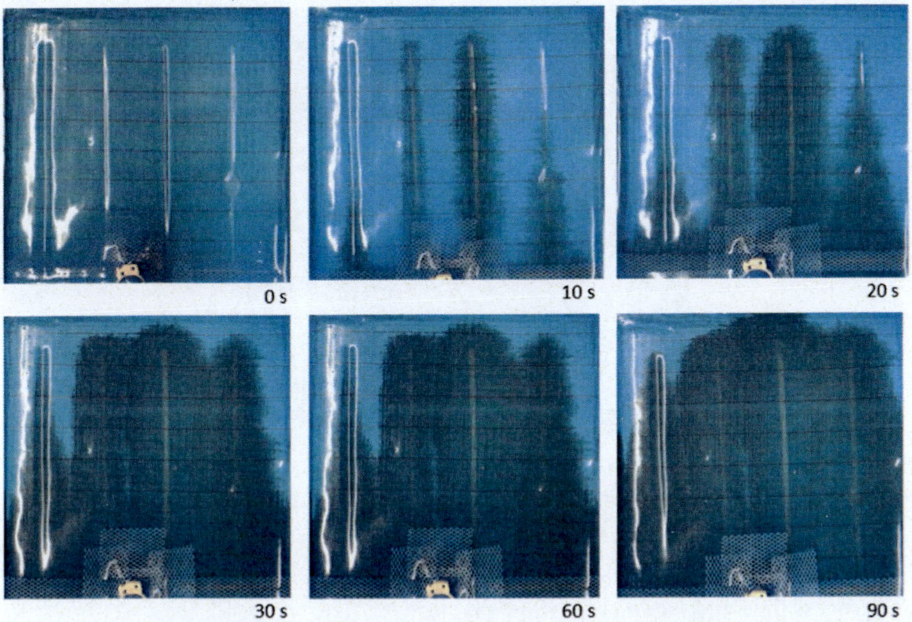

Fig. 3. Flow front curve during the VARI filling process with varying flow channels

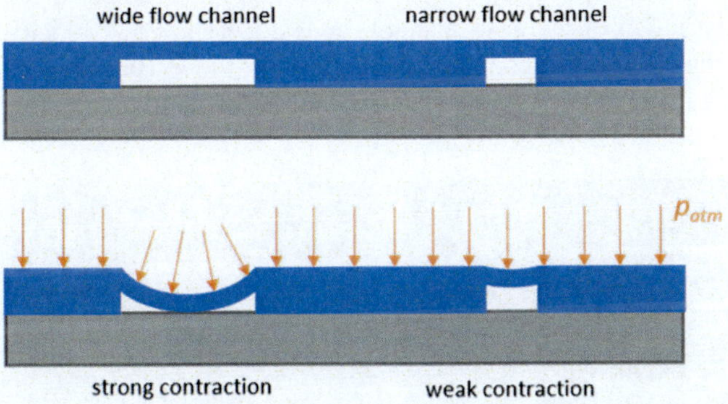

Fig. 4. Contraction of the silicone vacuum hood due to the negative pressure in the flow channel, leading to constriction of the flow cross-section

to defects (micro porosity) in the laminate in the affected areas (see Fig. 6). It can be seen from the test plate that weld lines occur to a much greater extent with flow fronts that meet in parallel than with flow fronts that run in a wedge shape to each other. In addition, a significantly worse impregnation quality of the reinforcement textile in the thickness direction can be observed at high flow rates. This is assumed to be due to a quick flow and spread of the resin on the whole area of the top layer of textile, before the resin can diffuse in the z-direction down to the lower layers. Thus, the quickest possible

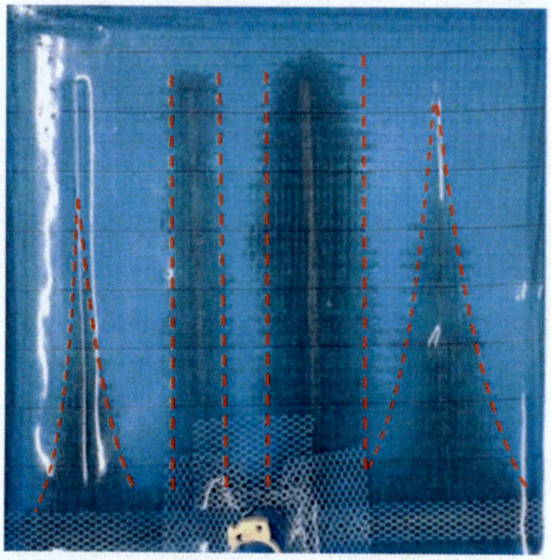

Fig. 5. Shapes of the flow front propagation as a function of the flow velocity in the impregnation channels 3 through 6

impregnation leads to low quality parts, and a compromise has to be found between a quick process time and the impregnation quality.

For a high-quality component with high strength, it is essential that no air pockets or micro porosity occur [10]. The best impregnation of the reinforcement textile is observed around the channel cross-section "3", so this geometry was selected for subsequent testing. The constriction of the flow channel slows down the flow velocity of the resin sufficiently, so that it has sufficient time to spread in the Z (thickness) direction. In addition, the flow front course is wedge-shaped, thus avoiding weld lines.

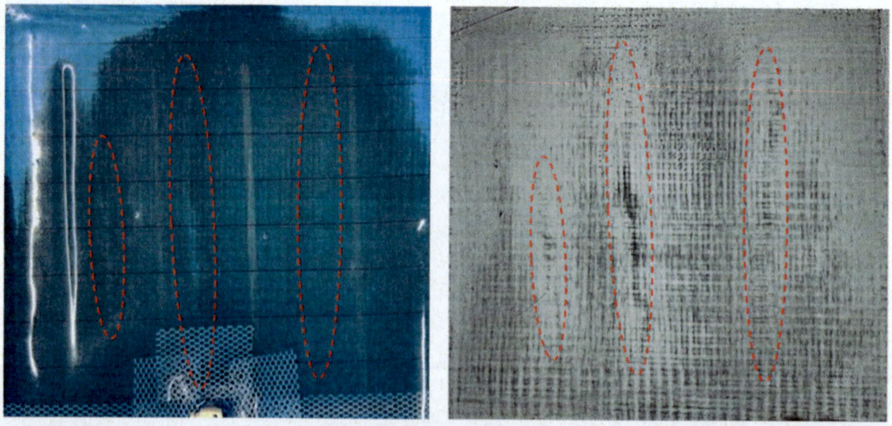

Fig. 6. Filling status after 90 s (left) and backlight image of the impregnated and cured test plate (right) with marked porosity on the weld lines due to trapped air. Darker regions between the weld lies (under the channels) indicate micro porosity due to insufficient impregnation in the z-direction.

2.3 Experimental Validation of Flow Channel Geometry on Complex Application Examples

Further VARI experiments were performed to validate whether the previously obtained findings regarding the good infiltration quality also apply to FRP structures with increased component size and geometric complexity. As an example, an existing mold was chosen with which glass fiber-reinforced plastic (GFRP)-door pillars for rail vehicles have already been produced using the conventional VARI process. A silicone vacuum hood with integrated infusion channels (9×1.5 mm) was designed and manufactured for this purpose (see Fig. 7 and Fig. 8). The channel geometries on the silicone vacuum hood are listed in Table 3. In comparison to the previously presented experiments, a so-called morphrunner system, which has a much larger volume for quick infusion times, was also used to realize the inlet sprue channels [11]. A VARI test was performed with the same parameters as before, listed in Table 2. The flow front profile and a demolded component are shown in Fig. 9 and Fig. 10.

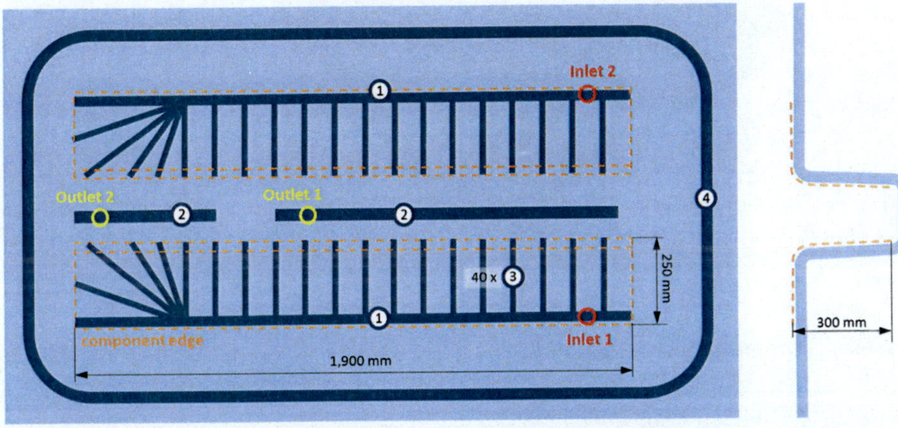

Fig. 7. Schematic representation of the experimental setup, silicone vacuum hood for complex application example

Observation of the flow front progression confirms that the goal of realizing a wedge-shaped resin spreading by the selected flow channel geometry is achieved. This is also reflected in the good quality in the impregnated areas on the component in question.

It can be seen that as the distance of the flow front from the inlet point increases, the flow velocity decreases, which can be explained by the pressure loss in the system. Due to the insufficient pot life of the resin, the test was stopped after 25 min, resulting in a non-impregnated area near the outlet “2”. In order to be able to avoid such process-related component defects in the future, suitable calculation methods for predicting the resin spread in the textile reinforcement structure must be determined for the process design.

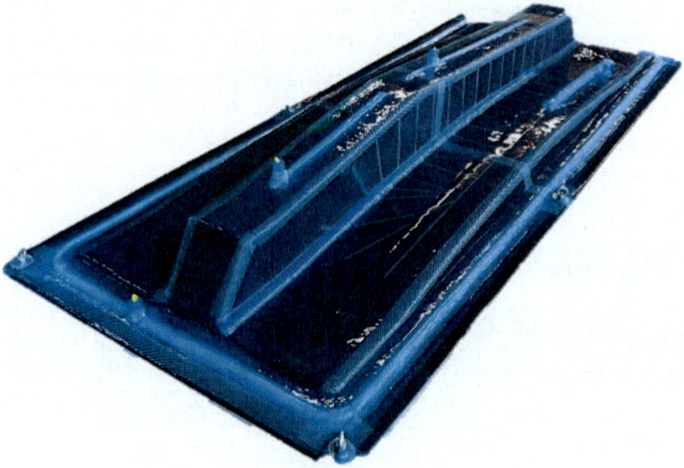

Fig. 8. Silicone vacuum hood for complex application example

Table 3. Channel geometries on silicone vacuum hood for complex application example

Number	Width [mm]	Height [mm]	Notes
1	10.0	8.0	Inlet Sprue channel with *morphrunner* in the first 600 mm
2	10.0	8.0	Outlet Vacuum channels, reinforced with fabric
3	1.5	1.5	40 x Flow channel
4	10.0	6.0	Vacuum channel for positioning and fixation, reinforced with fabric

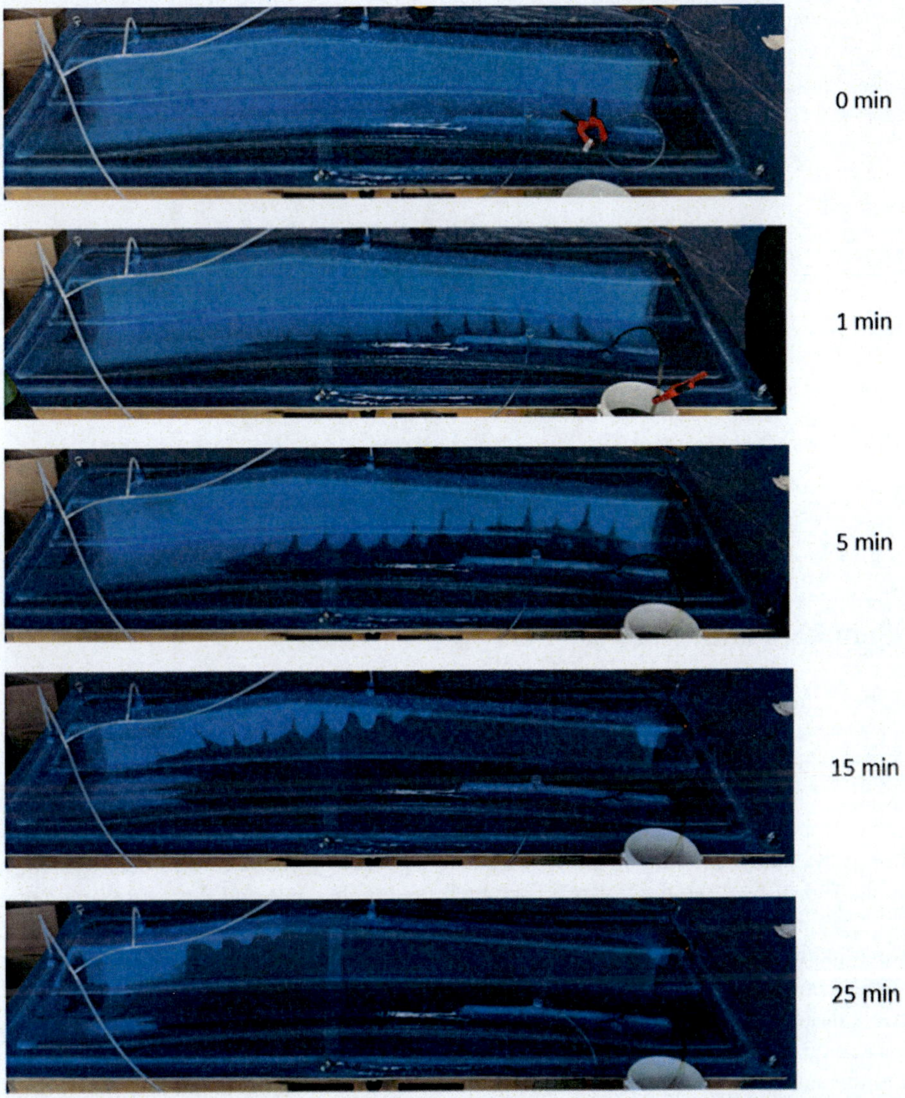

Fig. 9. Flow front curve during VARI filling process with complex application example

Fig. 10. Complex GFRP component manufactured using the FuVak process

3 Summary and Outlook

The study presented a novel improvement of the VARI process, the FuVak process, which represents an important technological step towards a faster and more reproducible manufacturing process. It is based on the use of reusable silicone vacuum hoods with integrated flow channel structures [3].

To determine design guidelines for functionally integrated silicone vacuum bags, a test setup was developed in which varying flow channel geometries were investigated. This has shown how the flow of resin on the infusion channels and their surroundings occur. The most important information are that while a very fast flow is possible, it leads to a low-quality part due to trapped air both in the lower layers of the fiberglass textile lay-up and on the "weld lines", where parallel resin flow fronts meet. A cross-section that allows a more controlled infusion was found, allowing a compromise between process speed and quality of the final laminated material. The lower flow speeds allow the resin to impregnate the lower layers of the fiberglass lay-up before any air is left trapped by the complete covering of the upper layers. Additionally the flow front takes a wedge-shape, instead of parallel expanding fronts. This avoids the trapping of air in weld lines, also leading to higher quality parts. Further-more, a channel with varying cross section, like a node point, leads to different resin flow speeds along its length, which can open new possibilities to control the resin flow in more complex scenarios.

The previously determined preferred variant for infusion channel design was validated on the mold of a real component used in the train industry. This application-typical component had increased size and geometric complexity. The FuVak process could successfully control the flow front on this part, even when the flow passed through sharp corners, leading to good impregnation on the majority of the part. Nevertheless, it was also shown that there is a need for further research in order to realize high-quality impregnation over the entire component structure as well as a reduction in cycle time. For this purpose, it is essential to have calculation methods for a realistic modeling of the infiltration process with a high predictive capability [7, 9]. These calculation methods have to be modified and validated to the proposed FuVak process, since no disposable flow aid layer is used and the channel cross-sections and density are very different from the conventional process. Such calculation methods are currently in development by the authors and will be the topic of future publications.

In addition, the component size and geometry complexity are to be further increased as part of further research activities. For this purpose, a silicone vacuum hood with integrated channels has already been designed, which will be used to produce an exemplary train front mask with a component area of approx. 11 m^2 (3.1 × 3.4 × 3.0 m^3) and to validate the FuVak process in greater depth (see Fig. 11).

If the further research activities on calculation methods and component complexity are successfully realized, the FuVak process will provide an improved VARI technology for the re-producible and cost-effective production of complex FRP structures with significantly lower wasted materials, as well as a suitable calculation approach for process design.

Fig. 11. Reusable silicone vacuum hood with integrated channels for the production of a GFRP train front mask using the FuVak process

Financing. This research was funded by the Sächsische Aufbaubank.

References

1. Malheiro, J.M., Nunes, J.P.: Simulation of vacuum assisted resin infusion (VARI) process for the production of composite material parts. In: Gaspar-Cunha, A., Periaux, J., Giannakoglou, K.C., Gauger, N.R., Quagliarella, D., Greiner, D. (eds.) Advances in Evolutionary and Deterministic Methods for Design, Optimization and Control in Engineering and Sciences. CMAS, vol. 55, pp. 319–340. Springer, Cham (2021). https://doi.org/10.1007/978-3-030-57422-2_21
2. Rajak, D.K., Wagh, P.H., Linul, E.: Manufacturing technologies of carbon/glass fiber-reinforced polymer composites and their properties: a review. Polymers **13**(21), 3721 (2021)
3. Scholz, S.: Vakuumhaube. German Patent Number DE 10 2016 120 799 B3 2017.12.28 (2017)
4. Klunker, F.: Aspekte zur modellierung und simulation des vacuum assisted resin infusion. Doctoral thesis, Technischen Universität Clausthal (2008)
5. Lionetto, F., Moscatello, A., Totaro, G., Raffone, M., Maffezzoli, A.: Experimental and numerical study of vacuum resin infusion of stiffened carbon fiber reinforced panels. Materials **13**(21), 4800 (2020)
6. Shimada, Y., Matsuzaki, R., Takahashi, A.: Numerical simulation of molding-defect formation during resin transfer molding. Adv. Compos. Mater. **25**, 17–32 (2016)
7. Fracassi, F.T., Donadon, M.V.: Simulation of vacuum assisted resin transfer molding process through dynamic system analysis. J. Compos. Mater. **52**, 3759–3771 (2018)
8. Matsuzaki, R., Shiota, M.: Data assimilation through integration of stochastic resin flow simulation with visual observation during vacuum-assisted resin transfer molding: a numerical study. Compos. A Appl. Sci. Manuf. **84**, 43–52 (2016)
9. Pillai, K.M.: Modeling the unsaturated flow in liquid composite molding processes: a review and some thoughts. J. Compos. Mater. **38**, 2097–2118 (2004)
10. Okabe, T., et al.: Multi-objective optimization for resin transfer molding process. Compos. A Appl. Sci. Manuf. **92**, 1–9 (2017)
11. Harper, A.: Resin transfer tool and method using an elastomeric insert to define a runner channel. International Patent Publication Number WO 01/02146 A1 (2001)

Challenges and Opportunities of Industry 4.0 at Mold Production Engineering and Management

Arminda Pata[1](✉) and Agostinho Silva[2]

[1] D. Dinis Higher Institute (ISDOM), Marinha Grande, Portugal
arminda.pata@isdom.pt
[2] CEI – Zipor, Companhia de Equipamentos Industriais, São João da Madeira, Portugal

Abstract. The performance of technology-enabled service ecosystems can be an alternative to the traditional system view. Service-Dominant Logic can offer a theoretical foundation that elucidates the performative nature of markets, technologies, and business models, overcoming boundless views of rationality in business model development through unification rather than division. Based on this theory, it is our intention to present a unification proposal for the mold-making industries. By transforming this sector of activity into Industry 4.0 with its integrated services. Unification enables them to respond quickly to the needs of the increasingly demanding and uncertain market. So, the aim of the paper is to propose a conceptual model, based on the Service-Dominant Logic theory that contributes to the development and innovation of the mold-making. In this paper, two pathways are proposed: Pathway 1: Private Connectivity Solution and Pathway 2: Network Connectivity Solution. The implementation of these two pathways integrated compose the teorical model that aims to contribute to increasing the responsiveness of mold-making, guaranteeing quality and delivery times. Value Creation Innovation Processes can positively improve the mold industries. In addition, each resource operating in a mold industry at the collaborative service's ecosystem-level can increase the company's annual revenue, compared to a company that has not started down the technology ecosystem path. Consequently, the companies' profits can increase.

Keywords: Industry 4.0 · Mold · Project · Production · Industrial engineering

1 Introduction

Products satisfy the daily needs of people in different fields of application. One product can be a service, a physical good or a mix of both [1]. Each physical good can result from the transformation of one or more materials (e.g., polymers, metals, ceramics, composites) and sometimes it needs one or more cut technologies (e.g., laser, plasma, water jet, electrical discharge machining, flame cutting) [2]. Different manufacturing processes can be used to obtain the shape of a physical good [3–5]. Manual processing

J. Machado et al. (Eds.): icieng 2022, LNME, pp. 13–23, 2023.
https://doi.org/10.1007/978-3-031-09382-1_2

(e.g., hand contact molding, vacuum, and pressure bagging, spraying or projection, autoclave molding), semi-automatic processes (e.g., resin injection, hot and cold pressing, compression molding) and automatic processes (e.g., injection molding). The transformation of the raw material into a physical good may only result from primary molding (e.g., rotational molding, injection molding, extrusion), or may also require secondary molding processes (e.g., thermoforming, blow molding). Most of the processes used require molds (e.g., sand, steel, wax). Steel molds are used to produce hollow, massive physical goods with or without intricate details [3, 4, 6–8]. Since the amount of plastic in the oceans is currently a concern, naturally the mold making industries, particularly those that produce plastic injection molds, must adapt to these new demands and design new solutions that continue to ensure the mass production of physical goods with alternative materials to plastic (e.g., bioplastic, magnesium), to remain competitive, having the need to adjust processes and thinking philosophies. So, the production of physical goods will continue for decades to satisfy the most diverse needs of the world's population. To satisfy these needs, it is necessary to produce large series of physical goods. That's why are used adequate processes to large series, like injection molding (e.g., bioplastic, plastic, magnesium) [9, 10]. Therefore, the production of molds for material injection is fundamental for guaranty the mass production of a huge variety of physical goods, namely for the aeronautic, aerospace, and automotive industries [11, 12].

Specifically, injection molding requires steel molds composed of a huge variety of systems and components to produce big series of physical goods. They are the most complex molds, with various systems and components (e.g., bi-material, different colors in simultaneous, different materials – plastic, bioplastic, magnesium sulfate injection).

Nowadays, a company specialized in the design, management, manufacturing, and optimization of injection molds is currently dealing with strong daily pressure. This pressure results from the current international market demands, high quality, and tight tolerances at components machining. And because to do one injection mold, it is necessary different enterprises specialized in different processes to do systems and mold components (e.g., hot channels; machining).

Each mold must be done in the shortest possible time. So, the unification of industries is extremely important for mold-making companies. To develop projects and manage to produce molds, it is not enough the transformation to industry 4.0, to acquire sophisticated machines, equipment, and software. It is indispensable to be able to take advantage of their functionalities and potentialities. The efficiency of industries of mold making should be improved.

The growth of these companies has resulted from the urgent demand for molds by international customers, and the need to design and produce more and better, with competitive prices. Naturally, the market urges companies to adopt work methodologies that allow them a comprehensive view of the project and production of each mold, so that they are prepared to make changes, without compromising delivery times. Such a requirement implies that they plan and control the production of each mold.

Industry 4.0 can allow the consultation of information stored in databases not only to plan and control the project and the production of each mold, internally, but also, the integration of information that allows companies to share some services that contribute to finish a project within the deadline established with the customer. The efficiency and mold-making performance can improve from technology-enabled service ecosystems.

2 Mold Production Engineering and Management

The mold is a hollow block that is filled with molten material in a liquid or malleable state (e.g., plastic, glass, metal), where the molten and homogenized material hardens inside the mold, adopting the shape of the physical good. In the process of molding physical goods, two-part molds or part molding can be used, creating only one section of the physical good. Molds are an increasingly used and indispensable type of tool in all branches of industrial activity, due to quality control and reliability, geometric tolerancing and their transportation. Among their great advantages are economic issues and the fact that they allow the construction of parts of extreme dimensions with great precision and accuracy. In production planning designing to order is suitable for design industries [13–15].

Designing and producing a mold for injection molding can be a very complex and demanding process, physically and mentally, depending on the shapes of the physical goods to be injected into the mold [6, 10, 16]. In this process there are a vast number of departments with a strict description of activities and functions to be performed. To minimize failures, and so that when there is a need to consult specific information of the mold to be produced, there is interconnection of departments and ease of communication of information and interpretation of work done and work to be done, network collaboration is fundamental.

The existence of production planning and control is extremely important in mold production systems. Production can be harmed individually or collectively. The lack of production planning and control, as well as the lack of efficiency and compliance with demands, tend to be increasingly harmful to Organizations. Verifying directly (e.g., financially) and indirectly (e.g., deadlines). Sometimes this indirect harmful form, tends to be the most serious in the sense that it is silent. Because it may not be as direct in the verification of loss of customer credibility, demotivation of teams, missed deadlines and in more critical cases loss of the customer/project itself, as well as increased psychosocial risks in the organization. Lean tools can be considered important for the improvement of planning and control of the mold production, as well as, in all the Organization in general. They can also contribute to the increase of profitability, improvements in meeting deadlines and the planning initially defined for each project, and the creation of workplaces duly safe and organized, and with a global efficiency of the mold production process of excellence (i.e., above 85%), the result of availability (90%), efficiency (95%), and quality (99%) [17, 18]. Nevertheless, the diversification of solutions destined to guarantee the workflow, through an assertive production planning and control, to obtain more profitability in the execution of all processes. It is also necessary to create work methodologies to check the acceptance capacity of new projects, plan their distribution internally and among the various partners. On the other hand, it is necessary to improve the planning, programming and control of the project and the production of the molds, in which it is possible to verify the internal capacity of the execution of the several injection molds processes. Therefore, the transformation of the mold sector into Industry 4.0 (e.g., robots exchanging tools) is urgent [19]. On the other hand, Augmented Reality (AR), the interaction between a virtual platform and the real environment, using a set of devices consisting of cameras, computer, screen, creates a junction between the real environment and the virtual environment, which will lead to a real-time interaction. The screen can be exchanged for AR glasses. AR, enables a simpler and immediate understanding of selected tasks even for less experienced operators, minimizing errors and delays in mold manufacturing and maintenance, ensuring greater predictability in the duration of critical tasks [20, 21]. It is also advocating the importance of multiple skills of the work team, that employers can have. Initiatives to improve the skills focus on data science and advanced analytics [12, 13], advanced simulation [22] and virtual mold modeling [23–25], network communication [26–28], human-machine interfaces [29], digital-to-physical transfer technologies (CAE, CAD/ CAM, CNC programing for machines, CIM) [30–33], management systems [4, 34, 35], real-time inventory and logistics systems [36–38], AR [20], Work-Related Musculoskeletal Disorders (WMSDs) [39] and teaching and learning infrastructures [40–43].

The conceptual view proposed gives an understanding of the dynamic interaction of entities and their complexity in the ever-widening space of objects, people and entities being interconnected. And a representation of the system that could be implemented at mold-making by the conceptual model. That model consists of concepts used to help people know, understand, or simulate the subject of that the conceptual model represents. It is also a set of concepts that can used to implement the propose presented.

3 Technological Platforms Applied to the Mold Industry

3.1 Functionality and Components Required

Today a close collaboration between all stakeholders of the mold making industry and the standardization bodies can ensure the needs and requirements of the sector in an efficient way. A close relationship for knowledge transfer about technological processes, laboratory studies, or even about regulatory and commercialization aspects. Thus, two pathways of technological connectivity are proposed to the mold-making: Pathway 1 – private connectivity solution and Pathway 2 – network connectivity solution. The first pathway refers to efficient internal communication, between departments. The second model represents the communication through industrial networks, to carry out joint work, technology-enabled service ecosystems. Both pathways should work simultaneously, internally and in a group. Below are the functionality and components needed for mold making industries to implement the internal connectivity technology. The private connectivity solution for mold industries 4.0 is shown in Fig. 1. A solution for private and autonomous deployment, within the premises of each industry. The main highlights are four work zones:

(1) comercial – in the first work zone, a specific need arises and the customer's request via email and respective delivery of the 3D model of the physical asset (e.g., component for the aircraft industry) for which the mold is desired. The customer is identified, as well as all the requirements and specifications required. A mold identification is assigned (QR Code). Next, the production capacity is analyzed (an

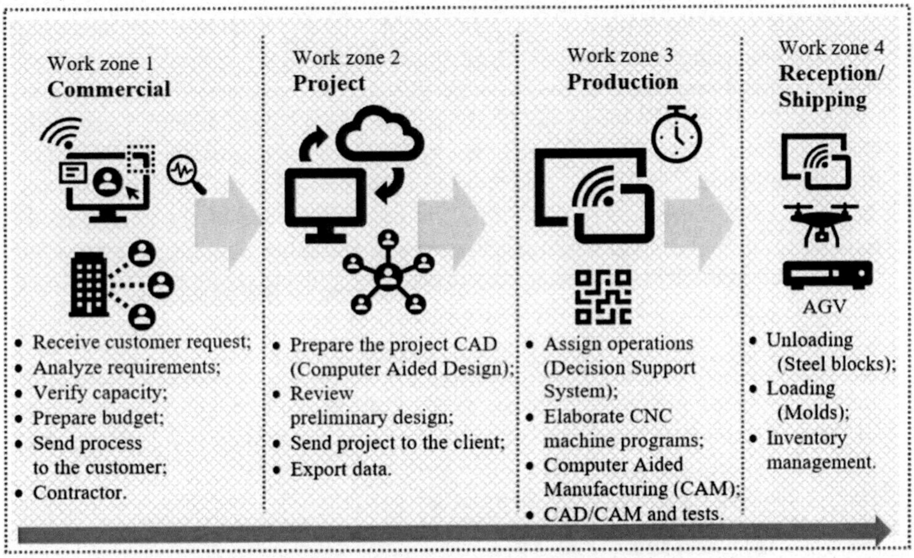

Fig. 1. Pathway 1: Private connectivity solution.

overall assessment is made of the eventual capacity to fulfill the project in the time-frame desired by the customer. It is also decided if there is necessary subcontracting of services (integrate to the proposed pathway 2). If the analysis is in favor of sub-contracting there is an evaluation of suppliers and a request for a budget. If there is internal capacity, the budget is made and, after internal approval, it is sent to the customer. Negotiations may or may not occur and the process is sent to the customer. After customer approval, the mold award is determined, external or internal execution.

(2) project – at the second work zone, if mold making in-house, the project manager analyzes with those responsible for the 3D sections of the physical asset that the customer intends to produce, or its 3D modeling is done at the software CAD – Computer Aided Design. At this stage there may be subcontracted services in some of the sections in order to avoid bottlenecks (integrate to the proposed pathway 2). This is followed by the 3D modeling of the awarded mold. During the design phase can be used the Lean philosophy (e.g., Poka-Yoke at the draw of the components). It is necessary planning a preliminary evaluation, sent to the customer where he can still introduce some changes to the mold, before its approval. Once the mold design is finished, automatically all sections will have access to the work orders.

(3) production – the detailed knowledge of the shop floor data in real time and with detail is a fundamental aspect in this cluster, allowing, not only that equipment with higher levels of intelligence can make decisions based on the state of the products they manufacture, but also so that decisions about the course of manufacturing can be made in time. What assumes a greater importance in this sector, considering the unit value of a mold, knowing also that each mold is unique.

At this stage there may or may not be CAD/CAM. If there is this need, the programming is done (integrate to the proposed pathway 2). For subsequent CNC machining and conventional milling and self-checking (integrate to the proposed pathway 2). It is not always necessary to heat treat before assembly and self-checking (integrate to the proposed pathway 2). At this work zone the final machining and self-checking is done (integrate to the proposed pathway 2). On the test date if all work orders have been fulfilled or deviations reported, the mold is ready for testing. It is checked for geometry, warping, and other possible defects. After this evaluation it may be necessary to rework the mold and perform a dimensional control (integrate to the proposed pathway 2).

(4) reception/shipping – in the final zone work, after the client's approval, the mold is shipped (integrate to the proposed pathway 2). Having resorted to subcontracting, a control is made internaly, in the sense of verifying if there is the need to make corrections, then tests are made (integrate to the proposed pathway 2). If the final control is in conformity, the mold is delivered to the customer.

3.2 Network for Mold Industries

The group connectivity solution for the 4.0 mold industries is shown in Fig. 2. A dynamic and networked solution use within the premises of each industry. The second pathway proposed results from the need to integrate and connect internal manufacturing orders (pathway 1) with outsourced services laboratory testing, scientific knowledge for research and development of new solutions, and planning and control of the design, production, and shipping of injection molds. Many companies work using obsolete conventional methods. Such methods cause disorder in the work processes, leading to lower quality products and higher prices. All these factors negatively influence the industry, leading to a loss of competitiveness and reputation in the marketplace.

It is true that these industries are equipped with (pathway 1) increasingly productive and sophisticated industrial machines, which allow to perform cuts with chip removal using conventional processes (e.g., turning, milling) and non-conventional processes (e.g., Electrical Discharge Machining, Laser, Plasma). In the last decades, the mold making industry has followed the technological evolution and integrated Computer Aided Design (CAD) with Computer Aided Manufacturing (CAM). As well as the integration of QR codes on the molds so that customers have access to mold-specific information. The transformations have been gradual. However, skilled labour is still not sufficient for all the current industry requirements (e.g. CNC programming, cience of materials, management shop floor). Since it is necessary to apply methodologies, technologies, and techniques to minimize waste and to consult updated information, in real time, to meet the deadlines for the manufacturing of a mold.

For the production and assembly of the systems and components of each mold to be efficient, the integration of hardware and software (e.g., CAD, CAM, planning and management) and a good organization in the infrastructures (e.g., Lean philosophy) is fundamental. Demanding multidisciplinary skills from the work team.

Contemporary mold industries need people with engineering skills in several areas of expertise intrinsic to the industrial production of molds. Namely, project management and industrial activity management, integrated with process technological innovation.

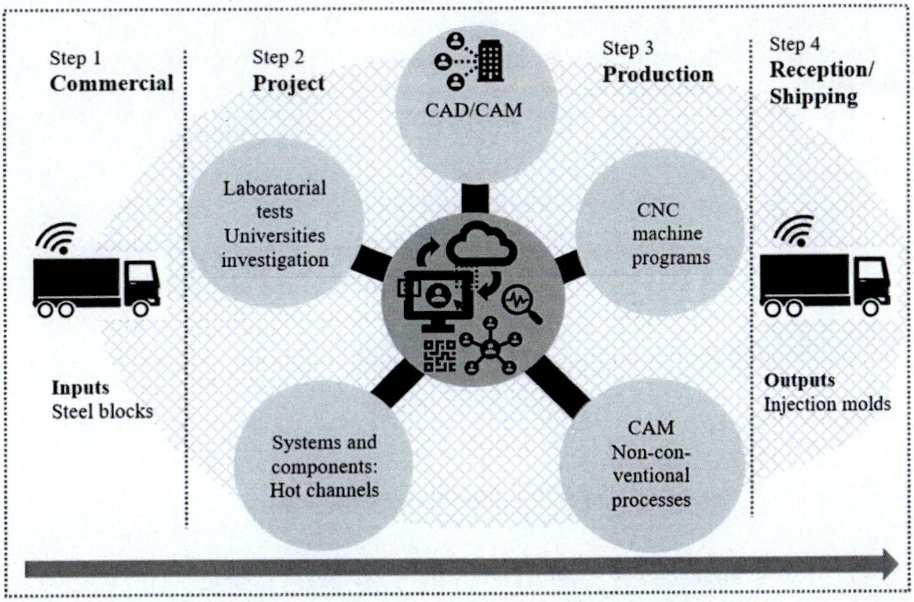

Fig. 2. Pathway 2: Network connectivity solution.

So, it's urgent, not only to implement Industry 4.0, but first of all, to increase the knowledge of the work teams in transdisciplinary areas, of the technologies applied from the project to the mold manufacturing. Updating the scientific, technical and practical domain of professionals and encouraging scientific integration with the work context. Promoting the transfer of knowledge and know-how of existing teams with future professionals, through partnerships with higher education institutions, by integrating classes in a real work context. Thus, it contributes to instill a permanent attitude of inquiry, experimentation, and teamwork, aligned with problem solving in the transformation of these industries into Industry 4.0. Simultaneously we are ensuring the training of people with the necessary skills in the mold-making, to prepare work teams for the mold industry, able to successfully perform their professional activities. Being autonomous and able to solve problems individually and in teams, essentially by the "know-how" integrated with scientific and practical competences and domains.

As a result of the partnerships the increase of interdisciplinary knowledge, both for those who already work and for those who intend to work in areas related to moldmaking. Such partnerships may also fill the current shortage of qualified labour in the mold industry, guaranteeing success in the implementation of contemporary industries (e.g., industry 4.0, industry 5.0), and so on.

3.3 Conceptual Model

The conceptual model (Fig. 3) represents the set of interdependence relations, regulated by the physical conditions that companies establish between themselves and with customer requests. To develop, test and validate innovative concepts, new products, and services, assuming itself as an open innovation ecosystem, with a set of available services that allow companies to meet customer needs while respecting the delivery time of each mold.

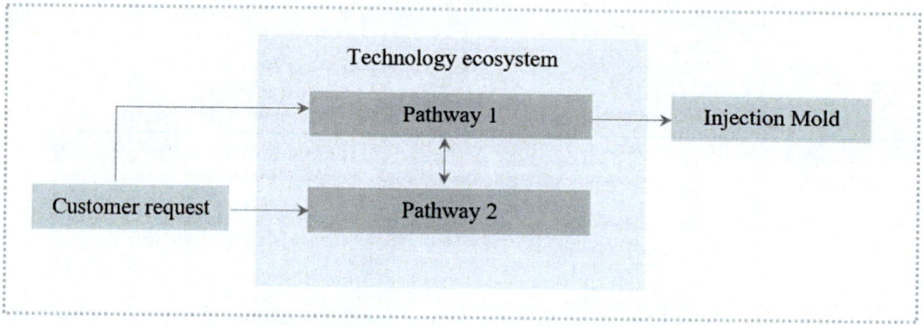

Fig. 3. Conceptual model

In short, the customer orders a mold directly from a company (pathway 1) or the entity that represents all the companies in the ecosystem (pathway 2). When the request is made directly to the company this company has verified its capacity and can request services outside (pathway 2). If the customer's request is made to the entity that represents the mold cluster, the entity checks the availability of the companies and awards the mold to the company with the highest overall process efficiency. The aim is to produce injection molds at the best price and in the least time possible.

4 Conclusion

The market has stimulated a competitiveness that grows exponentially. That has driving organizations in the search for cost reduction to remain active in the market, and to embrace the transformation of their companies into Industry 4.0.

How competitiveness grows exponentially and globally, it is necessary to have controlled and organized work processes at mold making. All possible improvements must be implemented. Whether they are considered small or large improvements in the organization.

Although big changes can start with small steps, for example, Lean tools are starting to be implemented in some mold industries, with the Poka-Yoke in mold design to facilitate the assembly of some mold components. It is important to ensure that the good results continue and increase as the Lean tools continue to be implemented, for example, with Lean methodologies at shop floor (e.g., SMED, 5S). Until a Lean Organization is achieved.

This is the only way to provide robust foundations for the implementation of new processes and investments at mold making. And to be aligned with the private connectivity solution or to implement a private convert solution (pathway 1). It is essential to the mold making grow. When there are weaknesses or threats at the implementation (pathway 1) in these intern improvements, should be overtaken to be prepared to the transformation to the network connectivity solution for mold industries (pathway 2).

The integration of technological processes (e.g., machining, CNC programming) from several companies can positively improve the offer of mold-making on a global scale. This way, each resource operating in a certain mold-making industry and belonging to the collaborative service ecosystem, that is contributing to a better use of existing and available resources. This can increase the companies' annual revenues because resource utilization is maximized compared to a company that does not belong to the technology ecosystem.

To keep up with this globalization era in which we live nowadays, it is important to create a new relationship between work, management, learning, and improve the ability of mold industries acting and collaborate for the growth of the organizations themselves. Future work should create the cluster based on the conceptual model presented.

References

1. Milberg, S.J., Park, C.W., McCarthy, M.S.: Managing negative feedback effects associated with brand extensions: the impact of alternative branding strategies. J. Consum. Psychol. **6**, 119–140 (1997)
2. Speranza, G., Liu, W., Minati, L.: Applications of Plasma Technologies to Material Processing. CRC Press, Boca Raton (2019)
3. Stokes, S.V.K.: Introduction to Plastics Engineering. Wiley, Hoboken (2020)
4. Goodship, V.: Injection Moulding: A Practical Guide. Walter de Gruyter GmbH & Co KG, Berlin (2020)
5. Gupta, R.S.: Principal of Structural Design: Wood, Steel, and Concrete. CRC Press, Boca Raton (2020)
6. Crawford, R.J., Martin, P.J.: Plastics Engineering. BH (2020)
7. Crawford, R.J., Kearns, M.P.: Practical Guide to Rotational Moulding. Elsevier, Amsterdam (2012)
8. Beaumont, J.P.: Runner and Gating Design Handbook Tools for Successful Injection Molding. Hanser Publications, Cincinnati (2019)
9. Thomas, S., Weimin, Y.: Advances in Polymer Processing 2020. Advances in Polymer Processing. Springer, Heidelberg (2009)
10. Kutz, M.: Applied Plastics Engineering Handbook. Processing, Materials and Applications. Elsevier, Amsterdam (2017)
11. Tornincasa, S.: Technical Drawing for Product Design Mastering. Springer, Cham (2021). https://doi.org/10.1007/978-3-030-60854-5
12. Mennig, G.: Mold-Making Handbook. Hanser Publications, Munich (2013)
13. Valero, J.R.L.: Plastics Injection Molding Scientific Molding, Recommendations and Best Practices. Hanser, Austin (2020)
14. Pruner, H.: Understanding Injection Molds. Hanser Publications, Austin (2013)
15. Kennedy, P.: Flow Analysis of Injection Molds 2e. Hanser Publications, Austin (2013)
16. Reyne, M.: Plastic Forming Processes. Wiley, Hoboken (2006)
17. Nicholas, J.: Lean Production for Competitive Advantage. Lean Production for Competitive Advantage. Routledge Taylor & Francis Group, Milton Park (2018)
18. Kennedy, R.K.: Understanding, Measuring, and Improving Overall Equipment Effectiveness: How to Use OEE to Drive Significant Process Improvement. Productivity Press; Taylor & Francis, CRC Press (2018)

19. Lalic, B., Majstorovic, V., Marjanovic, U., von Cieminski, G., Romero, D. (eds.): APMS 2020. IAICT, vol. 591. Springer, Cham (2020). https://doi.org/10.1007/978-3-030-57993-7
20. Zhylenko, T., Ivanov, V., Pavlenko, I., Martynova, N., Zuban, Y., Samokhvalov, D.: Mobile applications in engineering based on the technology of augmented reality. In: Machado, J., Soares, F., Trojanowska, J., Yildirim, S. (eds.) icieng 2021. LNME, pp. 366–376. Springer, Cham (2022). https://doi.org/10.1007/978-3-030-79168-1_33
21. Gronwald, K.-D.: Integrated Business Information Systems. Integrated Business Information Systems (2020)
22. Dey, N., Ashour, S., Fong, J.: Systems Simulation and Modeling for Cloud Computing and Big Data Applications. Elsevier, Amsterdam (2020)
23. Osswald, T.A., Hernandez-Orti, J.P.: Polymer Processing. Modeling and Simulation. Hanser Fachbuchverlag (2006)
24. Liu, X., et al.: Overview of the experimental trends in water-assisted injection molding. Macromol. Mater. Eng. **303**, 1–13 (2018)
25. Al Shalabi, E.W.: Low Salinity and Engineered Water Injection for Sandstone and Carbonate Reservoirs. Elsevier, Amsterdam (2017)
26. Takakura, Y., Yajima, T., Kawajiri, Y., Hashizume, S.: Application of critical path method to stochastic processes with historical operation data. Chem. Eng. Res. Des. **149**, 195–208 (2019)
27. Aneja, Y., Punnen, A.: Multiple bottleneck assignment problem. Eur. J. Oper. Res. **112**, 167–173 (1999)
28. Mossa, G., Boenzi, F., Digiesi, S., Mummolo, G., Romano, V.A.: Productivity and ergonomic risk in human based production systems: a job-rotation scheduling model. Int. J. Prod. Econ. **171**, 471–477 (2016)
29. Badri, A., Boudreau-Trudel, B., Souissi, A.S.: Occupational health and safety in the industry 4.0 era: a cause for major concern? Saf. Sci. **109**, 403–411 (2018)
30. Preston, E., Crawford, G., Coticchia, M.: CAD/CAM Dictionary. e-conversion - Proposal for a Cluster of Excellence. Mechanical Engineering (1985)
31. Pereira, R., Carvalho, A., Costa, V., Galvão, J., Matos, A.: The damping system in crutches: development of new model. In: Machado, J., Soares, F., Trojanowska, J., Ottaviano, E. (eds.) icieng 2021. LNME, pp. 371–387. Springer, Cham (2022). https://doi.org/10.1007/978-3-030-79165-0_35
32. Vukašinović, N., Duhovnik, J.: Advanced CAD Modeling: Explicit, Parametric, Free-Form CAD and Re-engineering. Springer, Cham (2019). https://doi.org/10.1007/978-3-030-02399-7
33. Davim, J.P., Ranjan, C., Kumar, K.: CNC Programming for Machining. Springer, Cham (2020). https://doi.org/10.1007/978-3-030-41279-1
34. Moayyedian, M.: Intelligent Optimization of Mold Design and Process Parameters in Injection Molding. Springer, Cham (2019)
35. Blokdyk, G.: Injection Mold Design Engineering Complete Self-Assessment Guide. 5STAR-Cooks (2018)
36. Leming-Lee, T.S., Polancich, S., Pilon, B.: The application of the toyota production system LEAN 5S methodology in the operating room setting. Nurs. Clinics **54**, 53–79 (2019)
37. Martínez, K.P., Morabito, R., Toso, E.A.V.: A coupled process configuration, lot-sizing and scheduling model for production planning in the molded pulp industry. Int. J. Prod. Econ. **204**, 227–243 (2018)
38. Glock, C.H., Grosse, E.H., Persona, A., Sgarbossa, F.: Reprint of "ergo-lot-sizing: an approach to integrate ergonomic and economic objectives in manual materials handling." Int. J. Prod. Econ. **194**, 32–42 (2017)
39. Pata, A., Moura, A.: Applying metaheuristics to minimize work-related musculoskeletal disorders. Int. J. Technol. Hum. Interact. **14**, 17–34 (2018)

40. Almeida, H.A., Vasco, J.C.: Progress in Digital and Physical Manufacturing. Lecture Notes in Mechanical Engineering, Springer, Cham (2019). https://doi.org/10.1007/978-3-030-29041-2
41. Sanghavi, D., Sahil, P., Raj, S.: Industry 4.0: tools and implementation. Manag. Prod. Eng. **10**, 3–13 (2019)
42. Tromp, J., Le, D., Le, C.: Emerging Extended Reality Technologies For Industry 4.0. Wiley, Hoboken (2020)
43. Ariffin, S., Karim, A.: Theoretical, Modelling and Numerical Simulations Toward Industry 4.0. Springer, Singapore (2021). https://doi.org/10.1007/978-981-15-8987-4

Influence of Mechanical Processing on the Durability of Parts in Additive Manufacturing Conditions

Viktoriya Pasternak[1], Oleg Zabolotnyi[1](✉), Kostiantyn Svirzhevskyi[1], Irina Zadorozhnikova[1], and José Machado[2]

[1] Lutsk National Technical University, 75 Lvivska Street, Lutsk 43018, Ukraine
volynasi@gmail.com

[2] Department of Mechanical Engineering, MEtRICs Research Centre, University of Minho, 4804-533 Guimarães, Portugal

Abstract. In this paper, the main problems associated with the mechanical processing of parts of various types and purposes are investigated. On the basis of this, a new technology for processing and strengthening the surface layer of parts made of high-speed steel of the P6M5 brand using the combined RTS effect is proposed. A block diagram of the main additive manufacturing processes, which are mainly used for manufacturing parts of various types of purposes, is presented. The results of the main stages of the technological process in the mechanical processing of parts are presented. The molecular diffusion of raw materials is studied. Based on the results obtained, the correlation between the components of the components was studied. Microstructural analysis revealed three different grain sizes: small grains, large grains, and molten grains. The homogenization process that occurs in the internal microstructure of the sample is justified. The maximum porosity, microparticle sizes, and total particle area were recorded. The main stages of the shear mechanism were investigated. Experimental results of studies on establishing the main regularities of changes in the structure of the surface layer of inhomogeneous materials are presented. A 3D model of the gear wheel is presented. Also, the average value of microhardness for gears in the conditions of additive production when performing the corresponding technological operations is predicted.

Keywords: Components · Cutting process · Feed rate · Microhardness · Microstructural analysis · Molecular diffusion · Shear mechanism · Surface layer · Technological processes

1 Introduction

The high quality of the surface layer of materials is one of the main factors that determines the durability of the final product and parts in general. Increasing the wear resistance of the working surfaces of any tools is one of the most important problems in the manufacture of products in mechanical engineering technology. Strengthening the surface layer

J. Machado et al. (Eds.): icieng 2022, LNME, pp. 24–35, 2023.
https://doi.org/10.1007/978-3-031-09382-1_3

of inhomogeneous materials by changing its structure is one of the most important and basic technological solutions in practice. At the same time, the strengthening of surface quality can be achieved through the main stages of the technological process (TP) of mechanical processing.

The main component of the technological process is technological operations. However, to obtain the surface layer of parts with better properties and stable physical and mechanical characteristics, it is advisable to use the combined RTS effect. The essence of this process is that it is necessary to pay more attention to the study of internal relationships between the main components of the electrostatic potential of structurally inhomogeneous materials, taking into account the influence of external factors (tool feed rate, workpiece diameter, cutting speed, material thickness, etc.). On the basis of which it is possible to predict the main indicators of quality and durability, as well as the microhardness of the surface layer of parts of any type and purpose, which is an important and basic indicator for any additive manufacturing and enterprise.

2 Literature Review

Research teams [1] were mainly engaged in problematic issues of additive manufacturing. This study is based on the production of additives with a high deposition rate. The detailed evolution of the microstructure that occurs in the general technical requirements ranges is presented, and the rate of heating and cooling of materials during mechanical processing is studied. In [2], the influence and speed of material removal during mechanical processing are highlighted. In this case, the main attention is focused only on the roughness of the surface layer, as well as the chemical etching process. In scientific works [3], the influence of external factors affecting the main parameters of materials during mechanical processing is considered. It should be noted that only the process of electrical discharge processing is studied in detail and the effect of hybrid aluminium composites is described. This does not take into account the main conditions for strengthening the surface layer of materials.

Research [4] examines the surface treatment and corrosion resistance of stainless steel by mechanical processing. It is important that the developed technology is widely used only for repairing cracks in some welded structures. In the works [5], automated milling and their influence on the accuracy of manufactured parts are investigated. It is revealed that the overlap parameter affects the flatness of the surface, and also takes into account the main average deviations and tolerances according to GOST. At the same time, the process of molecular diffusion of materials, which is important for conditions of any type of production, is poorly covered. Scientific experiments [6] show that only energy-intensive processing operations are prioritized. Measurement of the impact of electrical parameters, as well as improved quality indicators for power.

The team of authors [7] improved the processes of obtaining titanium alloys by mechanical processing methods. The phase relationship between the source materials is revealed. In the area of interaction of materials, the authors of the article presented a micrograph of the transition layer. We also found that the thickness of the surface layer practically does not change with increasing exposure time. It should also be noted that the researchers were able to increase the efficiency by 2–3% by optimizing the charge composition of TiAl components. In [8], the dynamic behaviour of processes during laser chemical treatment is elucidated. The authors of the paper claim that laser chemical treatment (LCM) is an unconventional processing method that allows smooth and precise microstructure of metal surfaces. Proceedings [9] consider the mechanical processing of the Ti-6Al-4V system using additive manufacturing. The authors improve the surface quality of materials and productivity with the intervention of laser processing (LAM), which limits the possibility of studying the combined effect of RTS. In scientific articles [10], the influence of the workpiece material on tool wear is investigated. The main focus is only on the performance and formation of tribofilm during mechanical processing. And also, for the rapid movement of the workpiece material to the surface layer, while not taking into account the main stages of the shear mechanism, which in turn limits the study of microhardness of parts of any type. The authors of [11] investigated the basic physical and mechanical properties of inhomogeneous Al-Ti materials, namely porosity. Presented the microstructure of the samples using application programs. We performed recognition of individual grains. An interpolation porosity diagram was constructed. It is important that the authors calculated the economic effect for individual parts based on the results of calculations of porosity and heat treatment modes.

In [12], the influence of thermal processes on the dynamics of drilling in the processing of deep holes was studied. The process of deep drilling and features of interaction of the cutting part of the drill with the part are considered. It was found that external vibrations lead to loss of exact dimensions of the part, as well as worsen the surface quality of materials. It should be noted that this study allows us to consider a narrow range of problems, since only the mechanical process for processing holes is considered in detail. The study [13] examines the effect of the electric pulse signal on the processed steel grade S235. This takes into account the cutting speed, feed rate and cutting depth during the electric turning process. It is worth noting that the technological process of mechanical processing is generally directed to the effects of electric pulse signals. Research teams [14] investigated the influence of various types of carbide materials by mechanical processing methods. The authors studied the chemical and phase composition of the initial components using transmission and scanning electron microscopy and X-ray dispersion spectroscopy. Based on the results obtained, the hardness and durability of TiAlN and AlTiN composites were predicted. In [15], the conditions of additive manufacturing are investigated, as well as the mechanical processing of parts made on the basis of titanium is considered. It should be noted that the elements of parts and structures are made only for the production of aircraft, which is an expensive moment and less practical for production. In the scientific work [16], there is an experimental study of the mechanical processing and quality conditions of an electric discharge that affect the part, and which are made of aluminium alloy of the Al5052 brand. It is important to note that in this paper, the authors were limited to certain boundary conditions that are not very supported by production. In materials [17], the mechanical processing of parts based on recycled aluminium alloys is considered. The authors of the work focus mainly on the mechanical properties of materials, while not taking into account the conditions and requirements according to GOST of any type of production.

Research works [18] improve the process of exposure to a temperature field on the surface of a layer made of aluminium alloy etched with a laser jet of water. At the same time, the processing of these compositions is studied using a laser installation, the main attention is directed only to the preparation of materials. It should be noted that this study generally analyses the basic law of temperature changes on the surface of a material during mechanical processing. Experimental papers [19, 20] argue for heat treatment on Ca and Sb. A mechanical process is performed for casting each type of Ca and Sb alloy. Basically, the samples were evaluated for hardness. The authors of this article did not take into account the influence of external factors, which made it impossible to obtain qualitative indicators of the final product in the conditions of additive production. Scientific results [21, 22] investigate the effect of the coolant pressure value on the surface layer of inhomogeneous materials. The dependence of roughness on pressing pressure in the processing of stainless steel is highlighted. The operation process of the STAR SR-32J centre and the outdoor unit of the HYTEK CHAV 160/150-AF-F-OL refrigeration unit is also described. This scientific study does not fully cover the main processes of additive manufacturing, and also partially takes into account the stages of the technological process, which limits the prediction of the durability of parts of any type. So, from the analysis of literature sources, it follows that research in the field of mechanical processing in production conditions is partially justified and requires further research. Namely, it is necessary to develop a technology for molecular diffusion of materials and study the homogenization process at the appropriate technological operations. This will allow us to determine the stable microhardness of parts and predict the durability of parts in additive manufacturing conditions.

3 Research Methodology

3.1 Basic Classification of Additive Manufacturing Processes

The purpose of the article is to develop a technology for molecular diffusion of structurally inhomogeneous materials, as well as to improve the homogenization process, on the basis of which the main regularities of changes in the structure of the surface layer of inhomogeneous materials were studied and to predict the average value of microhardness for parts of any type and purpose.

The basic classification of processes of any type of production at an enterprise or research laboratory is mainly complex, and also consists of several interrelated technological processes (TP). That is, it consists in developing the technological process of mechanical processing of manufacturing parts for any purpose. In particular, the technological process (TP) of mechanical processing of manufacturing parts in most cases consists of the following main components:

1) the workpiece part (production, selection, and preparation of the workpiece);
2) forming part (list of basic machining operations);
3) the thermal component (adjacent operations);
4) finishing (decorating).

Figure 1 shows a block diagram of the main additive manufacturing processes that are used for manufacturing parts of any type and for various purposes.

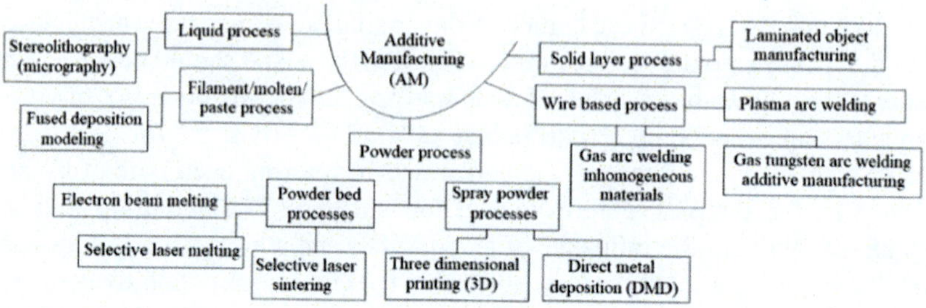

Fig. 1. Block diagram of the main additive manufacturing processes used for manufacturing parts for various purposes.

From the presented general block diagram, it can be seen that the mechanical processing of the main processes of additive production is mainly to analyse the initial data, provide technological control of the part, consists in choosing the types of production and workpiece, improves the basing process, develops the route of processing individual surfaces, provides the main type of equipment, calculates the allowances of intermediate and initial dimensions of the initial workpiece, and also provides technical rationing. It is also important to note that an assessment of the technical and economic indicators of the main additive manufacturing processes is a prerequisite.

3.2 Three-Dimensional Study of Molecular Electrostatics of Materials

To understand the greater interaction of internal bonds of high-speed steel grade P6M5, it is necessary to consider in more detail the molecular diffusion of these components. That is, to study the thermal motion of particles at temperatures above zero. We present a three-dimensional study of the molecular electrostatics of inhomogeneous materials of P6M5 steel. The colour scheme of this three-dimensional study indicates that yellow is responsible for the negative charge that passes in these structurally inhomogeneous materials. It should be noted that these negative charges of atoms can be expected from the temperature, viscosity of materials, or mass (size) of grains. The relationship of materials is shown in green. This is due to the fact that the movement of molecules does not stop and the process of molecular electrostatics of materials is automatically replaced by the process of self-diffusion. As a result, the initial components are gradually mixed until a uniform structure of the materials is obtained. It should be noted that while the molecules are in motion, the process of dynamic equilibrium occurs. The blue colour shows a small repulsion of certain external forces acting on particles, and the red spectrum aims to attract streams of molecules with a higher concentration.

Figure 2 shows a three-dimensional study of molecular electrostatics of materials.

The three-dimensional colour gamut of the presented results, which ranges from −5,605 to +5,605, shows the main interaction of structurally inhomogeneous materials of P6M5 Steel, which is formed in the internal structure of materials. That is, it characterizes the behavior of correlations between the main components of the electrostatic potential of materials.

Fig. 2. Three-dimensional study of molecular electrostatics of materials.

4 Results

4.1 Microstructural Analysis

Microstructural analysis performs a crucial function of the regularities of structure formation, as well as the dependence of the influence of the structure on mechanical properties, physical and mechanical properties, and electrical properties. Therefore, it is important to conduct this experimental study with maximum accuracy. When examining the sample, three different grain sizes were recorded in microstructural images: small grains, large grains, and molten grains that were formed as a result of 100% melting. In addition, it was found that large grains were reduced by increasing the temperature by tensile testing, which causes the process of homogenization (giving a uniform structure) in the internal microstructure of the sample. On the other hand, local grain sizes vary depending on the overall heat treatment, not just the high melting point. In addition, the larger size of the large (coarse) grain in the material leads to less tensile stress than in small grains.

Figure 3 shows the microstructure of samples of P6M5 materials. Where the grain size ranged from 1,5 mcm to 2 mcm, the maximum porosity reached $Q_{max} = 15\ldots18\%$, and the total area of grain particles was recorded, which was equal to 10126 mcm^2 and 12067 mcm^2.

The study is highly accurate. Because a small deviation (small difference) that occurred during the experiment can be attributed to the permissible limits of the boundary conditions. First of all, this is determined by the obtained mechanical characteristics of interatomic bonds, which mainly occur in the internal structure and are responsible for the strength of the material. Also, you can get additional various mechanical properties, namely: tension, compression and hardness.

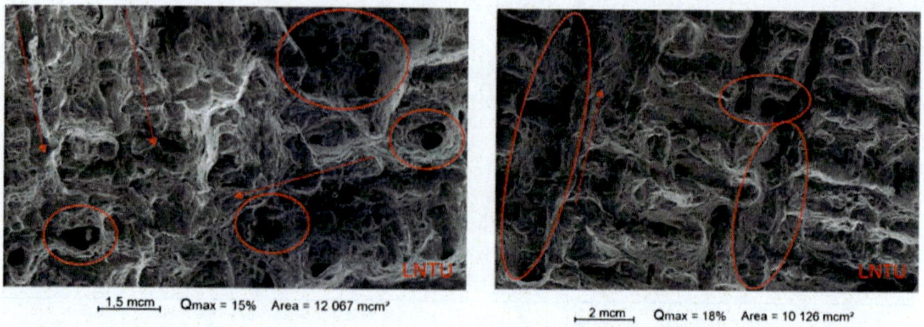

a) b)

Fig. 3. Microstructural analysis of grade steel samples P6M5, were: a) d = 1,5 mcm, Q_{max} = 15%, S = 12067 mcm^2; b) d = 2 mcm, Q_{max} = 18%, S = 10126 mcm^2.

4.2 Mechanism of Shear of Structurally Inhomogeneous Materials Under Additive Manufacturing Conditions

The shear mechanism occurs due to the sequential rotation of the workpiece when the tool is fed. This mechanism made it possible to obtain shavings of materials due to the low feed rate – 1,63 m/s. Also, to study the structure of materials with visible small cracks that occurred on the surface of samples of P6M5 materials, and which are formed mainly in the shear zone.

It should be noted that we have observed clear changes in the internal structure. Where two zones were recorded: upper and lower. The feed rate was 1,63 m/s. The resulting structure suggests that individual grains of materials undergo surface deformation due to their inhomogeneous mechanical properties. As a result, a shear band is observed (Fig. 4).

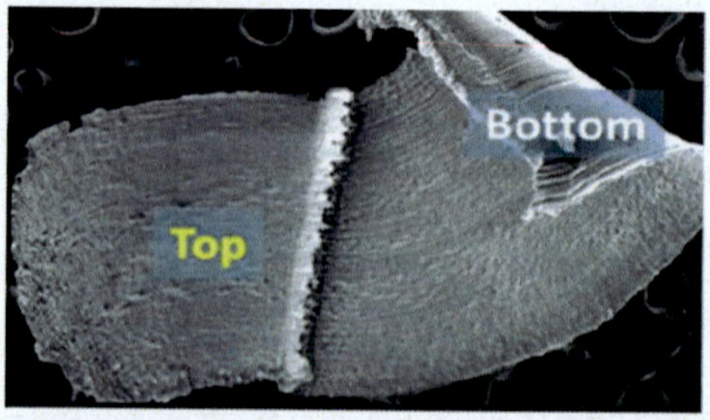

Fig. 4. Mechanism of shear of structurally inhomogeneous materials under additive manufacturing conditions.

It should also be noted that increasing the feed rate from 1,80 m/s to 2,50 m/s allows you to form a more lamellar structure and surface of materials. Because micro-cracks can be caused by a certain initiation of particle rupture, which subsequently delaminates into the material and disappears on the surface of materials, which is typical for observation during mechanical processing.

In addition, continuous chip formation at this feed rate from 1,80 m/s to 2,50 m/s is explained by the combined RTS effect. That is, the dependence of the feed rate on the maximum removal rate of materials. In this case, the chip thickness h_c of structurally inhomogeneous materials correlates with the tool feed rate, the workpiece diameter, and the cutting speed using formula 1, we obtain:

$$h_c = \frac{\pi \cdot 60 \cdot OD \cdot f}{V \cdot 1000} \tag{1}$$

where: OD – workpiece diameter, mm; f – tool feed rate, m/s; V – cutting speed, m/min.

We found that the removal rate of structurally inhomogeneous MRR material, measured in mm^3/min, also correlates well with the tool feed rate, workpiece diameter, and cutting speed. We get:

$$MRR = \frac{\pi^2 \cdot \left(OD^2 - ID^2\right) \cdot f_z \cdot N}{4 \cdot 1000} \tag{2}$$

where: OD^2 – outer diameter of the workpiece, mm; ID^2 – inner diameter of the workpiece, mm; f_z – tool feed rate, m/s; N – rotation, rev/min.

Thus, the results obtained fully ensure the desired length of contact of the workpiece with the tool. It is also important to note that with an increase in the feed rate, a more uniform structure of materials is provided, this is justified by the appearance of a shear mechanism and is explained by the combined RTS effect.

4.3 Microhardness of the Surface Layer of High-Speed Steel Grade Materials P6M5

Since, from the presented study in the previous section, it was revealed that the morphology of chips, that is, the mechanism of displacement of heterogeneous materials is mainly affected by the speed of tool feed, the diameter of the workpiece and the cutting speed, an important element here is to take into account the conditions of influence of various types of cutting processing on the microhardness of the surface layer of materials of high-speed steel grade P6M5. The maximum microhardness of the surface layer was 800 HV0,05. When turning with a cutting depth of t = 0,4 mm, the cutter was forcibly applied to the sample (while the heating temperature was up to 1000 °C). It should be noted that almost no changes were observed on the surface layer, while the maximum microhardness was 350…400 HV0,05 with an increase in the cutting depth to t = 0,8 mm, we observed a simultaneous interaction of two factors: the force index and thermal effects. The heating temperature was ±1200 °C. An important point here is that the temperature interaction was manifested only in certain zones on the surface layer of P6M5 steel materials. In this case, the plastic metal base was deformed as a result of the force factor of the cutter, and small particles of inhomogeneous materials moved along with the metal base. The maximum microhardness was 800 HV0,05.

In Fig. 5 shows the dependence of the microhardness of the surface layer of P6M5 steel materials on the cutting surface.

The drilling process during processing showed much less interaction on the test material of high-speed steel grade P6M5. The interaction of power and heat indicators

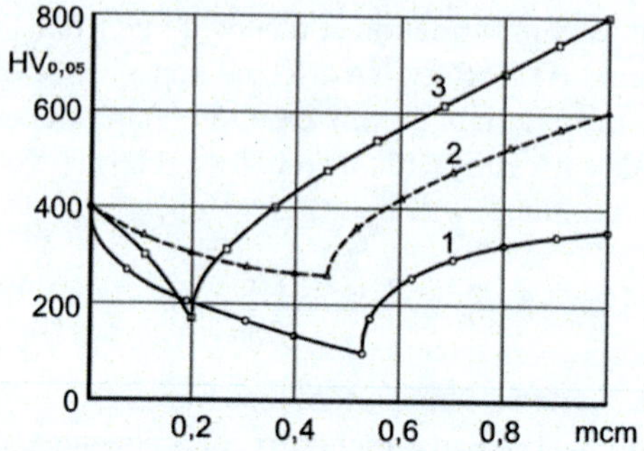

Fig. 5. Dependence of microhardness on the cutting surface, where: 1 – turning AT t = 0,4 mm; 2 – drilling; 3 – turning AT t = 0,8 mm.

was manifested to a lesser extent than during turning. It should be noted that the combined influence of force and heat factors extended to the cutting depth up to 0,3…0,5 mm, in the surface layer up to 0,7…0,8 mcm. At the same time, heating the sample from materials of the P6M5 brand during this drilling process did not eliminate the force interaction of the drill. And the microhardness ranged from 400 HV0,05 to 600 HV0,05.

4.4 Microhardness Research for Various Types of Parts

Based on a three-dimensional study of the molecular diffusion and homogenization process (correlation between the main components of structurally inhomogeneous materials), the average microhardness value HV = 2350 MPa was optimized for gear manufacturing. The level of accuracy obtained in 3D modelling of this part allowed us to predict the main properties of high-speed steel grade R6M5, namely: grain sizes, maximum porosity and total area of high-speed steel particles. The use of physical processing methods to optimize the main production processes of the gear wheel made it possible to determine the durability of the part and its reliability using calculated indicators (according to the wear criterion). Also, the systematic approach made it possible to comprehensively study and substantiate the main design patterns in the mechanical processing of parts (a gear wheel) for various purposes in the conditions of additive production.

In Fig. 6 shows a 3D simulation of a gear wheel. It should be noted that the results of the conducted experimental studies made it possible to establish the main regularities of changes in the structure of the surface layer of high-speed steel grade P6M5 and the teeth of gears during their mechanical processing at the corresponding technological operations.

And also predict the microhardness of the surface layer of gears, which is an important quality indicator for any production.

Fig. 6. 3D model of a gear wheel

5 Conclusions

Based on the combined RTS effect, a new technology for processing parts of any type and purpose, as well as strengthening their surface layer, is proposed. It should also be noted that the proposed technology of molecular diffusion of structurally inhomogeneous materials allows optimizing the basic physical and mechanical properties of inhomogeneous materials P6M5 (particle size, porosity, grain area), which allows taking into account the main technological requirements according to GOST when modelling specific products (gear wheel). The developed molecular diffusion technology makes it possible to carry out research on the following range of tasks:

1) consider in detail the correlation relationship (homogenization process) that is formed between the main components and components of structural and inhomogeneous materials;
2) to investigate the basic physical and mechanical properties of materials (high-speed steel grade P6M5);
3) perform reasonable particle microstructural analysis a steel grade P6M5;
4) substantiate the main consequences of the shear mechanism on the surface of samples of structurally inhomogeneous materials made of P6M5 steel.

It should also be noted that based on the improved homogenization process, the main regularities of changes in the structure of the surface layer of inhomogeneous materials were studied. This, in turn, made it possible to predict the average value of microhardness a gear wheel and for parts of any purpose in additive manufacturing conditions.

It is also worth noting that the optimization of microhardness (which is equal to HV = 2350 MPa), which is required for the manufacture of a gear wheel, allows us to recommend the durability of the part and its reliability using calculated indicators (according to the wear criterion).

References

1. Ho, A., Zhao, H., Fellowes, J., Martina, F., Davis, A., Prangnell, P.: On the origin of microstructural banding in Ti-6Al4V wire-arc based high deposition rate additive manufacturing. Acta Mater. **166**, 306–323 (2019)

2. Abbas, I., Abdulwahhab, A., Shabeeb, A.: Influence of FECL3 on material removal rate and surface roughness in chemical machining process. Kufa J. Eng. **10**(1), 44–55 (2019)
3. Amresh, K., et al.: Investigating the influence of WEDM process parameters in machining of hybrid aluminum composites. Adv. Compos. Lett. **29**, 1–14 (2020)
4. Pereira Rodrigues, J., Francisco do Nascimento, C., Rogerio de Freitas Batista, S., Vilarinho, L., Meneses Barros, J., Gonçalves, R.: Influence of the surface finish on corrosion resistance of stainless steels through machining and remelting on weld bead. Acta Scientiarum. Technol. **41**, 1–10 (2019)
5. Varga, J., et al.: The influence of automated machining strategy on geometric deviations of machined surfaces. Appl. Sci. **11**(2353), 1–15 (2021)
6. Sidhu, Ar., Singh, S., Kumar, R., Pimenov, D., Giasin, Kh.: Prioritizing energy-intensive machining operations and gauging the influence of electric parameters: an industrial case study. Energies **14**(4761), 1–39 (2021)
7. Pasternak, V., Zabolotnyi, O., Ilchuk, N., Cagáňová, D., Hulchuk, Y.: Improvement of processes for obtaining titanium alloys for manufacturing parts with design elements. In: Tonkonogyi, V., Ivanov, V., Trojanowska, J., Oborskyi, G., Pavlenko, I. (eds.) InterPartner 2021. LNME, pp. 323–333. Springer, Cham (2022). https://doi.org/10.1007/978-3-030-91327-4_32
8. Mehrafsun, S., Messaoudi, H.: Dynamic process behavior in laser chemical micro machining of metals. J. Manuf. Mater. Process. **2**(54), 1–18 (2018)
9. Woo, W.-S., Kim, E.-J., Jeong, H.-I., Lee, C.-M.: Laser-assisted machining of Ti-6Al-4V fabricated by DED additive manufacturing. Int. J. Precis. Eng. Manuf.-Green Technol. **7**, 559–572 (2020)
10. Yuan, J., Boyd, J., Covelli, D., Arif, T., Fox-Rabinovich, S., Veldhuis, S.: Influence of workpiece material on tool wear performance and tribofilm formation in machining hardened steel. Lubricants **4**(10), 1–10 (2016)
11. Sulym, H., Pasternak, Ia., Pasternak, V.: Boundary element modeling of pyroelectric solids with shell inclusions. Mech. Mech. Eng. **22**(3), 727–737 (2018). https://doi.org/10.2478/mme-2018-0057
12. Kondratenko, L., Mironova, L.: On the influence of thermal processes on the dynamics of a drill during deep hole machining. In: MATEC Web of Conferences, vol. 298, pp. 1–7 (2019)
13. Hameed, S., González Rojas, H., Perat Benavides, J., Nápoles Alberro, A., Sánchez Egea, A.: Influence of the regime of electropulsing-assisted machining on the plastic deformation of the layer being cut. Materials **11**(886), 1–10 (2018)
14. Kucharska, B., et al.: Influence of different types of cemented carbide blades and coating thickness on structure and properties of TiN/AlTiN and TiAlN/a-C:N coatings deposited by PVD techniques for machining of wood-based materials. Materials **14**(2740), 1–17 (2021)
15. Alonso, U., Veiga, F., Suárez, Al., Artaza, T.: Experimental investigation of the influence of wire arc additive manufacturing on the machinability of titanium parts. Metals **10**(24), 1–15 (2020)
16. Markopoulos, An., Papazoglou, Em., Karmiris-Obratański, P.: Experimental study on the influence of machining conditions on the quality of electrical discharge machined surfaces of aluminum alloy Al5052. Machines **8**(12), 1–13 (2020)
17. Mandatsy Moungomo, J., Kibouka, G., Nganga-Kouya, D.: Influence of heat treatments, parameters and machining conditions on machinability in the turning of recycled 6061 aluminum alloy. J. Manuf. Mater. Process. **5**(37), 1–16 (2021)
18. Chen, X., et al.: Study on the influence of surface temperature field of aluminum alloy etched by laser water jet composite machining. Materials **13**(3206), 1–21 (2020)
19. Náprstková, N., Kraus, P., Sviantek, J., Nguyen Van, T.: Influence of Ca, Sb and heat treatment on AlSi9CuMnNi alloy in frame of their properties from view of machining. In: MATEC Web of Conferences, vol. 244, pp. 1–7 (2018)

20. Cep, R., et al.: Influence of coolant pressure size on surface roughness when stainless steel machining. In: MATEC Web of Conferences, vol. 299, pp. 1–10 (2019)
21. Arrais-Castro, A., Varela, M.L.R., Putnik, G.D., Ribeiro, R.A., Machado, J., Ferreira, L.: Collaborative framework for virtual organisation synthesis based on a dynamic multi-criteria decision model. Int. J. Comput. Integr. Manuf. **31**(9), 857–868 (2018). https://doi.org/10.1080/0951192X.2018.1447146
22. Sousa, R.A., Varela, M.L.R., Alves, C., Machado, J.: Job shop schedules analysis in the context of industry 4.0. In: International Conference on Engineering, Technology and Innovation: Engineering, Technology and Innovation Management Beyond 2020: New Challenges, New Approaches, ICE/ITMC 2017 - Proceedings, pp. 711–717, January 2018. https://doi.org/10.1109/ICE.2017.8279955

ISO 2631 Vibration Analysis in Portuguese Alfa Pendular Trains

Patrícia Silva[1] (✉), Eurico Seabra[2], and Joaquim Mendes[2]

[1] School of Engineering, University of Minho, Guimarães, Portugal
patriciasilvaeb@gmail.com
[2] Faculty of Engineering, University of Porto, Guimarães, Portugal

Abstract. Trains are currently one of the most used transportation modes. Improving comfort is the key condition to attract more passengers. On train journeys, the comfort is highly influenced by vibration, which derives from train motion and results in whole-body vibration. This type of vibration is evaluated by the ISO 2631 standard.Following the international standard, the present study performs a ride comfort evaluation on the Alfa Pendular trains regarding the *Porto Campanhã – Lisbon Oriente* connection. Measurements were performed on the floor, seat surface and seatback, run on both seat classes. Regarding the ride comfort evaluation, it should be highlighted that low acceleration values were found for all journeys. Maximum discomfort peaks were obtained on the middle of train. The journeys were ranked as "little uncomfortable" or "not uncomfortable" during around 97% of its duration, concerning the seat surface evaluation. This value increased up to 99% for the seatback. It was also observed that when compared with the floor, the seat surface amplifies the total acceleration during the entire trip. This was the first study evaluating vibration comfort on Portuguese trains.

Keywords: ISO 2631 methodology · Vibration measurements · Ride comfort · Maximum discomfort · Alfa Pendular trains

1 Introduction

Trains are becoming a competitive transportation over air travel. Train passenger numbers are continuously increasing since 2013. From 2015 to 2019 an increase of 10% was found in Europe and, worldwide approximately 4000 billion passenger-kilometres were recorded [1–3].

A comfortable ride, mostly based on comfort, safety, and user conditions, is essential to obtain passenger satisfaction. Passengers evaluate the comfort journey based on the interaction of multiple parameters, such as seat type, noise, humidity, or air quality [4–11]. Among these, vibration is the key factor which affects passenger's comfort, performance and health, and can limit the user ability of carry out simple tasks, like reading and writing [12–14]. Depending on the transmission path, vibration can be classified as whole-body vibration (WBV), or localized hand-arm vibration. While the former is defined as the vibration transmitted into the body by means of a supported

J. Machado et al. (Eds.): icieng 2022, LNME, pp. 36–46, 2023.
https://doi.org/10.1007/978-3-031-09382-1_4

surface, the latter is characterized by a localized vibration transmitted to the hands and arms [13, 15]. Concerning a rail vehicle environment, vibration transmission occurs mostly due to the contact of the passenger with the floor, seat surface and seatback, being, this way, categorized as WBV [12, 13]. Besides affecting the comfort, this type of vibration may also cause fatigue and, in extreme cases, diseases. The human body has its natural vibration mode; when this mode match with an external induced vibration, resonance may occur which, if absorbed, can lead to tissues and organs physical stress [4, 12, 14, 16–19]. Therefore, it is crucial to assess the harmful consequences of vibration on train users. ISO 2631 standard is specially dedicated to the WBV evaluation for both, comfort and health and, thus, its methodology will be followed to fulfil the requirements of the present study [20].

This study purpose was to evaluate the WBV, concerning comfort approach, on the Portuguese Alfa Pendular trains operated at *Porto Campanhã – Lisboa Oriente* connection. This is a pioneer study on Portuguese trains.

1.1 WBV Evaluation – ISO 2631

ISO 2631 standard quantifies the WBV in relation to comfort, human health, and motion sickness. Once comfort and health are related in many ways, the evaluation method is the same for both. The standard defines the frequencies between 0.5–80 Hz for analysis, once this range of vibration affects the body as a whole. Moreover, this method is based on 3-axial acceleration measurements which should occur at the interface surfaces where the vibration transmission happens, namely the floor, seat surface and seatback. The root mean square (RMS) acceleration is calculated for each axis. As previously stated, the human body has its natural vibration mode, therefore, vibrations with identical intensities but different spectral content will induce different dynamic responses. In order to quantify this effect, the standard determines the application of weighting curves, which will assign different weights to the RMS acceleration depending on their impact on human body [18]. The weighting process is calculated according with Eq. (1):

$$a_w = \left[\sum (W_i a_i)^2\right]^{1/2} \tag{1}$$

where W_i represents the weighting frequencies and a_i the RMS accelerations. The application of the weighting curves is dependent on the measurement place and purpose. Finally, the total vibration (a_v) is obtained following Eq. (2):

$$a_v = \left(k_x^2 a_{wx}^2 + k_y^2 a_{wy}^2 + k_z^2 a_{wz}^2\right)^{1/2} \tag{2}$$

where a_w are the RMS accelerations for each axis and k is the multiplying factor corresponding to the measuring position. The comfort level is evaluated by a defined scale, Table 1 [20].

2 Materials and Methods

ISO 2631 approach was followed to perform the WBV evaluation during *Porto Campanhã – Lisboa Oriente* connection on Alfa Pendular trains. As mentioned, according

Table 1. ISO 2631 comfort evaluation scale. Adapted from [20].

$a_v (m/s^2)$	Ride comfort
≤0.315	Not uncomfortable
0.5–0.63	Little uncomfortable
0.63–0.8	Little uncomfortable to fairly uncomfortable
0.8–1.0	Fairly uncomfortable to uncomfortable
1.0–1.25	Uncomfortable
1.25–1.6	Uncomfortable to very uncomfortable
1.6–2.0	Very uncomfortable
2.0–2.5	Very uncomfortable to extremely uncomfortable
≥2.5	Extremely uncomfortable

with this standard, 3-axial measurements should occur at the floor, seat surface and seatback. The experiments run on Alfa Pendular of 4000 series trains, this is operated as a single unit, divided into 6 cars. The cars are organized according with two classes and bar facilities; therefore, the 1st and 2nd cars are classified as comfort class, on the 3rd car is placed the bar and, the 4th, 5th and 6th cars are designated as standard class. According with the train motion, the 1st car corresponds to the begging of the train, consequently the 6th car is placed at the opposite extremity, meaning the end of the train. The procedure comprised 9 full journeys following the location and class; thus, 3 journeys took place on the 1st car, 3 on the 4th car, representing the middle of the train, and 3 on the 6th car. Figure 1 presents the seat location where the measurements took place. The track takes around 2 h 50 m to complete and, presents a total length of 275 km, divided into 5 stations. The journeys run under regular operation conditions and passenger transportation.

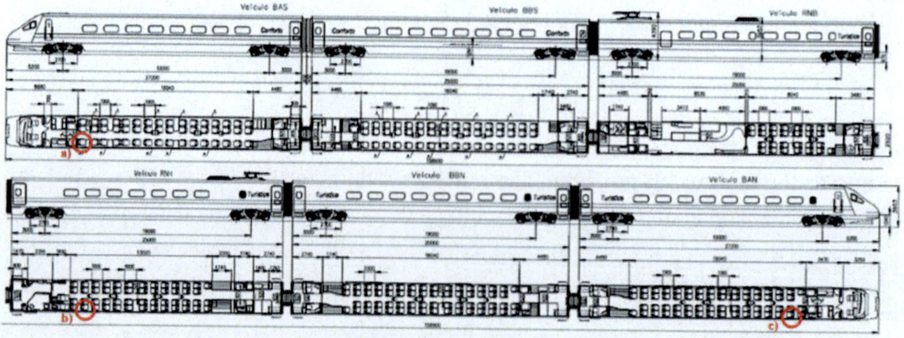

Fig. 1. Seat measurement location; a) 1st car; b) 4th car; c) 6th car.

2.1 Equipment

In-situ measurements were performed using three 3-axial accelerometers, specifically the PCE-VDL-24I Accelerometer [21]. Each accelerometer was fixed into a disc format, flexible, silicone seat pad and attached on the vibration transmission sources. The measurements happened synchronously, allowing the study of the vibration evolution from the floor to the seat. An illustration of the experimental setup on both classes can be found in Fig. 2.

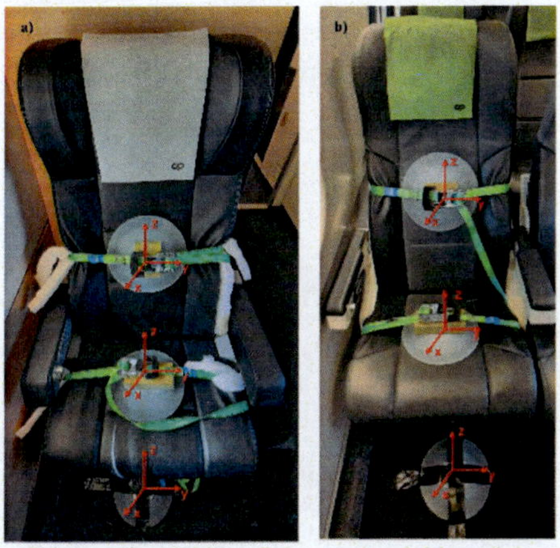

Fig. 2. Experimental setup: a) comfort class; b) standard class.

Concerning data analysis, a script was developed in MATLAB following the ISO 2631 references and applying the respective frequency weighting curves and multiplying factors indicated for comfort analysis, Table 2.

Table 2. Frequency weighting curves and multiplying factors defined by ISO 2631 for comfort analysis of a seated passenger.

	X - axis	Y - axis	Z - axis
Floor	W_k and $k_x = 0.25$	W_k and $k_y = 0.25$	W_k and $k_z = 0.40$
Seat surface	W_d and $k_x = 1.0$	W_d and $k_y = 1.0$	W_k and $k_z = 1.0$
Seatback	W_c and $k_x = 0.80$	W_d and $k_y = 0.50$	W_d and $k_z = 0.40$

A comfort analysis was obtained for each journey, then the mean total acceleration was calculated for each car. Also based on the same analysis, with the calculation of the total acceleration per second, it was possible to find the maximum discomfort presented during the journey. It was assumed that for each car the maximum discomfort was the higher discomfort presented by one of the three journeys. The instantaneous total acceleration allowed the comparison between the ride comfort of the seat surface and floor during the entire trip. Moreover, based on instantaneous comfort, it was possible to develop an observation concerning the acceleration percentage per comfort level according with the measurement location for each car.

3 Results

The following sections present the results according with the car number.

3.1 Measurement Results for Car 1

Relatively to the 1st car ride comfort evaluation, the mean comfort level was ranked as "not uncomfortable" for all measuring positions, Table 3.

Table 3. Total acceleration, ride comfort evaluation and maximum discomfort for car 1 measurements.

Position	Av	Ride comfort	Maximum discomfort	Ride comfort
Floor	0.0647	Not uncomfortable	0.4242	Little uncomfortable
Seat surface	0.2688	Not uncomfortable	2.2777	Very to extremely uncomfortable
Seatback	0.1701	Not uncomfortable	1.3026	Uncomfortable to very uncomfortbale

Besides being in the first discomfort level, it also should be pointed out the fact that the worst a_v results (0.2688 m/s^2) still 15%, below the limit of this level. Concerning the maximum discomfort, both seat surface and seatback presented high values, situated on the 8th and 6th discomfort levels respectively.

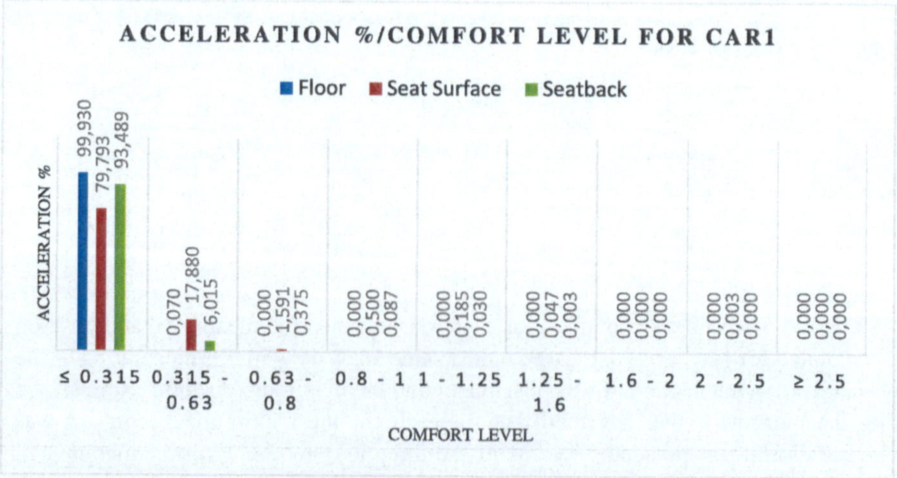

Fig. 3. Acceleration percentage per comfort level for car 1.

The acceleration percentage per comfort level, Fig. 3, showed that the journeys are considered 99.93% of the time as "not uncomfortable" on the floor location. This value decreases to 79.793% for the seat surface and, represents 93.489% of the journey considering the seatback position. Furthermore, for the seat surface, 17.88% of accelerations were obtained between 0.315–0.63 m/s^2, second discomfort level, which traduces on 97.673% of the journey ranked equally or below "little uncomfortable". The same comfort level was achieved by 99.504% for the seatback location.

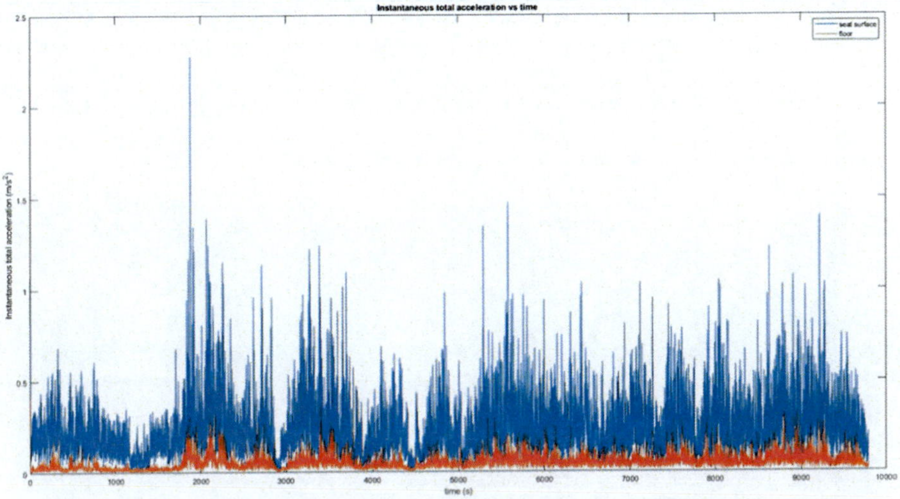

Fig. 4. Instantaneous total acceleration on the floor vs seat surface.

Observing Fig. 4, it is possible to conclude that the instantaneous total acceleration increases during the entire trip from the floor to the seat surface, with special attention on the floor peaks magnitude rising.

3.2 Measurement Results for Car 4

Regarding the 4th car results, Table 4, the journeys were also defined as "not uncomfortable", with a higher av value of 0.2566 m/s^2, presenting a gap of 0.0584 m/s^2 to the comfort level limit.

Relatively to the maximum discomfort, high peaks were found for both seat surface and seatback. While for the seat surface a maximum discomfort of 2.7525 m/s^2, corresponding to the "extremely uncomfortable" discomfort level was found. For the seatback a "uncomfortable to very uncomfortable", 1.5828 m/s^2 high value was obtained.

For the floor, 99.969% of the journey revealed to be "not uncomfortable", Fig. 5. Concerning the seat surface, 83.033% was ranked on the same evaluation and, a total of 97.179% of the journey was characterized below 0.63 m/s^2. The seatback presented similar comfort levels as the one previously found for car 1, as 93.797% of the journey as "not uncomfortable" and a total 99.415% of the journey was ranked equal or under the "little uncomfortable".

Table 4. Ride comfort evaluation and maximum discomfort for car 4.

Position	Av	Ride comfort	Maximum discomfort	Ride comfort
Floor	0.0548	Not uncomfortable	0.3717	Little uncomfortable
Seat surface	0.2566	Not uncomfortable	2.7525	Extremely uncomfortable
Seatback	0.1617	Not uncomfortable	1.5828	Uncomfortable to very uncomfortable

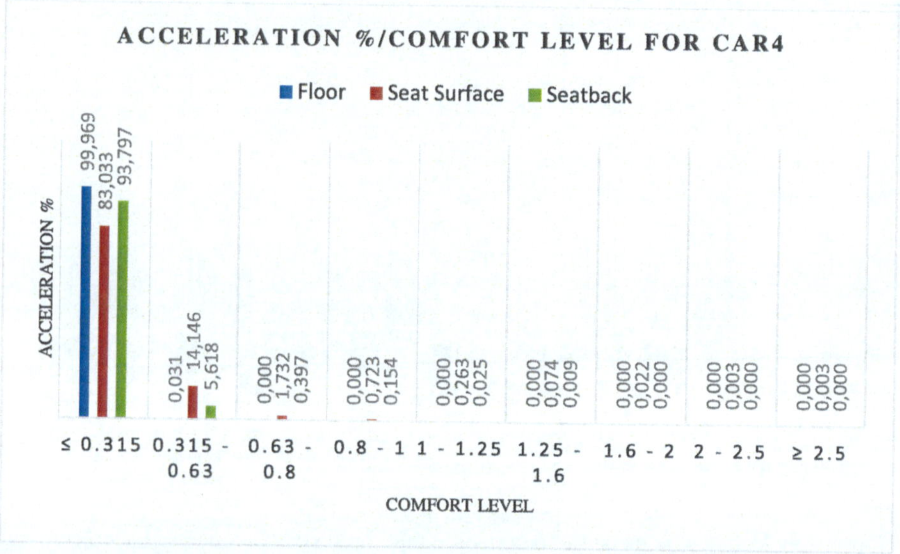

Fig. 5. Acceleration percentage rating concerning comfort levels on car 4.

Figure 6 presents the comfort vibration for both floor and seat surface. Like on the car 1 results, the vibration magnitude increases on the seat surface compared with the floor.

3.3 Measurement Results for Car 6

For car 6, the measurements concerning the WBV evaluation showed to be "not uncomfortable", Table 5. A 0.0357 m/s^2 gap to the comfort level limit was found for the worst case.

Maximum discomfort on levels "uncomfortable" and "very to extremely uncomfortable" were found for the seatback and seat surface respectively.

Figure 7 presents the acceleration percentage per comfort level. The floor revealed to be the most comfortable location with 99.954% of the journey ranked as "not uncomfortable". The seat surface presented 80.747% of the journey considered as "not uncomfortable" and, 97.031% presented acceleration values under 0.63 m/s^2. Concerning the

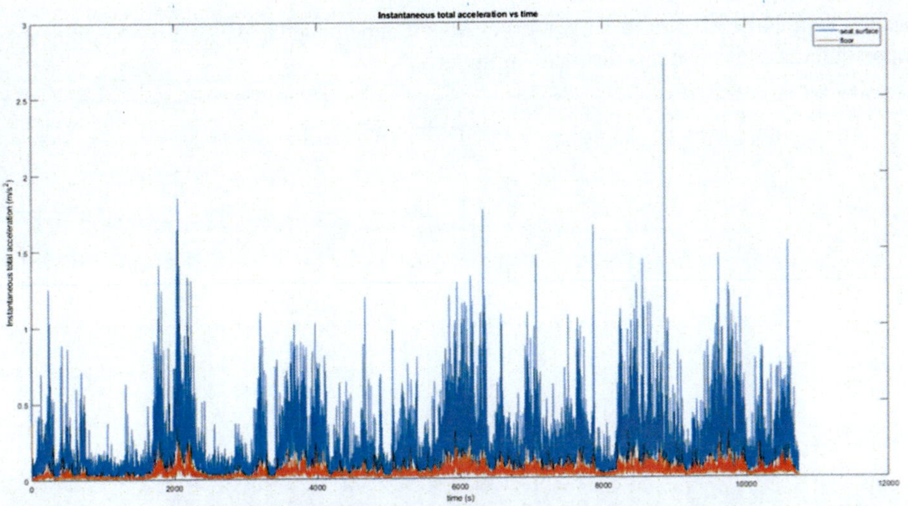

Fig. 6. Floor vs seat surface instantaneous acceleration.

Table 5. Total acceleration and maximum discomfort for car 6.

Position	Av	Ride comfort	Maximum discomfort	Ride comfort
Floor	0.0607	Not uncomfortable	0.5102	Little uncomfortable
Seat surface	0.2793	Not uncomfortable	2.0126	Very to extremely uncomfortable
Seatback	0.1572	Not uncomfortable	1.0793	Uncomfortable

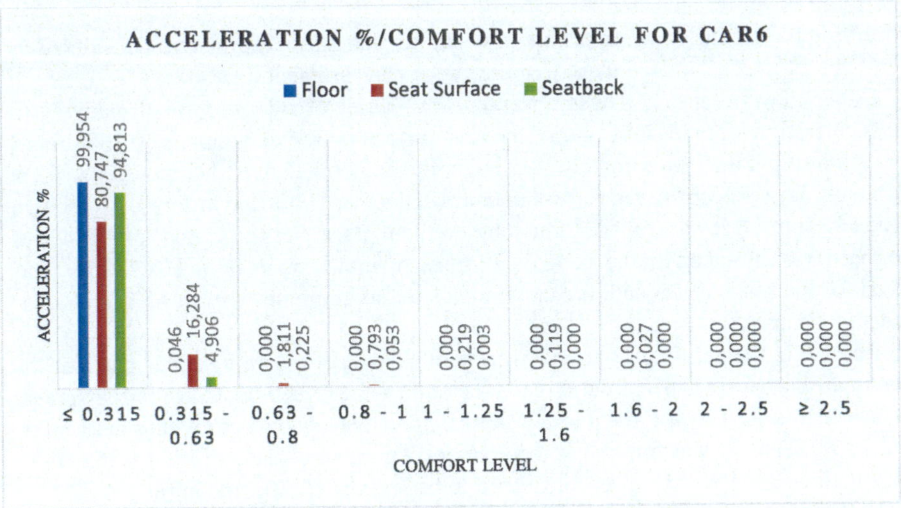

Fig. 7. Acceleration percentage per comfort level for car 6.

seatback, a 94.813% comfortable journey was found and a 99.719% of the journey was considered as "little uncomfortable" or under.

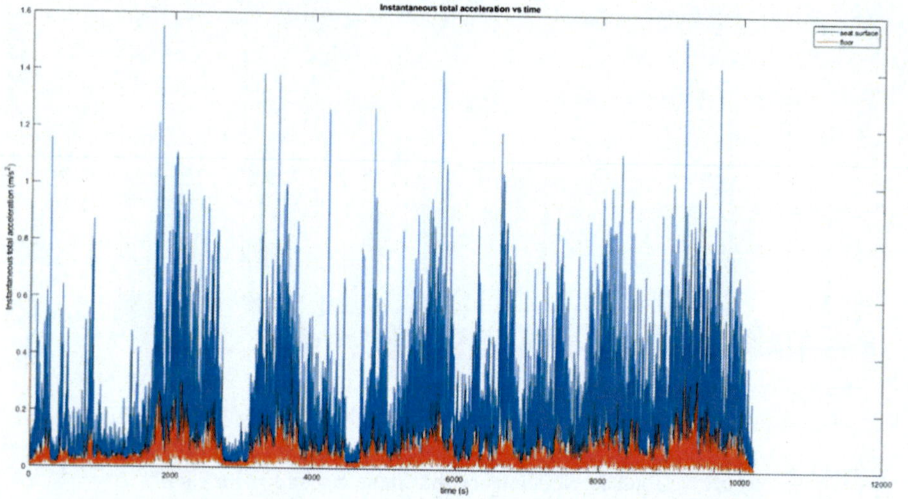

Fig. 8. Ride comfort evaluation on the floor vs seat surface.

Concerning car 6 results, Fig. 8, it was also possible to observe a rise of instantaneous total acceleration magnitude from the seat floor to the seat surface which matches with the previously findings for both car 1 and 4.

4 Discussion

Observing the results, it was possible to conclude that all journeys were considered as "not uncomfortable" independently of the car or measurement location.

Concerning the car 1, it was observed low values of ride comfort. Relatively to the acceleration worst-case scenario, the seat surface presented results of 0.2688 m/s^2, corresponding to an approximately 15% under the limit, 0.315 m/s^2. Moreover, 97.673% of the journey was considered as "not uncomfortable" or "little uncomfortable", the two lowest discomfort levels, for the same measurement place. For the seatback 99.504% of the journey was ranked on the same conditions. A significant peak of 2.2777 m/s^2 was found for the seat surface and, for the seatback a maximum discomfort of 1.3026 m/s^2 was also obtained.

Regarding the car 4, all journeys were ranked as "not uncomfortable" and, for the worst-case scenario a gap of around 18% was found to the level limit. Considering the seat surface results, 97.179% of the journey was calculated below 0.63 m/s^2, but a discomfort peak of 2.7525 m/s^2 was found. For the seatback a maximum discomfort of 1.5828 m/s^2 was obtained and 99.415% was considered to be less than "little uncomfortable".

Relatively to car 6, a difference of 11% under for the worst-case scenario to the first discomfort level pointed out the good ride comfort evaluation. This car presented the lowest discomfort peaks, with 2.0126 m/s^2 for the seat surface and 1.0793 m/s^2 for the seatback. Furthermore, 97.031% of the journeys were ranked as "little discomfort" or under for the seat surface and 99.719% of the journey matched the same rank for the seatback.

For the three cars an increase of approximately 0.2 m/s^2 was found on the total acceleration values comparing the floor to the seat surface. Based on Figs. 4, 6 and 8, it was possible to conclude that the increase between both locations happened during the entire journey and not only on localized points. These results are justified with the application of different weighting curves and multiplying factors on the floor and seat surface. Furthermore, vertical vibration transmission occurs from the floor to the seat surface, on the rail environment this is characterized by low frequencies which the foam cushion doesn't present the capability to absorb. To do that, the seat cushion will be too soft or too hard which will induce higher static discomfort levels [16–18].

Comparing the three cars' results, car 4 presented the worst results and seems to be more suitable for discomfort peaks. Cars 1 and 6 presented similar results, however car 1 presented a more comfortable journey percentage for the seat surface than car 6, and therefore, this can be considered the most comfortable place of the train. It should be highlighted the high comfort levels presented by all cars.

5 Conclusion

Trains present a complex vibration environment due to the interaction between rails, wheels, acceleration, braking and seats. WBV is transmitted to the users because of the contact with the seat and floor. Thus, it is crucial to assess the vibration exposure on train journeys. ISO 2631 presents the methodology to do it and to evaluate harmful situations for the passengers. Following this standard, the present work studied the comfort ride of the Portuguese customers on Alfa Pendular trains concerning the *Porto Campanhã – Lisboa Oriente* connection.

The measuring system developed for this purpose proved to be adequate as the experimental results showed precision and high resolution.

Results highlighted the comfort presented in 9 journeys performed on different locations. The beginning of the train (1st car) presented the best results while the middle (4th car) showed the worst. Moreover, car 4 location also seemed to be more suitable to discomfort peaks. More than 97% of the journeys time were considered as equal or under "little uncomfortable" for the seat surfaces. This value increases to 99% considering the seatback. Concerning the vibration transmission from the seat floor to the seat surface, an acceleration amplification was found, which agrees with previous studies.

This is a pioneer study in Portugal, that can provide a precious feedback for the operator to improve the comfort of the users.

The present study concluded that the passengers of the Alfa Pendular trains are travelling in a comfortable vibration environment regarding the *Porto Campanhã – Lisboa Oriente* connection.

Acknowledgements. The first author thanks to Fundação para a Ciência e Tecnologia (FCT) for a PhD scholarship under the project iRail (PD/BD/143161/2019). The authors would like to acknowledge the support of the project LAETA - UID/EMS/50022/2013.

References

1. Railway passenger transport statistics - quarterly and annual data. Eurostat - Statistics Explained (2021)
2. Railway passenger transport statistics-quarterly and annual data. Eurostat - Statistics Explained (2020)
3. Office of Rail and Road - Passenger Rail Usage 2020–21 Quarter 3 (2021)
4. Kim, Y.G., Kwon, H.B., Kim, S.W., et al.: Correlation of ride comfort evaluation methods for railway vehicles. Proc. Inst. Mech. Eng. Part F: J. Rail Rapid Transit **217**, 73–88 (2002). https://doi.org/10.1243/095440903765762823
5. Jiang, Y., Chen, B.K., Thompson, C.: A comparison study of ride comfort indices between Sperling's method and EN 12299. Int. J. Rail Transp. **7**, 1–18 (2019). https://doi.org/10.1080/23248378.2019.1616329
6. Khan, M., Sundström, J.: Vibration comfort in Swedish Inter-City trains—a survey on passenger posture and activities. In: ICA 2004, pp. 3733–3736 (2004)
7. Park, S.J., Min, S.N., Lee, H., et al.: Express train seat discomfort evaluation using body pressure and anthropometric data. J. Ergon. Soc. Korea **33**, 215–227 (2014). https://doi.org/10.5143/jesk.2014.33.3.215
8. Kelly, L.K.: Thermal comfort on train journeys (2011)
9. Maier, J., Zierke, O., Hoermann, H.J., Goerke, P.: Effects of personal control for thermal comfort in long-distance trains. Energy Build. **247**, 111125 (2021). https://doi.org/10.1016/j.enbuild.2021.111125
10. Sapena, J., Caminal, R.: Psychoacoustic evaluation of noises generated by passenger seats for high speed trains. In: Anderson, D., et al. (eds.) Noise and Vibration Mitigation for Rail Transportation Systems. NNFMMD, vol. 139, pp. 439–450. Springer, Cham (2018). https://doi.org/10.1007/978-3-319-73411-8_34
11. Zhang, J., Xiao, X.B., Sheng, X.Z., et al.: Characteristics of interior noise of a Chinese high-speed train under a variety of conditions. J. Zhejiang Univ. Sci. A **18**, 617–630 (2017). https://doi.org/10.1631/jzus.A1600695
12. Hostens, I.: Analysis of Seating during Low Frequency Vibration Exposure (2004)
13. Picu, L., Picu, M.: An analysis of whole-body vibration and hand-arm vibration exposure on the Danube ship crew. J. Phys. Conf. Ser. (2019)
14. Park, S.J., Subramaniyam, M.: Evaluating methods of vibration exposure and ride comfort in car. J. Ergon. Soc. Korea **32**, 381–387 (2013). https://doi.org/10.5143/jesk.2013.32.4.381
15. Huston, D.R., Johnson, C.C., Zhao, X.D.: A human analog for testing vibration attenuating seating. J. Sound Vib. **214**, 195–200 (1998)
16. Bovenzi, M.: Health effects of mechanical vibration. G. Ital. Med. Lav. Erg. **27**, 58–64 (2005)
17. Năstac, S., Picu, M.: Evaluating methods of whole-body-vibration exposure in trains. The Annals of "Dunarea de Jos" (2010)
18. Griffin, M.J.: Handbook of Human Vibration. Academic Press, London (1990)
19. Neto, M.F., Gomes, H.M.: Analysis of vibration levels in users of urban trains. Estudos Tecnológicos em Engenharia **12** (2018). https://doi.org/10.4013/ete.2018.122.02
20. International Standard Organization, ISO 2631 - Mechanical vibration and shock - Evaluation of human exposure to whole-body vibration (2001)
21. Instruments P. https://www.pce-instruments.com/espanol/

Noise Attenuation Analysis of a Multilayer Wall

Nicolae Adrian Niţu, Marcelin Benchea, Gelu Ianuş, and Carmen Bujoreanu(✉)

Technical University "Gheorghe Asachi" Iaşi, Iasi, Romania
nicolae-adrian.nitu@student.tuiasi.ro, marcelin.benchea@tuiasi.ro, {gelu.ianus, carmen.bujoreanu}@academic.tuiasi.ro

Abstract. Residential buildings, in which more and more people live, are overcrowded in terms of noise, which affects both their comfort and health. That is why there are regulations and standards that impose noise limits in the apartments of these buildings, which must be acoustically insulated, and the walls between the rooms must ensure the required attenuation. The paper aims to analyze the acoustic attenuation of a multilayer wall, both by calculations and by experiment. The experimental results obtained are close to those obtained by calculation, which makes us consider that a builder can first theoretically estimate the acoustic attenuation of an apartment, saving time and money, then he can build the wall. The analysis is of interest to civil engineers who can create multilayer wall structures by combining different acoustic materials. It results in a partition wall with improved acoustic properties, which complies with the noise limits imposed by the standards.

Keywords: Multilayer wall · Noise attenuation · Third octave analysis

1 Introduction

Noise level, both inside and outside buildings, are extremely important in ensuring acoustic comfort but also in the way it can affect people's health. The impetuous development of residential neighborhoods, inhabited by a large population, requires measures to reduce noise inside, so that the design and construction of partitions between rooms/apartments to ensure the acoustic comfort desired by civilized coexistence.

In construction there are different standards that regulate the allowed level of noise or impose measures to protect against noise: C125-2012 - Regulation on acoustics in construction and urban areas [1]; SR EN ISO 717-2000 - Acoustics. Evaluation of sound insulation of buildings and construction elements [2]; SR ISO 1996-2008 - Acoustics. Description, measurement and evaluation of environmental noise [3]. The terms, calculation methods and the methodology to measure the acoustic parameters are described in these standards. Thus, the parameters that define the performance of some subassemblies, elements and products for constructions are established, from the point of view of protection against noise. Our paper aims to perform the analysis and determination of noise attenuation due to a multilayer wall separating the rooms inside an apartment

J. Machado et al. (Eds.): icieng 2022, LNME, pp. 47–57, 2023.
https://doi.org/10.1007/978-3-031-09382-1_5

of a residential building. We obtained theoretical results close to the experimental ones, which demonstrates that the theoretical calculation of the noise attenuation in the construction area can be used successfully to characterize the required acoustic comfort, being helpful then in the effective realization of the partition walls between rooms from an apartment.

2 General Considerations

According to the C125–2012 regulation [1], in Table 1 are presented admissible values of the noise level, both inside and outside some buildings located in urban areas.

Table 1. Admissible values of noise levels.

Building type/functional unit	Indoor permissible limit, dB (A)	Outdoor permissible limit, dB (A)
Residential buildings (living rooms, bedrooms)	35	50
Office buildings (public offices)	45	65
Hospitals/polyclinics (rooms with more than 3 beds)	35	45

For human hearing, the perception of an increase in noise level is not directly related to the measured physical sound pressure level, which is summarized in Table 2 [4].

Table 2. Noise level increase perception.

Noise level increase, dB (A)	Subjective received noise
3	Almost noticeable
5	Noticeable difference
10	Noise doubling

The sounds we hear are a combination of different frequencies, but higher frequencies are more detectable by the human ear and therefore seem subjectively louder. Healthy people normally hear at frequencies from about 20 Hz to 20000 Hz. However, the human ear is more sensitive to sound in the frequency range of 1000 Hz to 4000 Hz compared to very low or high frequencies.

3 Determination of Noise Attenuation by Measurement

We take into consideration the acoustic behaviour inside an apartment where the rooms are separated through walls [4–6]. The walls are multilayer: bricks, insulated with plasterboard and mineral wool.

In this case, when propagating an acoustic field from one room to another, (see Fig. 1), the effective noise attenuation L_{ef} is calculated with the relation:

$$L_{ef}(f) = L_1(r_1) + R(f) - 10 \cdot lg(S/A_2) + L_2(r_2), \text{[dB]} \quad (1)$$

where: $L_1(r_1)$ and $L_2(r_2)$ are noise reduction corrections, dB, they are graphically determined [1]; S is area of the complete closure of the propagation path, m^2; A_2 is the equivalent area of sound absorption which belongs to the reception space, m^2; $R(f)$ is the sound attenuation index level of the complete closing path of the propagation path, dB.

The sound attenuation index for a construction element $R(f)$ can be calculated for each frequency from 1/3 octave with relation:

$$R(f) = L_1(f) - L_2(f) + 10 \cdot lg(S/A), \text{[dB]} \quad (2)$$

where: $L_1(f)$ and $L_2(f)$ are noise levels in emission and receiving rooms, respectively, measured in dB; S - sample area (wall), m^2; A is the equivalent absorption area in the reception room, m^2.

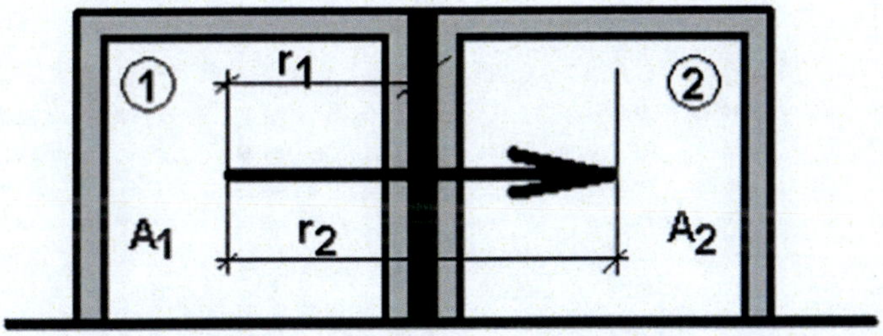

Fig. 1. Sound transmission from emission to receiving room.

For a residential building, between two rooms, from an apartment, separated by a wall of bricks and insulated with mineral wool and acoustic plasterboards, noise measurements were made in both rooms.

Noise sources used are: a circular saw that cuts a metal profile (called "noise 1"), and the white noise generated by a signal generator, respectively.

The measurements were performed with a professional sound level meter Svantek 979. Noise was measured in the emission room at a distance of 1 m from the noise source (near the partition wall), and in the receiving room at 2 m from the noise source. In the sound receiving room, three sets of noise measurements were made at a distance of 1 m from each other (center, right, left). The overall sound level (representing. the logarithmic value of the sound measured pressure relative to a value of reference) is denoted by L_w [4].

The data are centralized in Table 3. The first acoustic measurement was performed without an active noise source, normal quiet conditions, for which an overall noise level L_w = 29.28 dB was obtained, a value that falls within the limits allowed for residential buildings (living rooms). For the emission room (where the active sound source is located), for "noise 1" an overall sound level $L_w = 98.98$ dB was obtained, while the "white noise" generated a global sound level $L_w = 89.67$ dB. It can be seen that the noise level generated by the circular saw (noise 1) is higher than that generated by the white noise. Figure 2 depicts the third octave analysis of the measured sound level in the emission room, without an active noise source. It can be seen that the noise level is below 20 dB for the entire frequency range in third octave analysis, except for the 200 Hz frequency. We notice the low values of the noise level and the decreasing tendency in the most sensitive range for the human ear (1000 Hz–3150 Hz).

Table 3. Sound measurements.

Frequency [Hz]	without source [dB]	noise 1 [dB]	1 m center [dB]	1 m right [dB]	1 m left [dB]	white noise [dB]	1 m center [dB]	1 m right [dB]	1 m left [dB]
63	-5.04	5.37	-1.06	8.38	3.09	5.37	0.4	4.25	7
80	-0.27	11.08	3.99	11.08	-2.03	10.4	4.5	6.57	-1.06
100	5.56	30.12	0.4	6.72	-1.06	3.99	5.37	5.37	6.86
125	14.76	34.89	7.13	12.67	7.51	16.93	12.24	17.63	12.16
160	20.43	30.18	10.98	16.24	7.97	27.71	16.67	14.52	15.98
200	22.19	41.11	13.77	21.16	11.14	36.31	19.46	17.92	16.64
250	19.32	45.18	15.53	17.91	12.92	43.95	22.1	17.08	18.24
315	16.78	56.08	19.97	20.65	16.32	49.68	24.35	19.16	17.95
400	16.5	56.45	21.6	23.15	18.82	54.1	27.5	27.37	19.93
500	13.31	61.47	23.87	26.1	21.41	56.94	28.57	31.66	22.14
630	15.87	76.41	33.15	32.88	35.28	58.83	29.4	31.03	22.87
800	19.52	71.89	29.68	28.6	26.83	63.17	29.64	29.19	24.89
1000	16.89	70.51	32.51	34.98	33.59	69.14	30.75	29.15	26.21
1250	17.08	75.51	42.41	39.57	39.81	74.8	32.04	31.99	29.31
1600	16.03	83.07	45.51	45.19	44.48	78.25	33.55	32.91	32.65
2000	14.45	91.41	54.37	54.9	57.61	79.64	34.42	33.13	33.63
2500	14.07	90.02	52.33	54.27	54.4	79.29	33.98	31.49	32.14
3150	13.47	85.59	52.28	53.23	53.53	80.56	34.6	31.98	32.88
4000	12.24	87.85	50.93	51.15	50.45	82.83	33.88	31.78	33.26
Overall Lw	29.28	98.98	59.34	60.38	61.37	89.67	43.21	42.06	41.15

For the emission room, Fig. 3 shows the sound level using the third octave analysis for the two noise sources used in the measurements. It is obvious that in the whole frequency band the sound level of "noise 1" exceeds that generated by the white noise.

Figure 4 shows the noise level in third octave analysis in the receiving sound room, being performed three measurements at a distance of 1 m from each other, obtaining an overall noise level $L_w = \{59.34, 60.38, 61.37\}$ dB (for "noise 1", as we can see in Table 3), and $L_w = \{43.21, 42.06, 41.15\}$ dB (for "white noise" source). It can be seen that the level of noise generated by the circular saw is higher than that generated by the white noise, especially in the most noticeable range for the human ear (1000 Hz–3150 Hz), values that exceed the limits allowed for the interior of living rooms.

Table 4 summarizes the data calculated for each frequency of noise attenuation due to the partition wall between the two rooms, calculations made with the relationships described at the beginning of this paragraph.

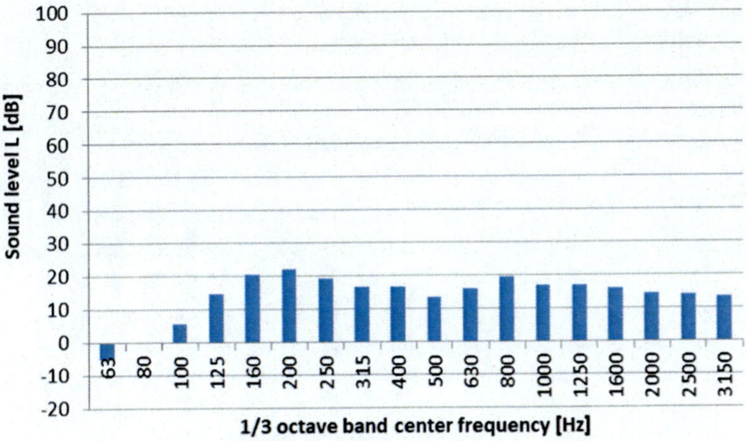

Fig. 2. Sound level without an active noise source (emission room)

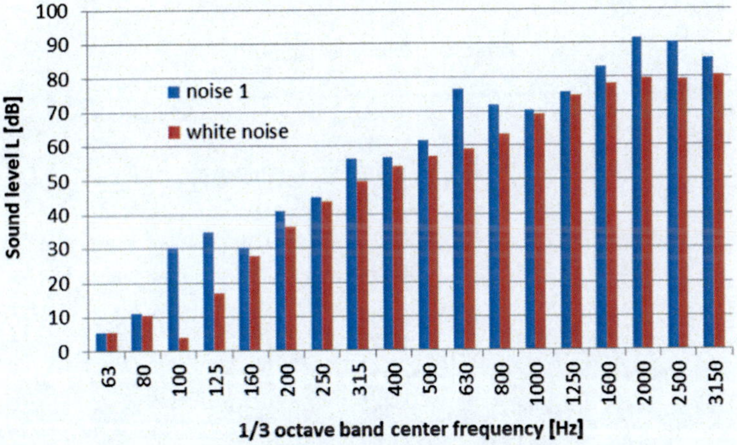

Fig. 3. Sound level for two noise sources (emission room)

The overall sound attenuation index is denoted by R_w and represents the difference of overall sound levels for a construction element [1].

For the two noise sources used in the measurements, the following values of the overall sound attenuation index R_w resulted:

- circular saw noise $R_w = \{46.64, 45.6, 44.61\}$ dB.
- white noise $R_w = \{50.46, 51.61, 52.52\}$ dB.

Figure 5 shows the effective noise attenuation L_{ef} for one of the measurements, noting that in the low frequency range (63 Hz–800 Hz) the attenuation has higher values for the noise generated by the circular saw.

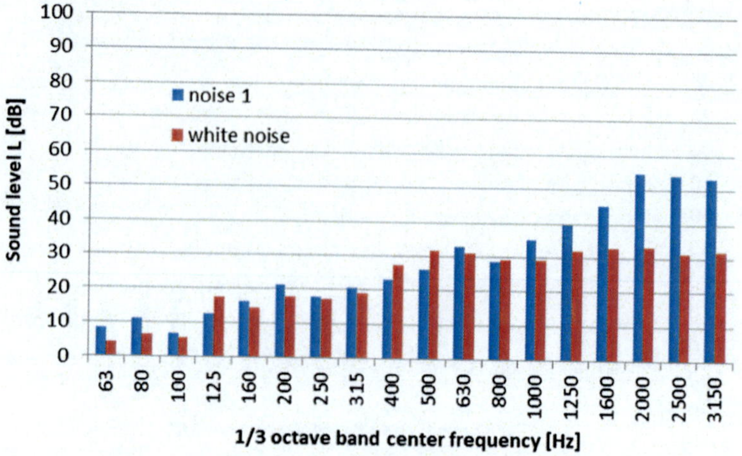

Fig. 4. Sound level for two noise sources (receiving room)

Table 4. Noise attenuation.

Frequency [Hz]	Noise 1[dB]			White noise [dB]		
	1 m center	1 m right	1 m left	1 m center	1 m right	1 m left
63	13.43	3.99	9.28	11.97	8.12	5.37
80	14.09	7	20.11	12.9	10.83	18.46
100	36.72	30.4	38.18	5.62	5.62	4.13
125	34.76	29.22	34.38	11.69	6.3	11.77
160	26.2	20.94	29.21	18.04	20.19	18.73
200	34.34	26.95	36.97	23.85	25.39	26.67
250	36.65	34.27	39.26	28.85	33.87	32.71
315	43.11	42.43	46.76	32.33	37.52	38.73
400	41.85	40.3	44.63	33.6	33.73	41.17
500	44.6	42.37	47.06	35.37	32.28	41.8
630	50.26	50.53	48.13	36.43	34.8	42.96
800	49.21	50.29	52.06	40.53	40.98	45.28
1000	45	42.53	43.92	45.39	46.99	49.93
1250	40.1	42.94	42.7	49.76	49.81	52.49
1600	44.56	44.88	45.59	51.7	52.34	52.6
2000	44.04	43.51	40.8	52.22	53.51	53.01
2500	44.69	42.75	42.62	52.31	54.8	54.15
3150	40.31	39.36	39.06	52.96	55.58	54.68
4000	43.92	43.7	44.4	55.95	58.05	56.57
Overall Rw	46.64	45.6	44.61	50.46	51.61	52.52

In the range of more noticeable frequencies for the human ear (1000 Hz–3150 Hz) the attenuation is higher for white noise, values that exceed 50 dB.

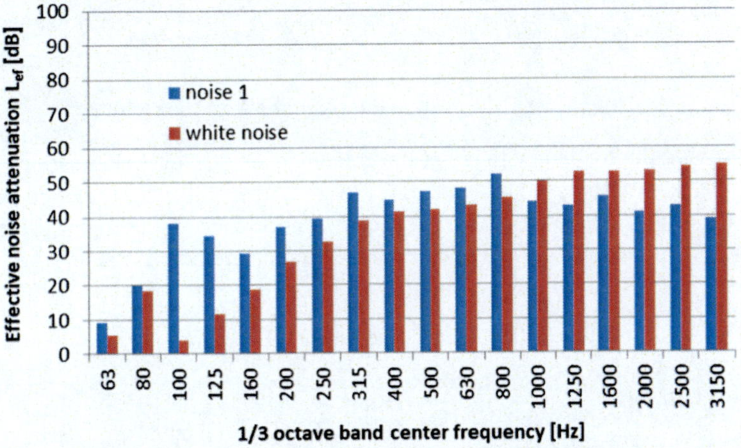

Fig. 5. Effective noise attenuation L_{ef} in third octave analysis

4 Determination of Noise Attenuation by Calculation

According to current standards, this involves plotting the $R_i(f)$ attenuation index curve for homogeneous closing elements, with simplified calculation methods for single layer or multilayer walls, such as the partition wall between the two rooms (consisting of brick pieces, acoustically insulated with mineral wool and plasterboards on both sides), as we can see in Fig. 6.

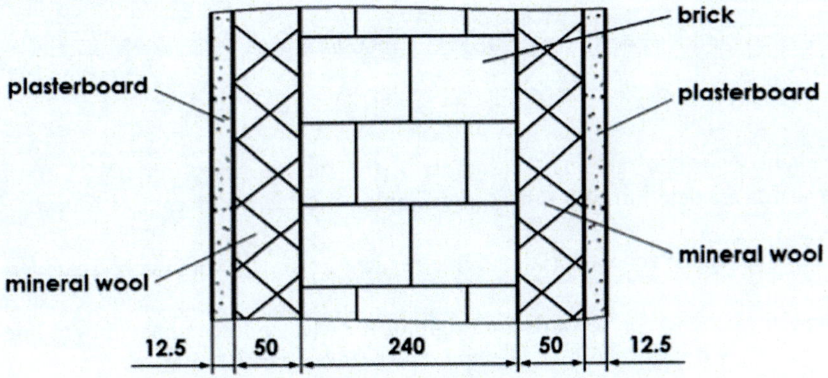

Fig. 6. Structure of the analyzed partition wall

The acoustic attenuation index curve $R_i(f)$ can be calculated as follows:

1. Determining the masses per unit area of the construction elements m_1 (brick with density ρ_1 and thickness d_1) and m_2 (plasterboard with density ρ_2, in kg/m^3, and thickness d_2, in m):

$$m_1 = d_1 \cdot \rho_1 = 0.24 \cdot 1800 = 432\ \text{kg/m}^2 \tag{3}$$

$$m_2 = d_2 \cdot \rho_2 = 2 \cdot 0.0125 \cdot 880 = 22\ \text{kg/m}^2$$

2. Determining the frequency range of the coincidence zone level attenuation (f_B - f_C) using the relationships from standard C125-2012 (see Table 5). It can be seen that the value of the attenuation index in the coincidence zone ($R_B = R_C$) is determined, depending on the material of which the construction element is made.

Table 5. The coincidence zone level attenuation.

Construction element	$R_B = R_C$ [dB]	f_B [Hz]	f_C [Hz]
Simple concrete, reinforced concrete	38	$\frac{19000}{m}$	$\frac{85000}{m}$
Brick	37	$\frac{17000}{m}$	$\frac{77000}{m}$
Autoclaved aerated concrete	29	$\frac{6700}{m}$	$\frac{43000}{m}$
Plasterboard	25	$\frac{5000}{m}$	$\frac{38000}{m}$
Wood elements	19	$\frac{2100}{m}$	$\frac{13600}{m}$

– for brick: $R_{B1} = R_{C1} = 37$ dB; $f_{B1} = 17000/m_1 = 17000/432 = 39.35$ Hz;

$$f_{C1} = 77000/m_1 = 77000/432 = 187.24\ \text{Hz}$$

– for plasterboard: $R_{B2} = R_{C2} = 25$ dB; $f_{B2} = 5000/m_2 = 5000/22 = 227.27$ Hz;

$$f_{C2} = 38000/m_2 = 38000/22 = 1727.27\ \text{Hz}$$

The curves of the attenuation indices $R_1(f)$ for the brick and $R_2(f)$ for the plasterboards are determined for these two simple constituent elements, as follows:

– In the coincidence zone a horizontal line segment is drawn, with the ordinate $R_B = R_C$;
– From the frequency f_B, towards the origin of the axes, a descending straight line is drawn, with a slope of 6 dB/octave, up to the frequency of 100 Hz;
– From the frequency f_C to the frequency $2 \cdot f_C$, therefore on an interval of one octave, an ascending line segment is drawn with a slope of 10 dB/octave;
– From frequency $2 \cdot f_C$ to the frequency of 3150 Hz, an ascending straight line with a slope of 6 dB/octave is drawn.

3. Construction of the curve $R(f) = R_1(f) + R_2(f)$;
4. Determining the final curve $R_i(f)$ with respect to:

$$R_i(f) = R(f) + R_a + R_{b1}(f) + R_{b2}(f) + R_c \text{ [dB]} \tag{4}$$

where:

R_a is the correction corresponding to the transmission of sound through collateral pathways; $R_{b1}(f)$ is the correction corresponding to the acoustic absorption of the space between the two simple constituent elements; $R_{b2}(f)$ is the correction corresponding to the stabilization of standing waves, in the space between the two simple constituent elements; R_c is the correction corresponding to the mechanical coupling of the two simple constituent elements. For the analyzed partition wall, the last two corrections have zero values, according to the data in the calculations, so the calculation method for the first two corrections is presented.

The correction R_a is determined by the relation:

$$R_a = -40 \cdot lg\left[\frac{P \cdot (\beta_1 \cdot m_1 + \beta_2 \cdot m_2)^{1.3}}{\sum_1^4 m_i' \cdot l_i} + 1\right], \text{[dB]} \tag{5}$$

where:

$m_{1,2}$ are the masses per unit area of each of the two simple constituent elements, kg/m^2; m_i' is the mass per unit area of the adjacent element "i" (joint with floor, ceiling and side walls), kg/m^2; P is the perimeter of the analyzed construction element, in meters; l_i is the length of the side "i" of the adjacent construction element, m; $\beta_{1,2}$ are the coefficients that take into account the way of attaching the simple constituent elements to the adjacent construction elements.

For the wall analyzed with sides of 2.65 × 4 m (height and length), with adjacent concrete elements (joint with floor, ceiling and side walls), considering the standard where $\beta_{1,2} = 0.8$, calculated correction R_a is of −16 dB.

The correction $R_{b1}(f)$ is determined by the relation:

$$R_{b1}(f) = -10 \cdot lg\left[\frac{1}{\alpha_m(f)}\right], \text{[dB]} \tag{6}$$

where $\alpha_m(f)$ is the average sound absorption coefficient of the interior surfaces of the space between the two simple constituent elements (mineral wool and plasterboard, whose values are taken from standard [1], at frequency f.

Table 6 summarizes the values obtained by calculations for plotting the attenuation index curve $R_i(f)$ of the analyzed wall. A global attenuation index $R_W = 55$ dB resulted.

Figure 7 shows the graph of attenuation indices $R_i(f)$ (green color) from which the increasing attenuation trend can be observed, values higher than 50 dB being registered in the range of frequencies more sensitive for the human ear (1000 Hz–3150 Hz), starting at 400 Hz.

The values obtained by measurement (curves for "noise 1" and "white noise") are close to those resulting from the calculations (curve $R + R_a + R_{b1}$) for an overall attenuation index $R_w = 55$ dB, as can be seen from Fig. 8, the attenuation indices measured being equal to those calculated on the frequency range 100–400 Hz. However, for the

Table 6. Attenuation index curve data.

Frequency [Hz]	Brick R_1	Plasterboard R_2	R_1+R_2	$R+R_a$	Standard Plasterboard absorption coefficient	Standard Mineral wool absorption coefficient	Average sound absorption coefficient	R_{b1}	$R+R_a+R_{b1}$
63	37	21.5	58.5	42.5	0.01	0.07	0.04	-13.9794001	28.520599
80	37	22	59	43	0.01	0.07	0.04	-13.9794001	29.020599
100	37	22.5	59.5	43.5	0.02	0.08	0.05	-13.0103	30.4897
125	37	23	60	44	0.02	0.08	0.05	-13.0103	30.9897
160	37	23.5	60.5	44.5	0.02	0.12	0.07	-11.5490196	32.9509804
200	39	24	63	47	0.02	0.16	0.09	-10.4575749	36.5424251
250	42.5	25	67.5	51.5	0.02	0.2	0.11	-9.58607315	41.9139269
315	45	25	70	54	0.02	0.35	0.185	-7.32828272	46.6717173
400	48	25	73	57	0.02	0.4	0.21	-6.77780705	50.2221929
500	49	25	74	58	0.03	0.65	0.34	-4.68521083	53.3147892
630	50	25	75	59	0.03	0.7	0.365	-4.37707136	54.6229286
800	51	25	76	60	0.03	0.82	0.425	-3.7161107	56.2838893
1000	52	25	77	61	0.03	0.94	0.485	-3.14258261	57.8574174
1250	53	25	78	62	0.03	0.96	0.495	-3.05394801	58.946052
1600	54	25	79	63	0.03	0.98	0.505	-2.96708622	60.0329138
2000	55	28	83	67	0.04	1	0.52	-2.83996656	64.1600334
2500	56	30	86	70	0.04	0.96	0.5	-3.01029996	66.9897
3150	57	33	90	74	0.04	0.98	0.51	-2.92429824	71.0757018

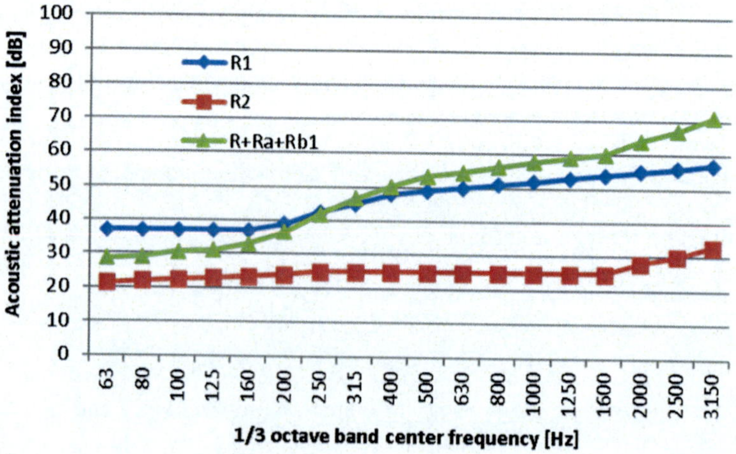

Fig. 7. Graph of the acoustic attenuation indices $R_i(f)$ for the analyzed partition wall

analyzed case, the apartment is unfinished, there are no doors, and noise attenuation is expected to be higher when the construction will be completed.

5 Conclusions

The paper presents a study of acoustic attenuation in an apartment of a residential building. Between the two rooms there is a partition wall composed of brick, mineral

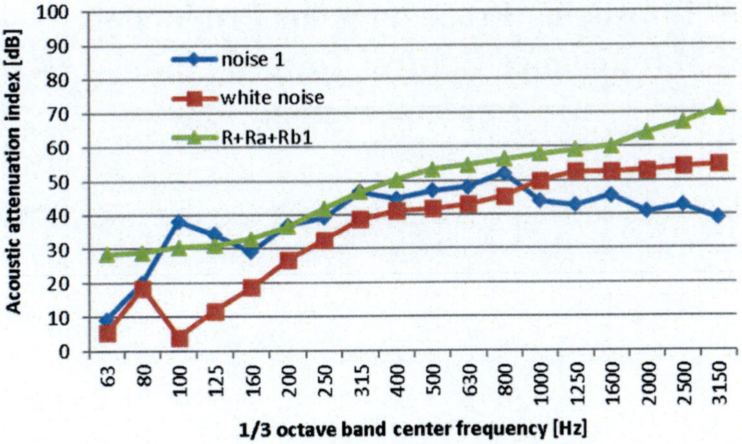

Fig. 8. Acoustic attenuation curves

wool and plasterboard. In accordance with the actual standards, we performed both measurements and calculations to assess the level of acoustic attenuation due to the partition wall. The measurement results are close to those calculated. This leads us to the idea that the manufacturer can establish the optimal materials to achieve the desired acoustic attenuation only by theoretical modeling. It is a preliminary action, which helps the civil engineer to then develop an experimental model that meets the desired noise attenuation requirements. It saves money and time. Future research aims to make other combinations of materials in the structure of a partition wall, so that the acoustic attenuation is improved, ensuring the tenants an increased comfort and a better state of health.

References

1. C125-2012: Regulation on acoustics in construction and urban areas
2. SR EN ISO 717-2000: Acoustics. Evaluation of sound insulation of buildings and construction elements
3. SR ISO 1996-2008: Acoustics. Description, measurement and evaluation of environmental noise
4. Vér, I.L, Beranek, L.L: Noise and Vibrations Control Engineering: Principles and Applications, p. 501. Wiley (2006)
5. Yang, T., Kang, J: Sound attenuation and reverberation in sequential spaces: an experimental study. Appl. Acoust. **182**, 108248 (2021)
6. Chmelík, V., et al.: Methodology for development of airborne sound insulation descriptor valid for light-weight and masonry walls. Appl. Acoust. **160**, 107144 (2020)

Improving Procedures for Production and Maintenance Control Towards Industry 4.0 Implementation

Rúben D. F. S. Costa[1], Marta L. S. Barbosa[1], Francisco J. G. Silva[1,2](✉), José C. Sá[1,2], Luís P. Ferreira[1,2], and Bebiana Pinto[1]

[1] ISEP - School of Engineering, Polytechnic of Porto, Porto, Portugal
{fgs,cvs}@isep.ipp.pt, luispintoferreira@eu.ipp.pt, bebiana.pinto@ficosa.com

[2] INEGI - Instituto de Ciência e Inovação em Engenharia Mecânica e Engenharia Industrial, Porto, Portugal

Abstract. The automotive industry is constantly stimulated by strong competitiveness in terms of product quality, cost control and commitment to delivery times. In the last two decades, competitiveness has been increasing, demanding from organizations greater control over production and surrounding processes, in order to reduce production time and, consequently, reducing costs. However, this production time of a product goes beyond what is its construction, going through validations, controls and records, among others. In an organization where the range of production is varied and the profit margin is relatively low, it is necessary to reduce all processes additional to the production of the products and increase its control.

This work presents a proposal for the implementation of production control systems according to the requirements of the 4th Industrial Revolution, in a Bowden cables manufacturing company. These systems use a MES (Manufacturing Execution System) tool with different technologies segmented by the needs of each process, both at the shop floor level and non-productive processes. The objectives proposed by the authors of this article are to increase control over production and automate processes, making them more effective, efficient and reliable, reducing resources and increasing the value of the product, by reducing costs allocated to non-productive processes.

Keywords: Industry 4.0 · Production control · Automation · Production programming · Maintenance · Information system · Mapex · Automobile industry

1 Introduction

Over the last few decades, there has been a clear evolution of the industrial sector in terms of technological modernization and increased competitiveness, having the automotive component industries also followed this trend [1]. The production of components for the automotive industry has been one of the fastest-growing sectors in recent

J. Machado et al. (Eds.): icieng 2022, LNME, pp. 58–67, 2023.
https://doi.org/10.1007/978-3-031-09382-1_6

years [2, 3], as the sale of these same components is based on very low prices, which makes this market very competitive [4]. The evolution and improvement of methods and technology monitoring must be a constant to obtain maximum profit in their production [5, 6]. Thus, production scheduling and control has a significant influence on the performance of production systems [7], which is even greater when production is under variable conditions [8]. The various production scheduling and control problems are usually defined by their high complexity and dynamics [9]. The case of this paper is no exception to this, as the production control is manual and has communication issues and information loss, a major problem in this industry. In this context, with the increasing use of collaborative machines equipped with sensors that allow the collection of data on the current state of the system in real time, perfectly updated virtual representations of the production systems are achieved [10]. Therefore, it is important to integrate Industry 4.0 concepts into production control systems, thus increasing control and decreasing response time, in general. In this study, it is presented the used methodology by this research team, which may be used by the decision-makers in the industry in order to solve their similar problems.

For this implementation, there are implications to consider, which are listed below:

(a) Big Data Collection and Analysis – connections of all resources on the shop floor, sharing information and behaviours, and giving rise to the 5V's (volume, variety, velocity, veracity and value) [11];
(b) MES/SCADA (Supervisory control and data acquisition) – Monitoring of the shop floor with real-time data collection, using SCADA systems and remote production control, considering the restrictions of the MES system [12];
(c) Cloud computing - is a technology which offers high performance and low cost [13], increasing storage capacity and data handling [14];
(d) Flexibility – programming in I4.0 must be generic enough to be applied in different scenarios and using different protocols [15];
(e) Cyber-physical systems – induced improved interconnectivity [16] between the physical components and the software is the highest level of integration [17], with constant interactions in countless ways, which adapt to the context [18];
(f) IoT and related technologies – platforms that can be considered a global network infrastructure [14], able to support sensors, communication, network, wireless and data processing technologies [19];
(g) Information security and privacy protection – with the integration, information security problems become increasingly serious [20].

Several studies have been developed in order to increase the production dynamics using Industry 4.0 concepts. Witzel et al. [21] developed a software that allowed managers to control overall fish and crop production, over factors such as efficiency, waste, and energy use. The basis of the system is the connection between the planning and the production systems, allowing to recognize abnormal states, or critical deviations in the production process, reducing the response time. Altaf et al. [22] present the RFID (Radio Frequency Identification) system as a data acquisition module. This raw data goes through a parameter estimation algorithm to extract a model from the data, calculating the time at each workstation and the production performance in real time. Caterino et al. [23] present a system capable of improving the performance of a car assembly line, focusing on times, quality and accuracy. After an initial diagnosis, a mathematical model has been developed to calculate the saturation of each worker, the OEE (Overall Equipment Effectiveness) of the line and the index of perfection, all in real time. This allows to decide if there is a need for corrective measures or not. Schleipen et al. [24] studied the cooperation of a robot gripper and an operator, in order to control the production, where an artificial vision camera makes it possible to monitor the operator's safe areas of movement. The control achieved in this system allows the safe interaction of an operator and a robot gripper working at the same workstation on an assembly line. Costa et al. [25] studied a system with the purpose to be integrated into precision lathes that would allow the analysis of stresses and vibrations which could indicate subsequent quality problems, allowing the combination of maintenance intervention with production scheduling. Barbosa et al. [26] developed an integrated machining system, capable of collecting the necessary tools and jigs, as well as carrying out the desired machining operations, without human intervention and with direct connection to the previous and next workstations, providing production information in real time. Pinto et al. [27] developed an integrated Bowden cables production system that allows real-time production counting, issuing a label for the packaging at the end of the batch and also an order completion notice for internal and external logistics services. All these cases prove how researchers have promoted the development of integrated systems that, through adequate means of sensing and communication, facilitate the production and control of manufactured products.

2 Methodology

In order to start the discussion and implementation of production control, it is important to survey the processes involved in the assembly lines of control cables.

In assembly lines, the flow is continuous, starting with the materials that go to the assembly line and ending with the final product. In the same production line, the products can also be diverse, ending up in different stations.

The production control, in the assembly lines, is performed through sensors present in each station, as well as through the existing consoles. The information taken from the console is only the production registered in each hour of the current shift. Afterwards, this information is registered on the production control board, done per shift, in a manual way.

Then, the procedure is to caption the production board's numbering, in order to understand the information registered every hour. This data includes the following information:

- Reference of the parts;
- Number of people working on the line;
- Objective;
- Real production;
- Stops associated with defects;
- Time taken;
- Description of the problem encountered.

The information is analysed by the APC (Autonomous Production Cell) managers every two hours and registered in a follow-up sheet that is presented to the APG (Autonomous Production Group) and the APU (Autonomous Production Unit) managers, who also make a round of all the APU lines. At the end of the shift, the presences in the assembly line, existing stops and production data are recorded in a logbook by the line manager. Afterwards, the APC transfers the information to the PC (Production Control), which is the ERP (Enterprise Resource Planning) of this company's Group.

As the amount of information placed on the production control boards is high and manual, its transfer to logbooks and PC causes several problems, such as the existence of varied records, altered production, high resources used and late response.

The information on the board is manually entered by the tuners and maintenance technicians, being then registered in the MAC (Maintenance Assisted Control) software to register the requests and interventions made in the assembly line, documentation relating to equipment and management of entry and exit of material. However, as this management is done manually in the production boards and in the software, additional problems of control and monitoring of the equipment arise. This is the case of the incoherence with the data presented by the production, lack of waiting times registering, orders of older requests that are not fulfilled, lack of communication between the passage of WO (Work Order) from the tuner for the technician, digital register done only by the technicians and difficulty in analysing systematic malfunctions in the equipment.

This way, it was decided to implement the MES system, in order to increase production control, data reliability between production and maintenance, but also the reduction of resources allocated to support tasks of production control and management.

Before moving on to the choice of software, a SWOT (Strengths, Weaknesses, Opportunities and Threats) analysis was established, as presented in Fig. 1, enabling strategic planning for the organisation and its decision making for implementation.

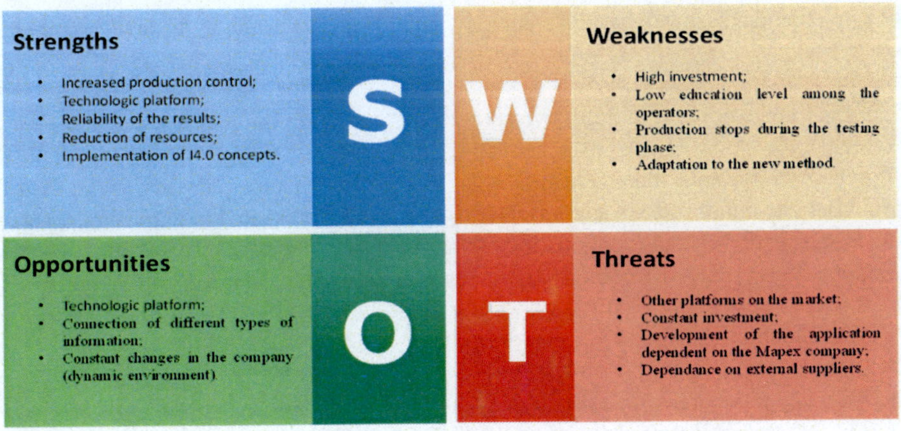

Fig. 1. SWOT analysis

The Mapex software was then chosen to implement in the assembly lines, namely in its PC (Production Control) and MM (Maintenance Management) aspects. Mapex PC allows the management of:

- Productivity of machines, lines or manual stations;
- Setup time, production, stops, failures and maintenance;
- Operators and work teams;
- Work orders, operations, phases and components;
- Waste, rejection or scrap;
- Analysis of the most common KPI's (Key Performance Indicators): OEE, availability, effectiveness, efficiency, quality, etc.;
- Documentation management: procedures, instructions, plans, etc.

On the other hand, Mapex MM guarantees:

- Detailed control of all technicians and intervention schedules;
- Control and management of spare parts (stocks, requests, delivery notes, costs, among others);
- Preparation of the preventive maintenance map;
- Management of corrective maintenance;
- Direct connection with the machines, machine cycles and counters in real time;
- Automatic creation of WOs;
- Direct connection with the production module;
- Checklist creation.

3 Discussion and Results

The implemented software has different viewing modes, according to the level assigned to each user:

- For the line level on shop floor, two screens are presented: the operator's and the TV screen;
- For line operational management levels, in addition to those already mentioned, SCADA and detail screens are presented. These can be presented in graphic or report form, presenting dynamic reports in both cases;
- For configuration levels, configuration menus are also presented for, among others:
  - Work centres;
  - Operators;
  - Shift and meal break times;
  - Masters for importing data from BPCS (Business Planning and Control System), the company's ERP;
  - Documentation.

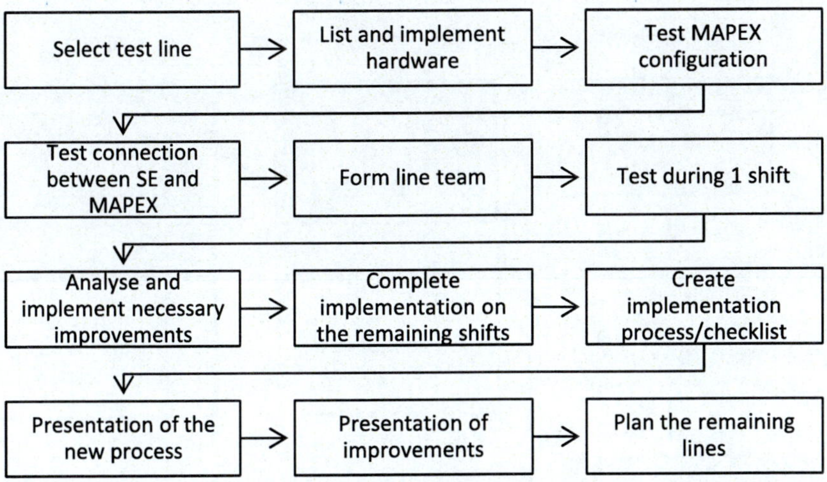

Fig. 2. Mapex implementation strategy

Thus, an implementation strategy is defined and presented in Fig. 2. With the implementation of MAPEX, the reduction of involved processes was visible, evidenced in Fig. 3.

In addition, in the old process, the information was transmitted from the Machine Counters to the Production Board, following later to the Logbook and ending at the PC, all through manual methods. In the new process, Mapex receives the information from the machine counters automatically, forwarding it directly to the PC. Not only does this new method require less time to operate, but it also saves manpower.

On the other hand, the implementation strategy for the MM module is presented in Fig. 4.

With the implementation of the maintenance module, the improvements were evident, allowing all the problems detected to be solved, in addition to the increase in the records made by the teams. In order to prove the improvements, a comparison was made of the number of registrations between the period of one year before and after the implementation of Mapex MM. In the previous period, approximately 25 000 WOs were detected and after the registration, there were 40 391. This increase demonstrated to the organization:

- The daily problems existing in the equipment;
- The workload of the tuners and maintenance technicians;
- The lack of records previously made;
- The possibility to analyse reliable data automatically.

The last bullet point was crucial for the organisation, as, due to the process becoming uniform and similar for any intervention, the data analysis was made automatic, by presenting important KPI's to the organisation.

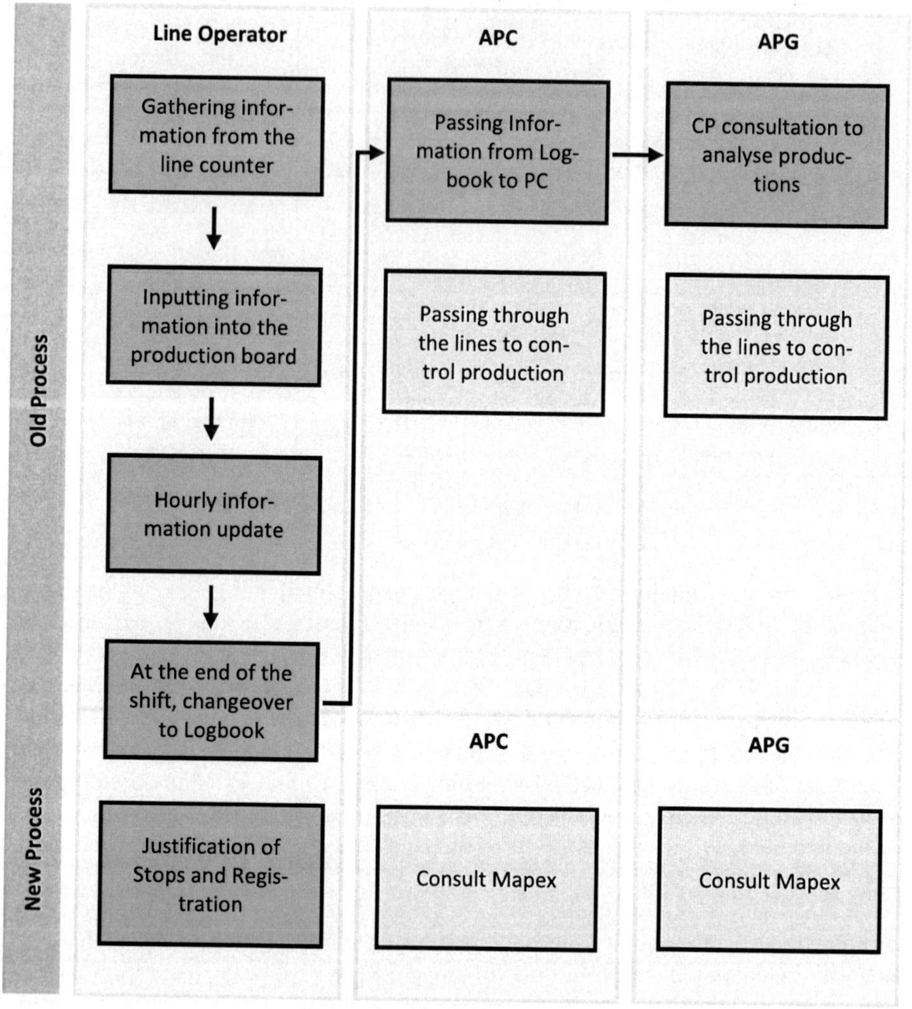

Fig. 3. Production control and registration steps: Old vs. New Process

The results obtained with the implementation of a MES system capable of controlling the maintenance and production areas allowed both qualitative and quantitative gains. With an investment per line in the maintenance and production modules of around 2750 M.U. (Monetary Units), the value is recovered in just one year, as shown in Table 1.

All these gains make the process more profitable for the company, as it is no longer necessary for the managers of the assembly lines to pass by regularly to intervene, to present such a detailed status report or to use production boards and respective accessories, also saving physical workspace.

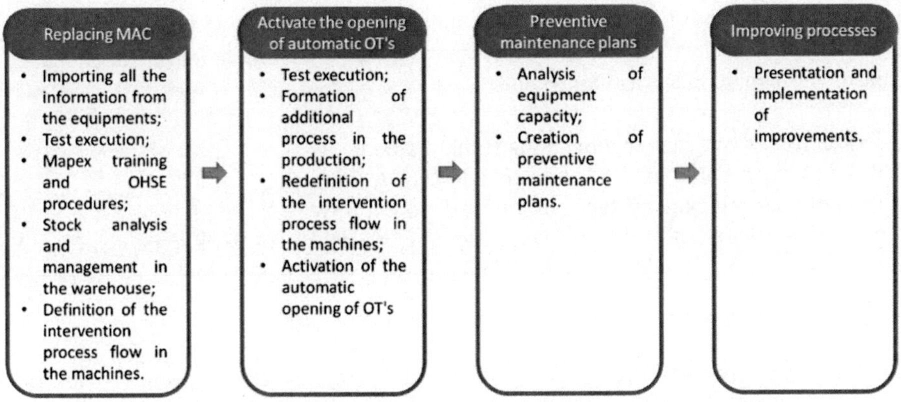

Fig. 4. Description of the implementation phases of Mapex MM

Table 1. Identification of the gains in the Mapex implementation

	Identification of the Gain	Value (year)
Team-leader gains	Time spent on the control round	1293.75 M.U
	Time spent passing the information to the CP	431.25 M.U
Supervisor gains	Time spent on the control round	658.85 M.U
	Time spent passing the information to the UAP Manager	107.80 M.U
Operator gains	Time spent filling in the board	1257.80 M.U
	Time spent filling in the Logbook	587.00 M.U
Maintenance gains	Time spent registering on the production board and MAC	1357.25 M.U
	Time used to analyse breakdowns and workload	1799.35 M.U
Physical resources	Production boards	550.00 M.U
	Reduction of materials	474.05 M.U
	TOTAL	**8517.15 M.U**

In addition to the monetary gains, it was possible with the implementation:

- To reduce repetitive processes;
- To eliminate the filling out of documentation in relation to the control of production by shift;
- To reduce errors made in the passing of information;
- To increase the reliability of results;
- To access information in real time;
- To access information anywhere;

Besides the information being able to be compared and analysed with the other departments, Mapex MM has become a key tool for resource management for the maintenance department management, as it allows:

- To identify where each technician is at the moment;
- To know the number of WO's allocated to each technician;
- To obtain reports with all types of analysis from the WO's;
- A lower workload of each employee, allowing other types of work to be distributed.

4 Conclusion

The automotive industry is highly demanding, translating this competitiveness to the factories that depend on it, one of which is the components industry. In these organisations, all projects and process improvements that are idealised, analysed and implemented are challenging, with the aim of increasing control, decreasing production costs, increasing product value and, above all, making processes reliable.

In this article, problems have been identified on the shop floor, due to the need to increase information and control over the production and maintenance processes, allowing the integration of Industry 4.0 concepts in the organisation.

The systems presented have links with different information, allowing in a single system, Mapex, the consultation of information and elimination of communication loss. This way, the decision-makers in the industry should privilege a system which englobes all the information about production control in a digital format, with the objective for it to always be updated and accessible in real time.

It is expected for the authors and the organisation to integrate other areas such as logistics, quality, metrology and industrialisation so that the increase of the flow and speed of information update is perceptible. All these systems have information available online, which can be consulted anywhere and by anyone, eliminating manual processes.

References

1. Araújo, W.F.S., Silva, F.J.G., Campilho, R.D.S.G., Matos, J.A.: Manufacturing cushions and suspension mats for vehicle seats: a novel cell concept. Int. J. Adv. Manufac. Technol. **90**(5–8), 1539–1545 (2016). https://doi.org/10.1007/s00170-016-9475-6
2. Figueiredo, D., Silva, F.J.G., Campilho, R.D.S.G., Silva, A., Pimentel, C., Matias, J.C.O.: A new concept of automated manufacturing process for wire rope terminals. Procedia Manuf. **51**, 431–437 (2020)
3. Moreira, B., Gouveia, R.M., Silva, F.J.G., Campilho, R.: A novel concept of production and assembly processes integration. Procedia Manuf. **11**, 1385–1395 (2017)
4. Silva, F.J.G., et al.: A novel automated system for the handling of car seat wires on plastic over-injection molding machines. Machines **9**, 141 (2021)
5. Martins, M., Godina, R., Pimentel, C., Silva, F.J.G., Matias, J.C.O.: A practical study of the application of SMED to electron-beam machining in automotive industry. Procedia Manuf. **17**, 647–654 (2018)
6. Costa, R.J.S., Silva, F.J.G., Campilho, R.D.S.G.: A novel concept of agile assembly machine for sets applied in the automotive industry. Int. J. Adv. Manufac. Technol. **91**(9–12), 4043–4054 (2017). https://doi.org/10.1007/s00170-017-0109-4

7. Nyhuis, P., Mayer, J.: Modelling the influence of setup optimized sequencing on lateness and productivity behaviour of workstations. CIRP Ann. Manuf. Technol. **66**(1), 421–424 (2017)
8. Stricker, N., Pfeiffer, A., Moser, E., Kádár, B., Lanza, G., Monostori, L.: Supporting multi-level and robust production planning and execution. CIRP Ann. - Manuf. Technol. **64**(1), 415–418 (2015)
9. Branke, J., Nguyen, S., Pickardt, C.W., Zhang, M:. Automated design of production scheduling heuristics: a review. IEEE Trans. Evol. Comput. **20**(1), 110–124 (2016)
10. Monostori, L., et al.: Cyber-physical systems in manufacturing. CIRP Ann. **65**(2), 621–641 (2016)
11. Wamba, S.F., Akter, S., Edwards, A., Chopin, G., Gnanzou, D.: How 'big data' can make big impact: Findings from a systematic review and a longitudinal case study. Int. J. Prod. Econ. **165**, 234–246 (2015)
12. Dalenogare, L.S., Benitez, G.B., Ayala, N.F., Frank, A.G.: The expected contribution of Industry 4.0 technologies for industrial performance. Int. J. Prod. Econ. **204**, 383–394 (2018)
13. Mitra, A., Kundu, A., Chattopadhyay, M., Chattopadhyay, S.: A cost-efficient one time password-based authentication in cloud environment using equal length cellular automata. J. Ind. Inf. Integr. **5**, 17–25 (2017)
14. Porter, M.E., Heppelmann, J.E.: How smart, connected products are transforming competition. Harv. Bus. Rev. **92**(11), 64 (2014)
15. Peres, R.S., Rocha, A.D., Leitão, P., Barata, J.: IDARTS – towards intelligent data analysis and real-time supervision for industry 4.0. Comput. Ind. **101**, 138–146 (2018)
16. Nayak, A., Levalle, R.R., Lee, S., Nof, S.Y.: Resource sharing in cyber-physical systems: modelling framework and case studies. Int. J. Prod. Res. **54**(23), 6969–6983 (2016)
17. Chen, H.: Theoretical foundations for cyber-physical systems: a literature review. J. Ind. Integr. Manag. **2**(03), 1750013 (2017)
18. Gürdür, D., El-Khoury, J., Seceleanu, T., Lednicki, L.: Making interoperability visible: data visualization of cyber-physical systems development tool chains. J. Ind. Inf. Integr. **4**, 26–34 (2016)
19. Mao, J., et al.: A hybrid reader tranceiver design for industrial internet of things. J. Ind. Inf. Integr. **2**, 19–29 (2016)
20. Kusiak, A.: Smart manufacturing. Int. J. Prod. Res. **56**(1–2), 508–517 (2018)
21. Witzel, O., Wilm, S., Karimanzira, D., Baganz, D.: Controlling and regulation of integrated aquaponic production systems – an approach for a management execution system (MES). Inf. Process. Agric. **6**(3), 326–334 (2019)
22. Altaf, M.S., Bouferguene, A., Liu, H., Al-Hussein, M., Yu, H.: Integrated production planning and control system for a panelized home prefabrication facility using simulation and RFID. Autom. Constr. **85**, 369–383 (2018)
23. Caterino, M., et al.: Simulation techniques for production lines performance control. Procedia Manuf. **42**, 91–96 (2020)
24. Schleipen, M., Gilani, S.S., Bischoff, T., Pfrommer, J.: OPC UA & Industry 4.0 - enabling technology with high diversity and variability. Procedia CIRP **57**, 315–320 (2016)
25. Costa, S., Silva, F.J.G., Campilho, R.D.S.G., Pereira, T.: Guidelines for machine tool sensing and smart manufacturing integration. Procedia Manuf. **51**, 251–257 (2020)
26. Barbosa, M., Silva, F.J.G., Pimentel, C., Gouveia, R.M.: A novel concept of CNC machining center automatic feeder. Procedia Manuf. **17**, 952–959 (2018)
27. Pinto, B., Silva, F.J.G., Costa, T., Campilho, R.D.S.G., Pereira, M.T.: A strategic model to take the first step towards industry 4.0 in SMEs. Procedia Manuf. **38**, 637–645 (2019)

Industry 4.0 Technologies for Maintenance Management – An Overview

Małgorzata Jasiulewicz-Kaczmarek[1] and Katarzyna Antosz[2](✉)

[1] Faculty of Management Engineering, Poznan University of Technology, Rychlewskiego 2, 60-965 Poznań, Poland
malgorzata.jasiulewicz-kaczmarek@put.poznan.pl

[2] Faculty of Mechanical Engineering and Aeronautics, Rzeszow University of Technology, Al. Powstancow Warszawy 12, 35-959 Rzeszów, Poland
katarzyna.antosz@prz.edu.pl

Abstract. Few decades ago, maintenance was mainly seen as something that was difficult to manage. Nowadays, taking into account the impact of maintenance on other company processes (e.g. production, logistics, occupational health and safety management, environment), the maintenance process is recognized as an important business function. The economic impact of improper designed maintenance processes is related to costs, downtime, breakdowns, defects, additional inventory, etc., which affect product quality and the company's performance. Maintenance technology is about finding and applying cost-effective ways to avoid or overcome degradation in machine performance and availability. The main goal of the paper is to present an overview of Industry 4.0 technologies in maintenance on the newest paradigm in the maintenance management. By the using of the advanced Industry 4.0, technologies, it is possible to monitor the state of the manufacturing equipment, diagnose faults, predict and solve problems before they occur. Moreover, the digitization of maintenance requires the mastery of both new technologies and organizational issues related to their planning, implementation and maintenance.

Keywords: Industry 4.0 · Maintenance 4.0 · I4.0 technologies

1 Introduction

Nowadays production involves data-driven business processes on various levels which result in the development of different manufacturing patterns. "Changes" are termed as the fourth industrial revolution in the literature considering production, frequently called as "Smart manufacturing" (mostly in the USA), "Industry 4.0" (mostly in Germany/Europe) and "Smart Factory" (mostly in Korea/Asia/Europe) [1]. This new economy model is based on digital factories, capable of combining information technology and machines with intelligent products [2, 3]. The development of Industry 4.0 (I4.0) is aimed at achieving a high level of operational efficiency, productivity and automation of production systems. However, despite the existence of a common goal, there is no consensus in the literature for its definition. According to [4] I4.0 can be defined as "a set of

J. Machado et al. (Eds.): icieng 2022, LNME, pp. 68–79, 2023.
https://doi.org/10.1007/978-3-031-09382-1_7

technologies, devices and processes capable of operating in an integrated manner at several stages of the production process and at several levels of the supply chain [5] which allows for self-sufficient production, integrated operations, decentralized decisions, minimal human intervention". However, according to [6] I4.0 is a transformational journey where changes can be both evolutionary and revolutionary. Evolutionary as some I4.0 assistive technologies, such as sensors and robots, are the result of a long-term innovation that began with the third industrial revolution [7–9]. Revolutionary as I4.0 can affect the core business of manufacturers, shifting it from manufacturing and selling products to delivering solutions [10, 11].

One of the priority areas of Industry 4.0 is the interdisciplinary cooperation between production and maintenance in order to obtain fast and efficient service, enabling companies to implement a cost-effective production system [12]. In the context of maintenance, Sufix "4.0" should be understood as another step towards making this area of the company's operations more effective. According to [13] the implementation of predictive and proactive maintenance strategies in place of a reactive and preventive approach should be one of the first steps of company management to be carried out in the context of Industry 4.0.

The aim of the article is to describe the latest trends in the field of maintenance management in the context of Industry 4.0 (I4.0) technology. The structure of the paper is as follows. Sections 2 provides a short literature review on Industry 4.0 technologies. Section 3 presents the possibility of applications I4.0 in maintenance activities. Finally (Sect. 4), the conclusion are presented. This paper is a continuation of the previously undertaken work presented in [14–16].

2 Industry 4.0 Technologies

According to [17] I4.0 refers to nine pillars of Industry 4.0: Autonomous/Collaborative Robots, Big Data, Cloud Computing, Internet of Things, Horizontal and vertical system integration, Simulation and Digital Twin, Virtual, Augmented reality, Additive manufacturing and Cyber-security (Fig. 1).

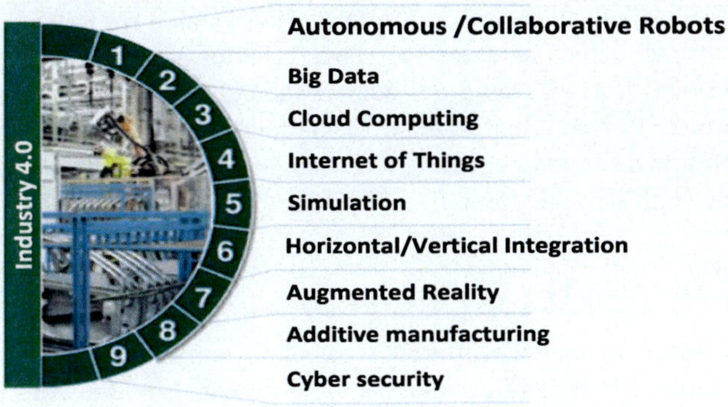

Fig. 1. Nine pillars of Industry 4.0

2.1 Autonomous/Collaborative Robots (A/CR)

Manufacturers in many industries have long used robots to perform complex tasks. However, modern robots are evolving towards even greater usability. They become more autonomous, flexible and are able to communicate and cooperate with each other and work hand in hand with people and learn from them [18]. Autonomous robots have the ability to obtain information about their environment and work for a long time without human intervention. Simpler autonomous robots use infrared or ultrasonic sensors to observe obstacles, more advanced robots use stereo vision to see their surroundings; cameras give them the perception of depth, and the software allows them to locate and classify objects in real time, without interaction with the operator [19].

2.2 Big Data (BD)

The concept of Big Data can be defined as the use of a "mass" amount of data supporting the decision-making process [20]. Most often, this pillar of Industry 4.0 is characterized by capital letters "V". The big 'Vs' have evolved from the classic three - Volume Velocity and Variety - to the two newer additional 'Vs' - Veracity and Value [21].

Analysts of the American research institute Gartner in their annual report "Gartner Top Strategic Technology Trends for 2012" listed Big Data in 7th place among strategic technological trends, indicating that big data is "collections of information about large volume, high variability or great variety that require new forms of processing to support decision making, discover new knowledge and optimize processes" [22].

2.3 Cloud Computing (CC)

According to the American National Institute of Standard and Technology (NITS), cloud computing is a new model for the provision and use of IT resources, such as computing resources, data centers, network bandwidth and applications (software). This model is characterized by high flexibility in access to computing power, tariffication of services (pay-as-you-use) and wide access to the network from virtually any device connected to the Internet, regardless of where you are. The most important components of cloud computing are: applications, platform and infrastructure, which form the basis of three service models: infrastructure as a service (IaaS), Platform as a service (PaaS) and software as a service (Software as a Service - SaaS) [23]. Analyzes of the Gartner Institute in the report "Gartner Top Strategic Technology Trends for 2012" listed Cloud Computing in tenth place among strategic technological trends [22]. A year later, this pillar I4.0 was the sixth strategic trend in the ranking [24].

2.4 Internet of Things (IoT)

IoT describes the ability to network different types of objects, not just computers. These objects can work intelligently and exchange data with other devices with little human involvement. According to [25], "The Internet of Things allows people and things to connect anywhere, anytime, with anything and with anyone, preferably via any path/network and any service" In 2012, the International Telecommunication Union (ITU) defined the Internet of Things as "a global infrastructure for the information society, enabling advanced services by connecting (physical and virtual) things based on existing and developing, interoperable information and communication technologies" [26]. IoT is one of the main sources of big data and the basic technology of cyber-physical systems (CPS) [27]. In 2013, the IoT was listed in fifth place in the "Gartner Top Strategic Technology Trends for 2013" ranking [24].

2.5 Horizontal and Vertical System Integration (HVI)

Horizontal integration is integration between enterprises [28]. As part of vertical integration, enterprises interact and create an interconnected ecosystem within the same value creation network [29]. On the other hand, a vertically integrated enterprise maintains the entire internal supply chain, starting with the concept and design of products, and ending with production, marketing, sales and distribution [30]. Vertical integration in I4.0 enables all areas in the enterprise to be connected: production, maintenance, logistics, etc.

2.6 Simulation and Digital Twin (S and DT)

For the successful implementation of digital production, an indispensable tool is computer simulation, which enables a better understanding of the dynamics of business systems [31], dealing with the complexity of systems and uncertainties that cannot be solved by ordinary mathematical models. Simulation modeling helps reduce costs, shorten development cycles and improve product quality [32]. New possibilities of approaching simulation were opened by technologies supporting Industry 4.0 (BD, CC, IoT). This new simulation approach generally applies to the development of DT [33]. In a production context, DT can be used for monitoring, control, diagnostics and forecasting, not just for simulation. According to [34] a digital twin is about creating a virtual model of a physical object in digital form, to simulate its behavior, monitor its current state, recognize internal and external complexities, detect abnormal patterns, reflect system performance, and predict future trends. Since 2016, more than 7,000 have appeared in the WIPO database. any applications related to this technology. In 2017, analysts of the American research institute Gartner in their annual report "Gartner Top Strategic Technology Trends for 2017" listed the digital twin among strategic technological trends in 5th place [35]. A year later, in 2018, DT technology was already in fourth place [36].

2.7 Virtual and Augmented Reality (VAR)

According to [37] AR as a system that: combines the real world and virtual reality, is interactive in real time and allows freedom of movement in three dimensions. Many publications emphasize the benefits of using AR in enterprises, including: improving work safety, environmental safety, quality management, in maintenance.

Zhang and Kwok [38] developed an AR interface for additive manufacturing, Lai et al. [39] proposed an intelligent instructional system using AR and deep-learning tools to improve employee performance through auxiliary instruction. In the report "Gartner Top Strategic Technology Trends for 2017", virtual and augmented reality were listed fourth among strategic technological trends [35].

2.8 Additive Manufacturing (AM)

One of the most frequently mentioned technologies supporting the implementation of the I4.0 challenges is a group of technologies collectively referred to as "Additive Manufacturing (AM)". As defined by ASTM International [40], AM is "the process of combining materials to create objects from 3D model data, typically layer by layer, as opposed to subtractive manufacturing methodology". Kim et al. [41] define AM as a process of creating a 3D object based on the layer-by-layer or drop-by-drop deposition of materials in a computer-controlled system. AM technologies can also be referred to other synonyms, such as rapid prototyping, layer manufacturing, digital manufacturing or 3D printing [42].

The main advantage of AM technology is the ability to create almost any possible shape. AM is most often used for modeling, prototyping and instrumentation. Thanks to the evolution of rapid prototyping technology, it has become possible to obtain parts representative for mass production in a very short time. Potential benefits of using AM include [43–46]: manufactured parts directly from CAD data files, shortening the design and introduction cycle of new products, production of complex shapes (some geometries cannot be obtained in conventional processes), greater personalization without additional tools and associated costs, maximization use of materials ("zero waste").

2.9 Cyber-Security (CS)

For the dissemination of Industry 4.0 technology in enterprises, it will also be crucial to ensure adequate cyber-security for the data sent there and to ensure safe, digital procedures for accessing devices to authorized users and other devices [47]. Machines become social cyber-physical systems, acting as real-time IoT systems of systems, creating a cyber threat related to the transferred data. CS is a technology whose task is to protect, detect and respond to attacks/threats [48].

The I4.0 pillars described above offer many opportunities for enterprises that can use them in their strategic plans to improve quality, productivity, reduce machine downtime and faster adaptation of products and services to customer requirements.

3 I4.0 Technologies Applications in Maintenance Management

In the recent years, the progress in technology has caused the increase in machines complexity as well as in their playing a more and more crucial role in reliability and availability [49, 50]. As a result, the nature of maintenance management has changed from reactive (Maintenance 1.0) to preventive (Maintenance 2.0) at first, and subsequently to the condition-based maintenance (Maintenance 3.0) and, finally, to the latest Maintenance 4.0 predictive and prescriptive maintenance (Fig. 2).

Fumagalli, et al., [52] with regard to Industry 4.0, implied that the role of maintenance is commonly defined as Maintenance 4.0, Smart Maintenance or e-Maintenance.

As a result of the application of innovative analytical methods in relation to the big data (such as technical condition, exploitation, environment, maintenance background, comparable equipment in different places as well as any type of data that can be associated with the asset performance), Maintenance 4.0 may prevent the occurrence of upcoming failures in assets and, in due course, the most efficient preventing means can be prescribed.

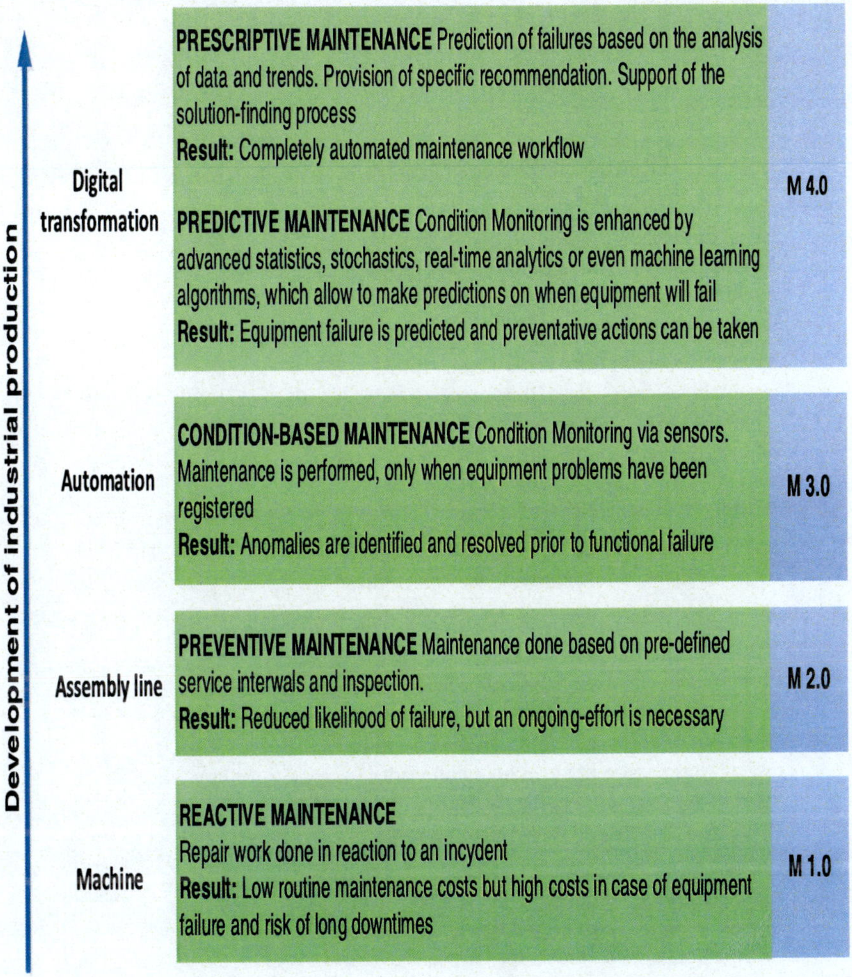

Fig. 2. Development of industrial maintenance [51]

Many authors suggest that the technologies of the I4.0 such as: Internet of Things (IoT), Cloud Computing (CC), Big data analytics (BDA), Digital Twin (DT), Additive Manufacturing (AM), Augmented Reality (AR), Artificial Intelligence (AI), blockchain and simulation are key-drivers in pursuit of maintenance activities to become more value added [33, 53, 54]. In Table 1 the potential applications of I4.0 technologies in maintenance processes are presented.

Literature analysis shows that through the application of I4.0 technologies in maintenance have a various advantages including: increased safety, improved reliability, increased manufacturing equipment availability and reduced life cycle costs [55, 83–86].

Table 1. I4.0 technologies applications in maintenance management

I4.0 T	Role in maintenance	References
IoT	IoT makes it possible to monitor manufacturing equipment and to provide timely information through IoT actuators, sensors and hardware device	[5, 53, 55–59]
CC	CC proposes maintenance services to store and collect the data from the various IoT devices, microcontrollers, sensors, etc	[5, 58–63]
BDA	BDA, uses mathematical methods to obtain information from data in order to optimize maintenance processes and support managers in the decision-making	[5, 59, 63–68]
HVI	VI is necessary in the operational and organizational processes of production and maintenance to measure and monitor the efficiency of each production unit, verify adopted maintenance policies and identify critical areas and elements for improvements	[5, 59, 63]
S&DT	Simulation and DT solutions are designed to optimize inspection and maintenance activities	[5, 32, 59, 63, 70–73]
AM	AM is referred as a group technology, which enables manufacturing complex and personalized products without any specialized tool requirements	[41, 43, 44, 46, 59, 63, 74, 75]
AR	AR offers maintenance staff to interact with the digital world while performing maintenance task and helping in reducing the staff error, enabling a new way of educating maintenance workers	[44–46, 59, 63, 76]
A/CR	Autonomous/collaborative robots can be used in most processes carried out in a smart factory, helping operators and technical services in their tasks and interactions with other cobots. Robots also acquire data from control and maintenance activities, thus enabling factual decision-making	[59, 77–82]
CS	Cybersecurity refers to preventative methods used to protect information from being stolen, compromised or attacked	[12, 59]

4 Conclusions

All industries are currently facing the challenges related to digital transformation. Maintenance management standards have not changed - maintenance managers still need to balance cost and risk while increasing efficiency, making optimal use of resources, avoiding incidents and coping with profit targets and constantly changing legal regulations. What has changed is the need to manage and use the enormous amount of data generated every day. Digitization can support maintenance processes. By the using of the advanced Industry 4.0, technologies, it is possible to monitor the state of the manufacturing equipment, diagnose faults, predict and solve problems before they occur. Thanks to this, the decisions made will be more responsible and will support the implementation of continuous improvement in production processes. The digitization of maintenance requires the mastery of both new technologies and organizational issues related to their planning, implementation and maintenance. From the technological point of view, each of the nine pillars of I4.0 can play an important role in keeping machines and devices in the required technical condition. In addition, the use of modern I4.0 technologies significantly affects the maintenance processes and slowly leads to more intelligent and more responsible (environmental, social and economic requirements of sustainable development) realization of maintenance function in the company.

References

1. Thoben, K.D., Wiesner, S., Wuest, T.: Industrie 4.0 and smart manufacturing–a review of research issues and application examples. Int. J. Autom. Technol. **11**, 4–16 (2017)
2. Gola, A.: Economic aspects of manufacturing systems design. Actual Probl. Econ. **156**(6), 205–212 (2014)
3. Kotowska, J., Markowski, M., Burduk, A.: Optimization of the supply of components for mass production with the use of the ant colony algorithm. In: Burduk, A., Mazurkiewicz, D. (eds.) Intelligent Systems in Production Engineering and Maintenance – ISPEM 2017, vol. 637, pp. 347–357. Springer, Cham (2018). https://doi.org/10.1007/978-3-319-64465-3_34
4. Castelo-Branco, I., Cruz-Jesus, F., Oliveira, T.: Assessing industry 4.0 readiness in manufacturing: evidence for the European union. Comput. Ind. **107**, 22–32 (2019)
5. Bai, C., Dallasega, P., Orzes, G., Sarkis, J.: Industry 4.0 technologies assessment: a sustainability perspective. Int. J. Prod. Econ. **229**, 107776 (2020)
6. Calabrese, A., Dora, M., Levialdi Ghiron, N., Tiburzi, L.: Industry's 4.0 transformation process: how to start, where to aim, what to be aware of. Prod. Plann. Control **33**(5), 492–512 (2020)
7. Reischauer, G.: Industry 4.0 as policy-driven discourse to institutionalize innovation systems in manufacturing. Technol. Forecast. Soc. Change **132**, 26–33 (2018)
8. Arrais-Castro, A., Varela, M.L.R., Putnik, G.D., Ribeiro, R.A., Machado, J., Ferreira, L.: Collaborative framework for virtual organisation synthesis based on a dynamic multi-criteria decision model. Int. J. Comput. Integr. Manuf. **31**(9), 857–868 (2018)
9. Burduk, A., Musiał, K.: Genetic algorithm adoption to transport task optimization. In: Graña, M., López-Guede, J.M., Etxaniz, O., Herrero, Á., Quintián, H., Corchado, E. (eds.) SOCO/CISIS/ICEUTE -2016. AISC, vol. 527, pp. 366–375. Springer, Cham (2017). https://doi.org/10.1007/978-3-319-47364-2_35
10. Strozzi, F., Colicchia, C., Creazza, A., Noe, C.: Literature review on the 'Smart Factory' concept using bibliometric tools. Int. J. Prod. Res. **55**(22), 6572–6591 (2017)

11. Gangala, C., Modi, M., Manupati, V.K., Varela, M.L.R., Machado, J., Trojanowska, J.: Cycle time reduction in deck roller assembly production unit with value stream mapping analysis. In: Rocha, Á., Correia, A.M., Adeli, H., Reis, L.P., Costanzo, S. (eds.) WorldCIST 2017. AISC, vol. 571, pp. 509–518. Springer, Cham (2017). https://doi.org/10.1007/978-3-319-56541-5_52
12. Rodseth, H., Schjolberg, P.: Data-driven predictive maintenance for green manufacturing. In: IWAMA 2016, Manchester, UK, 10–11 November 2016, pp. 36–41 (2016)
13. Mosyurchak, A., Veselkov, V., Turygin, A., Hammer, M.: Prognosis of behaviour of machine tool spindles their diagnostics and maintenance. MM Sci. J. **5**, 2100–2104 (2017)
14. Antosz, K., Jasiulewicz-Kaczmarek, M., Paśko, Ł., Zhang, C., Wang, S.: Application of machine learning and rough set theory in lean maintenance decision support system development. Eksploatacja i Niezawodnosc Maintenance Reliab. **23**(4): 695–708 (2021)
15. Antosz, K., Pasko, L., Gola, A.: The use of artificial intelligence methods to assess the effectiveness of lean maintenance concept implementation in manufacturing enterprises. Appl. Sci. **10**, 7922 (2020)
16. Antosz, K., Pasko, L., Gola, A.: The use of intelligent systems to support the decision-making process in lean maintenance management. IFAC-PapersOnLine **52**(10), 148–153 (2019)
17. Rüßmann, M., et al.: Industry 4.0: the future of productivity and growth in manufacturing industries. Boston Consult. Group **9**(1), 54–89 (2015)
18. Kamble, S.S., Gunasekaran, A., Gawankar, S.A.: Sustainable Industry 4.0 framework: a systematic literature review identifying the current trends and future perspectives. Process Saf. Environ. Protect. **117**, 408–425 (2018)
19. Ben-Ari, M., Mondada, F.: Robots and their application. In: Ben-Ari, M., Mondada, F. (eds.) Elements of Robotics, pp. 1–20. Springer, Cham (2018). https://doi.org/10.1007/978-3-319-62533-1_1
20. Davenport, T.H.: How strategists use "big data" to support internal business decisions, discovery and production. Strategy Leadersh. **42**(4), 45–50 (2014)
21. Wamba, S.F., Gunasekaran, A., Akter, S., Ren, S.J.F., Dubey, R., Childe, S.J.: Big data analytics and firm performance: effects of dynamic capabilities. J. Bus. Res. **70**, 356–365 (2017)
22. Report "Gartner Top Strategic Technology Trends for 2012". https://www.datamation.com/trends/gartners-top-10-strategic-technology-trends-for-2012
23. Kapeliński, W.: Kierunki rozwoju chmury obliczeniowej według założeń komisji europejskiej. In: Nowicki, A., Jelonek, D. (eds.) "Technologie informacyjne w kreowaniu przedsiębiorczości", Sekcja Wydawnictw Wydziału Zarządzania Politechniki Częstochowskiej, Częstochowa (2014)
24. Report "Gartner Top Strategic Technology Trends for 2012". https://www.businesstoday.in/industry/it/story/gartner-top-10-strategic-technology-trends-for-2013-36258-2012-10-24
25. Guillemin, P., Friess, P.: Internet of Things strategic research roadmap. European Commission - Information Society and Media DG (2009)
26. ITU: Overview of the Internet of Things. In: Series Y: Global Information Infrastructure, Internet Protocol Aspects and Next-Generation Networks—Frameworks and Functional Architecture Models (2012)
27. Atlam, H.F., Wills, G.B.: Technical aspects of blockchain and IoT. In: Advances in Computers, vol. 115, pp. 1–39 (2019)
28. Suri, K., Cuccuru, A., Cadavid, J., Gerard, S., Gaaloul, W., Tata, S.: Model-based development of modular complex systems for accomplishing system integration for Industry 4.0. In: MODELSWARD, pp. 487–495 (2017)

29. Salkin, C., Oner, M., Ustundag, A., Cevikcan, E.: A conceptual framework for Industry 4.0. In: Salkin, C., Oner, M., Ustundag, A., Cevikcan, E. (eds.) Industry 4.0: Managing the Digital Transformation, pp. 3–23. Springer, Cham (2018). https://doi.org/10.1007/978-3-319-57870-5_1
30. Brettel, M., Friederichsen, N., Keller, M., Rosenberg, M.: How virtualization, decentralization and network building change the manufacturing landscape: an industry 4.0 perspective. FormaMente **12**, 47–62 (2017)
31. Rodič, B.: Industry 4.0 and the new simulation modelling paradigm. Organizacija **50**(3), 193–207 (2017)
32. Jasiulewicz-Kaczmarek, M., Bartkowiak, T.: Improving the performance of a filling line based on simulation. In: IOP Conference Series: Materials Science and Engineering, vol. 145, no. 4, p. 042024. IOP Publishing (2016)
33. Negri, E., Fumagalli, L., Macchi, M.: A review of the roles of digital twin in CPS-based production systems. Procedia Manuf. **11**, 939–948 (2017)
34. Qi, Q., Tao, F.: Digital twin and big data towards smart manufacturing and industry 4.0: 360 degree comparison. IEEE Access **6**, 3585–3593 (2018)
35. Report "Gartner Top Strategic Technology Trends for 2012". https://www.gartner.com/smarterwithgartner/gartners-top-10-technology-trends-2017
36. Report "Gartner Top Strategic Technology Trends for 2012". https://www.gartner.com/en/newsroom/press-releases/2017-10-04-gartner-identifies-the-top-10-strategic-technology-trends-for-2018
37. Azuma, R.T.: A survey of augmented reality. Presence Teleoperators Virtual Environ. **6**(4), 355–385 (1997)
38. Zhang, Y., Kwok, T.H.: Design and interaction interface using augmented reality for smart manufacturing. Procedia Manuf. **26**, 1278–1286 (2018)
39. Lai, Z.H., Tao, W., Leu, M.C., Yin, Z.: Smart augmented reality instructional system for mechanical assembly towards worker - centered intelligent manufacturing. J. Manuf. Syst. **55**, 69–81 (2020)
40. ASTM F2792, Standard Terminology for Additive Manufacturing Technologies, ASTM (2014)
41. Kim, Y., Cha, M., He, F., Mun, D.: Robust generation of the delta volume for the damaged area of a part using the marching cubes algorithm to support additive manufacturing–based part maintenance. Int. J. Adv. Manuf. Technol. **117**(5–6), 1473–1489 (2021). https://doi.org/10.1007/s00170-021-07828-3
42. Chong, L., Ramakrishna, S., Singh, S.: A review of digital manufacturing-based hybrid additive manufacturing processes. Int. J. Adv. Manuf. Technol. **95**(5–8), 2281–2300 (2017). https://doi.org/10.1007/s00170-017-1345-3
43. Contreras, J.D., Garcia, J.I., Gomez, J.: Maintenance management of an additive manufacturing system based on the I4.0 model. In: Figueroa-García, J.C., Díaz-Gutierrez, Y., Gaona-García, E.E., Orjuela-Cañón, A.D. (eds.) WEA 2021. CCIS, vol. 1431, pp. 299–311. Springer, Cham (2021). https://doi.org/10.1007/978-3-030-86702-7_26
44. Ceruti, A., Marzocca, P., Liverani, A., Cees, B.: Maintenance in aeronautics in an Industry 4.0 context: the role of augmented reality and additive manufacturing. J. Comput. Des. Eng. **6**(4), 516–526 (2019)
45. Siew, C.Y., Ong, S.K., Nee, A.Y.: A practical augmented reality-assisted maintenance system framework for adaptive user support. Robot. Comput. Integr. Manuf. **59**, 115–129 (2019)
46. Knofius, N., van der Heijden, M.C., Zijm, W.H.: Consolidating spare parts for asset maintenance with additive manufacturing. Int. J. Prod. Econ. **208**, 269–280 (2019)
47. Radanliev, P., et al.: Future developments in cyber risk assessment for the Internet of Things. Comput. Ind. **102**, 14–22 (2018)

48. Piedrahita, A.F.M., Gaur, V., Giraldo, J., Cardenas, A.A., Rueda, S.J.: Virtual incident response functions in control systems. Comput. Netw. **135**, 147–159 (2018)
49. Loska, A., Paszkowski, W.: Geometric approach to machine exploitation efficiency: modelling and assessment. Eksploatacja i Niezawodnosc **24**(1), 114–122 (2022)
50. Gola, A., Nieoczym, A.: Application of OEE coefficient for manufacturing lines reliability improvement. In: AEBMR-Advances in Economics Business and Management Research, vol. 31, pp. 189–194 (2017)
51. Jasiulewicz-Kaczmarek, M., Legutko, S., Kluk, P.: Maintenance 4.0 technologies—new opportunities for sustainability driven maintenance. Manag. Prod. Eng. Rev. **11**(2), 74–87 (2020)
52. Fumagalli, L., Macchi, M., Colace, C., Rondi, M., Alfieri, A.: A smart maintenance tool for a safe electric arc furnace. IFAC-PapersOnLine **49**(3), 19–24 (2016)
53. Compare, M., Baraldi, P., Zio, E.: Challenges to IoT-enabled predictive maintenance for industry 4.0. IEEE Internet Things J. **7**(5), 4585–4597 (2019)
54. Zonta, T., da Costa, C.A., da Rosa Righi, R., de Lima, M.J., da Trindade, E.S., Li, G.P.: Predictive maintenance in the Industry 4.0: a systematic literature review. Comput. Ind. Eng. **150**, 106889 (2020)
55. Gbadamosi, A.Q., et al.: IoT for predictive assets monitoring and maintenance: an implementation strategy for the UK rail industry. Autom. Constr. **122**, 103486 (2021)
56. Calabrese, M., et al.: SOPHIA: an event-based IoT and machine learning architecture for predictive maintenance in industry 4.0. Information **11**(4), 202 (2020)
57. Ayvaz, S., Alpay, K.: Predictive maintenance system for production lines in manufacturing: a machine learning approach using IoT data in real-time. Expert Syst. Appl. **173**, 114598 (2021)
58. Daji, D., Ghule, K., Gagdani, S., Butala, A., Talele, P., Kamat, H.: Cloud-based asset monitoring and predictive maintenance in an industrial IoT system. In: 2020 International Conference for Emerging Technology (INCET), pp. 1–5. IEEE (2020)
59. Ansari, F., Glawar, R., Nemeth, T.: PriMa: a prescriptive maintenance model for cyber-physical production systems. Int. J. Comput. Integr. Manuf. **32**(4–5), 482–503 (2019)
60. Bouzidi, Z., Zerhouni, N.: A new approach of PHM as a service in cloud computing. In: 2016 4th IEEE International Colloquium on Information Science and Technology (CiSt), pp. 610–614. IEEE, October 2016
61. Cachada, A., et al.: Maintenance 4.0: intelligent and predictive maintenance system architecture. In: 2018 IEEE 23rd International Conference on Emerging Technologies and Factory Automation (ETFA), vol. 1, pp. 139–146. IEEE, September 2018
62. Peng, Z.: An operation and maintenance strategy of intelligent building information model data based on cloud computing. Ingénierie des Systèmes d'Information **25**(4), 461–467 (2020)
63. Paschou, T., Rapaccini, M., Adrodegari, F., Saccani, N.: Digital servitization in manufacturing: a systematic literature review and research agenda. Ind. Mark. Manag. **89**, 278–292 (2020)
64. Razali, M.N., Jamaluddin, A.F., Jalil, R.A., Nguyen, T.K.: Big data analytics for predictive maintenance in maintenance management. Prop. Manag. **38**(4), 513–529 (2020)
65. Sahal, R., Breslin, J.G., Ali, M.I.: Big data and stream processing platforms for Industry 4.0 requirements mapping for a predictive maintenance use case. J. Manuf. Syst. **54**, 138–151 (2020)
66. Kumar, A., Shankar, R., Thakur, L.S.: A big data driven sustainable manufacturing framework for condition-based maintenance prediction. J. Comput. Sci. **27**, 428–439 (2018)
67. Yu, W., Dillon, T., Mostafa, F., Rahayu, W., Liu, Y.: A global manufacturing big data ecosystem for fault detection in predictive maintenance. IEEE Trans. Ind. Inf. **16**(1), 183–192 (2019)

68. Li, Z., Hao, J., Li, X., Li, K.: Research on integrated data engineering of equipment maintenance support in the big data era. In: 2021 IEEE 4th Advanced Information Management, Communicates, Electronic and Automation Control Conference (IMCEC), vol. 4, pp. 447–451. IEEE (2021)
69. Ansari, F., Glawar, R., Sihn, W.: Prescriptive maintenance of CPPS by integrating multimodal data with dynamic Bayesian networks. In: Beyerer, J., Maier, A., Niggemann, O. (eds.) Machine Learning for Cyber Physical Systems. Technologien für die intelligente Automation, vol. 11, pp. 1–8. Springer, Heidelberg (2020). https://doi.org/10.1007/978-3-662-59084-3_1
70. Errandonea, I., Beltrán, S., Arrizabalaga, S.: Digital Twin for maintenance: a literature review. Comput. Ind. **123**, 103316 (2020)
71. Lu, Q., Xie, X., Parlikad, A.K., Schooling, J.M.: Digital twin-enabled anomaly detection for built asset monitoring in operation and maintenance. Autom. Constr. **118**, 103277 (2020)
72. Aivaliotis, P., Georgoulias, K., Chryssolouris, G.: The use of Digital Twin for predictive maintenance in manufacturing. Int. J. Comput. Integr. Manuf. **32**(11), 1067–1080 (2019)
73. Tao, F., Zhang, M., Liu, Y., Nee, A.Y.: Digital twin driven prognostics and health management for complex equipment. CIRP Ann. **67**(1), 169–172 (2018)
74. Cardeal, G., Höse, K., Ribeiro, I., Götze, U.: Sustainable business models-canvas for sustainability, evaluation method, and their application to additive manufacturing in aircraft maintenance. Sustainability **12**(21), 9130 (2020)
75. Navas, M.A., Sancho, C., Carpio, J.: Disruptive maintenance engineering 4.0. Int. J. Qual. Reliab. Manag. **37**(6/7), 853–871 (2020)
76. Espíndola, D.B., Fumagalli, L., Garetti, M., Pereira, C.E., Botelho, S.S.C., Henriques, R.V.: A model-based approach for data integration to improve maintenance management by mixed reality. Comput. Ind. **64**(4), 376–391 (2013)
77. Pei, F., Tong, Y., He, F., Li, D.: Research on design of the smart factory for forging enterprise in the industry 4.0 environment. Mechanics **23**(1), 146–152 (2017)
78. Pedersen, M.R., et al.: Robot skills for manufacturing: from concept to industrial deployment. Robot. Comput.-Integr. Manuf. **37**, 282–291 (2016)
79. Dogar, M., Spielberg, A., Baker, S., Rus, D.: Multi-robot grasp planning for sequential assembly operations. Auton. Robot. **43**(3), 649–664 (2018). https://doi.org/10.1007/s10514-018-9748-z
80. Müller, J.M., Kiel, D., Voigt, K.I.: What drives the implementation of Industry 4.0? The role of opportunities and challenges in the context of sustainability. Sustainability **10**(1), 247 (2018)
81. El Makrini, I., et al.: Working with walt: how a cobot was developed and inserted on an auto assembly line. IEEE Robot. Autom. Mag. **25**(2), 51–58 (2018)
82. Koch, P.J., et al.: A skill-based robot co-worker for industrial maintenance tasks. Procedia Manuf. **11**, 83–90 (2017)
83. Turner, C.J., Emmanouilidis, C., Tomiyama, T., Tiwari, A., Roy, R.: Intelligent decision support for maintenance: an overview and future trends. Int. J. Comput. Integr. Manuf. **32**(10), 936–959 (2019)
84. Franciosi, C., Di Pasquale, V., Iannone, R., Miranda, S.: Multi-stakeholder perspectives on indicators for sustainable maintenance performance in production contexts: an exploratory study. J. Qual. Maint. Eng. **27**(2), 308–330 (2020)
85. Tortorella, G.L., Fogliatto, F.S., Cauchick-Miguel, P.A., Kurnia, S., Jurburg, D.: Integration of industry 4.0 technologies into total productive maintenance practices. Int. J. Prod. Econ. **240**, 108224 (2021)
86. Sousa, R.A., Varela, M.L.R., Alves, C., Machado, J.: Job shop schedules analysis in the context of industry 4.0. In: International Conference on Engineering, Technology and Innovation: Engineering, Technology and Innovation Management Beyond 2020: New Challenges, New Approaches, ICE/ITMC 2017 - Proceedings, January 2018, pp. 711–717 (2018)

Design Concept of a Non-invasive Tagging Device for Blue Sharks

José Azevedo[1], Tiago Bartolomeu[2], Senhorinha Teixeira[3], and José Teixeira[4](✉)

[1] Department of Mechanical Engineering, University of Minho, 4800-058 Guimarães, Portugal
[2] CEiiA, Av. D. Afonso Henriques, 1825, 4450-017 Matosinhos, Portugal
[3] ALGORITMI Research Centre, University of Minho, 4800-058 Guimarães, Portugal
st@dps.uminho.pt
[4] MEtRICs Research Centre, University of Minho, 4800-058 Guimarães, Portugal

Abstract. This work describes the design concept process taken to develop a non-intrusive tagging device for blue sharks. The identification of the more important aspects involved in its accomplishment gave origin to the establishment of the design requirements, which served as guidelines for the design and validation of the generated concepts. By excluding any procedure that causes any, permanent or temporary, lesion to the tagged animal and limiting the resistance offered by the tag to the shark's motion assures minimal influence on the gathered information and, at the same time, high quality and relevance of the data from every mission. The design requirements for a non-intrusive tag were established as well as the equipment selection process after market research. Finally, the conceptualization of a shell for the equipment was developed. An optimization process can now be implemented to find the final product.

Keywords: Blue sharks · Product design · Tags

1 Introduction

The hydrosphere represents more than seventy percent of the Earth's surface. However, very little is still known to man about what occurs below its surface. The potential around this ecosystem is high and a wide variety of industries are growing around it.

To ensure that this process occurs smoothly and sustainably it is important to have extensive background knowledge concerning ocean activity, in particular, the existing life within this environment.

Tagging sharks presents a cheap, fast, and sustainable opportunity for grasping an insight into these apex predators' routines. This knowledge may itself be used to infer about lower chain level zoological species, as well as their surrounding environmental conditions.

Blue sharks are ideal primary study targets, mainly to understand the rapid decline noticed in the last years, but also due to their broad presence across all the oceans and their wide range of motion, which promotes an increased coverage of each tagging

J. Machado et al. (Eds.): icieng 2022, LNME, pp. 80–90, 2023.
https://doi.org/10.1007/978-3-031-09382-1_8

mission. Blue sharks are one of the most widely spread species of sharks and also one of the most fished (Queiroz et al 2012). Although there are many unknowns about their existence, distribution, and habits, they are considered as "near threatened" (Rigby et al. 2019) on the extinction scale.

Blue sharks can be identified by their long and fusiform bodies with long pectoral fins (Szpilman 2004). Its name relates to its color: a gradient from deep indigo-blue on its dorsal region up to a white shade on its pectoral area. Its size may vary among individuals and between sexes. Despite this fact, an average fork length will be between 166 and 221 cm for males and 183 and 218 cm for females (Colmenero et al. 2017). A long-tail fin provides the power for its movement, which is essential to the sharks' respiratory system.

Another important sensorial mechanism is called the Ampullae of Lorenzini. Sharks use this mechanism to sense electric fields. They get highly stimulated and even disorientated by sources of magnetic fields. Blue sharks not only face the risk of fishing activities but are also prey for bigger shark species and killer whales. Hence, their furtiveness and speed represent a critical element for survival.

In the past, monitoring wildlife was an extremely difficult, if not impossible, task. In the beginning, it was carried out by mere visual observation. Taking into consideration the high mobility range of some animals, the tendency of some to predominantly perform nocturnal activities, and the dangerous habitats they live in, one can foresee huge obstacles for the execution of this monitoring. Placing traditional tags on sea animals opened a new insight into the understanding of underwater biology. Presently, with remote tracking methods, it is possible not only to verify where animals were tagged and found but also to pinpoint the intermediate locations where they passed (Arai and Okuyama 2012).

Many different types of tags are used on sea animals. Their applications are many and the number of possible configurations is vast. This happens mainly because these devices are almost invariably configured by their users to better suit their needs. However, it is possible to classify tags by the philosophy of gathering and transmitting data they employ. Transmission tags are systems of telemetry which emit ultrasound and make use of receiver boats or stations (on the coast or submersed) to monitor passing animals (Arai and Okuyama 2012). They can track fine-scale movements (Grusha and Patterson 2005) in real-time, but are usually labor-intensive methods. They are highly used by fishing crews and in areas of high marine interest (Arai and Okuyama 2012). At archival tags, the information is stored in the tag by the sensors to be later recovered, with the retrieval of the tag. They are mainly used for fish that seldom surfaces and for longer deployments (Grusha and Patterson 2005). The retrieval of this type of tag highly depends on fishery practices. They have low precision determining position, hence are only used to follow ocean scale movements (Grusha and Patterson 2005). Hybrid tags store information for pre-determined periods. After this period, it detaches from the animal and floats to the surface. Once at the surface, it emits a resume of the mission via satellite. When and, if recovered, more detailed information may be collected (Grusha and Patterson 2005). The technological advancements of telemetry tags, as well as the progress in microelectronics during the last few decades, are progressively allowing for the miniaturization of devices and improved power efficiency, along with the incorporation of more sophisticated sensors, such as Fastloc GPS technology, accelerometers and camera systems (McIntyre 2014; Pavlov and Rashad 2012; Zhang et al. 2019).

All the different categories of tags will contemplate different combinations of sensors according to the monitoring intentions. Examples of these sensors include accelerometers, magnetometers, gyroscopes, pressure sensors, temperature sensors, and video cameras (Baeyaert et al. 2016; Witt et al. 2016). Each sensor will capture a different piece of information which may be used directly or to infer about other parameters.

Most of these parameters present low associated errors. As such, the results are meaningful in the context of the studies. The same may not always be true regarding geolocation. This is the most relevant of the parameters to monitor (Arai and Okuyama 2012) and it is usually the one of lowest precision, especially when monitored remotely. However, the emergence of new technologies brought a wider range of options. The most representative of these are Ambient light; Argos Satellite System; Global Positioning System (GPS) and fast GPS (Witt et al. 2016).

Concerning tagging sharks, it is important to discuss how the tag will stay connected to the animal for the intended period of operation. A certain trade-off will be necessary between the degree of intrusion and the maximum duration of the research. Reviews on shark tagging (Hammerschlag et al. 2011) point that intrusive methods are the most frequently used. This happens mainly to assure long periods of attachment or to overcome the difficulties of capturing and restraining these large predators (Gleiss et al. 2009). Non-intrusive methods are also being used, although on a smaller scale. Research is being conducted, with some success (Chapple et al. 2015), with the purpose of optimizing such attachment methods in different species of animals.

In order to assure the data will be representative of the animal's normal behavior and can be extrapolated for the whole species, it is crucial to assure that the tag will not disturb its carrier (Hazekamp et al. 2010). For this, biologists main effort has been the development of ever more hydrodynamic tags which are adapted for their application (Gleiss et al. 2009). Certain details, such as color, shape, tethers, and energy management, have been highlighted by different authors and present important aspects when designing a tag for marine applications.

Although tagging animals allows scientists to understand the behavior of sharks, this process is still related to significant costs and setbacks. Problems associated with this practice are still common. Even though efforts have been, and are still being made to minimize these limitations, it is still not possible to disregard them.

Most studies report two different groups of problems when using tags. On one hand, technical problems associated with the operability of tags must be considered. On the other hand, presenting a much more significant risk is the impact such equipment has on the physiology and behavior of the tagged sharks. In a conclusion, even though shark tagging is already a common practice among biologists, due to its degree of intrusion and the associated increase in the animal's energy consumption, there are several negative consequences arising from this practice which may severely limit the validity of the collected data. This paper describes the design and optimization process of a non-intrusive tagging device for blue sharks, leading to a hydrostatic stable design.

2 Design Requirements and Equipment Selection

2.1 Product's Requirements Identification

Having established a foundation of knowledge regarding the main topics related to shark tagging activities it is important to organize it in a favorable way for the understanding of the problem and to enlighten the conceptualization of possible solutions.

For engineering purposes, this task is of high importance. It presents an opportunity to outline the main design requirements as well as to evaluate the interferences and synergies of the intervening requisites.

Given the intricacies of this project and the number of parameters to consider it seemed wise to invest some effort in cautious planning. For this reason, it was decided to use a tool originally developed from the Quality Function Deployment (QFD) technique: House of Quality.

This tool is, in its essence, a matrix that allows the designer to adapt the product to the clients' needs. To achieve this matrix a required number of ordered steps, designed to lead to those which are the most important design requirements. A detailed description of this tool may be found in Ullman (2010).

To deliver a satisfying product it is crucial, in a preliminary phase, to identify those who are the "target clients". Subsequently, their idealized requirements ought to be gathered and organized. For the present project, the clients identified were, on one hand, the scientific community, which has an interest in the data collected by the tag, and, on the other hand, the sharks which will carry the tag while in usage.

As for their requirements, they were organized by the operation stage: attachment, usage, and recovery. In the ensuing phase, the relative importance between each of the requirements and each of the target clients should be established. Figure 1 summarizes the outcome of these processes.

Probably, the most relevant and sensitive stage requires the designer to compose a list of measurable parameters or engineering requirements. These specifications serve as a base for the design targets and should be accompanied by the units they are measured in, as well as their sense of improvement. Figure 2 exemplifies the process for two of the requirements.

Having identified and developed the applicable requirements, their degree of correlation with the clients' requirements was evaluated and presented in the main area of the house of quality. The subsequent stage involves the computation of the absolute and relative importance of each of the engineering specifications on the overall project, which output has the most relevant requirements for the project.

Figure 3 displays the outcome of this process showing the 10 most important requirements.

		Scientific Interest	Sea Life Preservation
Attachment	Easy to attach	3	3
	Safe to attach	4	2
	Can be employed to different animals	2	1
Usage	Determines precisely shark's horizintal and vertical geolocation in time	5	1
	Operate remotely	3	1
	Allow real.time tracking	1	1
	Relates shark's location with its activity	4	1
	Gives a perspective of the shark's view	4	1
	Operates for the require period	3	1
	Having the tag does not require significant extra energy	3	5
	The tag does not affect swiming motion	2	4
	The tag will remain free of fouling and paresites	1	3
	Tag will remain undetectd to prey and predators	1	5
	Operative time can be optimized	3	1
Recovery	Easy to collect	4	2

Fig. 1. Product's interested parties and respective requirements.

User Requirements	Engineering Requirements	Units
Easy to attach	Time spent on the attachment process	s
	Tools required to the attachment	-
	Steps in order to attach	-
Determines precisely sharks' horizontal and vertical geolocation time	Latitude accuracy of the positioning system	m
	Longitude accuracy of the positioning system	m
	Depth accuracy of the positioning system	m
	Percentage of accepted data points of the positioning system	%

Fig. 2. Generation of engineering requirements.

2.2 Equipment Selection

The list of requirements obtained through the house of quality serves as a starting point for the entire design process. As such, also the equipment to include ought to be chosen based on it.

For this selection, a variety of components and technologies, commonly used or suggested by the scientific community, were grouped accordingly to their purpose, namely attachment type, positioning technology, supplementary sensors, and other features.

The influence of each piece of equipment considered in complying with every requirement was evaluated and the overall impact of its implementation was estimated. This process is summarized in Fig. 4.

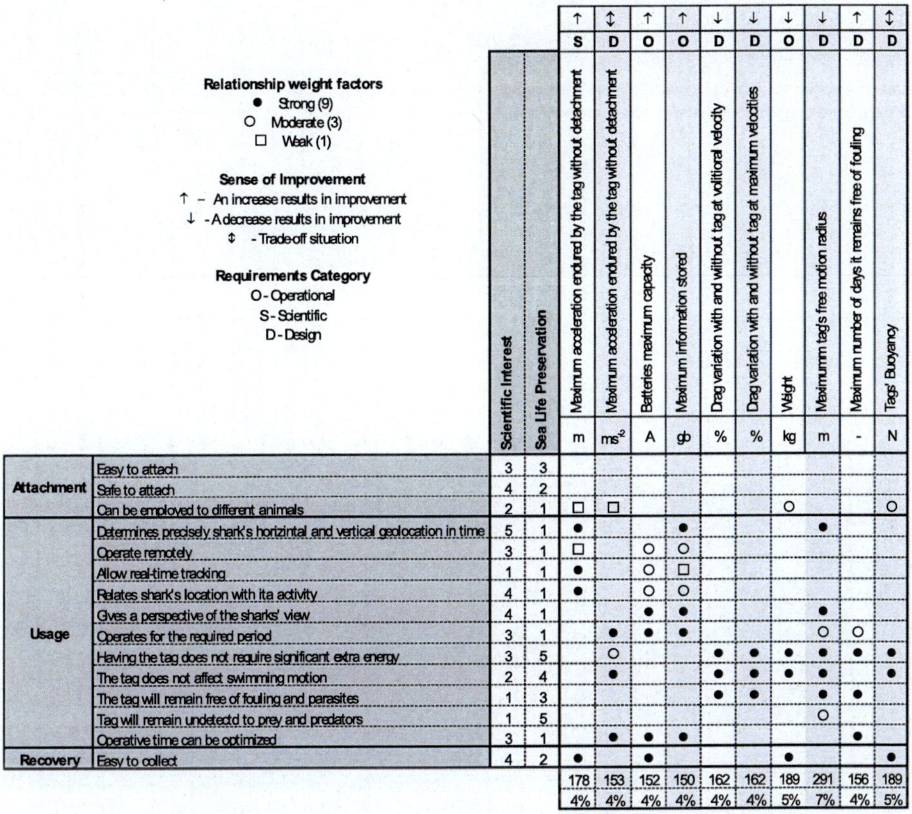

Fig. 3. The most relevant requirements for the project.

Through this analysis, it was possible to identify the loop attachment method as the most reliable in terms of deployment time and easiness of attachment. Despite the higher drag, when compared with the other methods, the expected increase in position locks increases the overall value of each deployment, since the tether allows the tag to break water even when the shark is at a small depth.

Relatively to geolocation, GPS Fastloc technology leads the comparison by coupling its high precision, to the high percentage of valid gathered positions and the convenience of its passive use. Nonetheless, the need to include supplementary geolocation systems was verified, aiming to track the animal while underwater and fine-tracking the tag in the recovery stage. As such, acoustic and Very High Frequency (VHF) transmitters were considered important elements for this system.

Regarding other sensors, both pressure and temperature gathering equipment were considered valuable assets not only in complementing the other data but also to use as filter and validation criteria for obtained positions by comparison with meteorological data.

Finally, despite the null value for the video camera, its added scientific interest would be significant. For this reason, it was decided to invest time in searching for alternatives

		Maximum tads' free motion radius	Weight	Tags' buoyancy	Precision of passive acquired horizontal position	Drag variation with and without tag at volitional velocity	Drag variation with and without tag at maximum velocities	Maximum number of days it remains free of fouling	Maximum acceleration endured by the tag without detachment	Batteries maximum capacity	Maximum information stored	Regenerative capacity	Time elapsed locking one position point	Tether's length	Time spent on the attachment process	Signil emitting radius at surface	Depth accuracy of the positioning system up to 1,000 m	Maximum number of connected receivers to accept data	Active tracking system's location precision	Tags' colour wave length in visible spectrum	Angle of sight recorded	Frames per second recorded	Percentage of accepetd locations of the positioning system among all transmitions	Σ
Attachment Types	Fin clamp	2		Tradeoff requirement					-1				-1	Tradeoff requirement	-1		1			Tradeoff requirement				**0**
	Suction cups	2							-2				-1		-2		1							**-2**
	Body loop	-1					-1	-1	2				-1		2		-1							**-1**
Positioning	Light level interference									-1			-2										-2	**-5**
	Argos		-1		1					-1	-1		1			1		1					1	**2**
	GPS		-1		2					-2	-1		2					2					2	**4**
	Acoustic									-1			-1			2	1	-1	2				1	**3**
Sensors	Accelerometers				1					-1													1	**1**
	Magnetometers									-1														**-1**
	Gyroscopes				1					-1														**0**
	Pressure				1					-1							2						2	**4**
	Termometer				1					-1													2	**2**
Others	Anti-fouling paint					1	1	2																**4**
	Video camera		-2							-1	-1										2	2		**0**
	Solar panels									1		2				1								**4**
	Micro turbine					-1	-1			1		2												**1**

Fig. 4. Equipment importance matrix.

that could soften the negative impact it has on weight and drag. On the other hand, energy generative equipment, such as solar panels, would prove useful for long deployments. However, as this system is intended to operate for a few weeks, the power supply was not a concern for this specific application.

Manufacturers of this equipment were identified, and their products analyzed and compared, yielding the selection of the ensuing devices. Concerning the GPS transmitter, a market analysis was conducted through the three companies that currently own commercial rights of Fastloc® technology. Wildlife Computers™ ("Wildlife Computers," n.d.) provided the best solution through their SPLASH 10-P324. Coupling an antenna interface between the tag and the GPS satellites with embedded pressure and temperature sensors to lightweight and 2,000 m depth rated equipment was simply the best offer in the market.

Several companies manufacture acoustic and radio simple transmitters. However, the research was limited to the brands already in use by national operators of such devices, which already own several compatible receptors and, as such, do not require new investment. Among the considered options, the final choice fell on VEMCO V13P-1H ("VEMCO," n.d.) for acoustic and ATS F1835B+ ("ATS," n.d.) for radio signals.

2.3 Materials Selection

Although the majority of the equipment is obtained off-the-shelf, two components ought to be manufactured in-house: a) the tag's body and b) the ballast weights used to balance its stability. As such, the material selection process was necessary for both.

As far as the weight is concerned, the choice was simple. A very dense material was needed so that with a minimum volume of material strategically placed, one obtains a strong impact on mass distribution. It was also required a relatively inexpensive material and easy to manufacture. For these reasons, the most evident choice was lead.

Conversely, the chosen material for the tag must allow the use of a significant amount of volume to provide buoyancy. For this purpose, its density must be smaller than that of the water. Also, this is the outermost component of the tag, providing not only protection for the remaining equipment of the tag, but also plays a major contribution to the flow development around it. Therefore, it must also resist deformation at the high pressures observed at higher depths.

For these reasons, the most commonly used material in similar applications is Syntactic foam. As such, this was also the choice, more specifically, BMTI®'s VLD2000 ("BMTI," n.d.), which is rated for a maximum depth of 2,000 m and has a density of 400 $kg.m^{-3}$.

3 Conceptual Design

After an extensive description of what is required from the tag was given, the focus shall be on making use of this knowledge as a basis for the creation of concept designs with the potential to meet these requirements.

3.1 Approach for Concept Generation

Knowing the equipment to include in the assembly, the time arises to conceptualize the actual platform to carry them. The main objective in this process is to create a shell around the components that not only assembles them, but also works as an active element to meet the desired requirements of buoyancy, equipment orientation, and minimum drag.

The approach designed to obtain the concepts will follow three design stages and one optimization phase, which, together constitute an iterative process of concept creation and refinement. The first design stage comprehends the layout of the selected equipment, respecting its requirements. Also in this stage, the initial concept for the hydrodynamic shell morphology is idealized. The second design stage consists of the assessment of the product's submerged hydrostatics characteristics, detailing the required adjustments to improve any deviation from the desired outcome. The last design stage, much like the previous, is an assessment of the tag's hydrostatic stability, when surfaced. Once this stage is cleared, the concept is considered hydrostatically validated. The final stage, the optimization phase consists of the appraisal of the hydrodynamic behavior of the tag. This paper addresses the first stage of the design process.

3.2 Concept Creation

The first step of the conceptual design is the actual creation of concepts, defined to the extent in which they allow for the study of the physical principles and mechanisms dictating their behavior once materialized.

For this work's analysis, the concepts require the definition of mass distribution, operative requirements, and external geometry. To answer these needs, this process is subdivided into two processes: the equipment layout process and the initial morphology definition process.

Designing a hydrodynamic shell around the actual monitoring equipment demands careful thought on where to place them, optimizing mass distribution, and respecting their operational requirements.

Generally, for marine devices, the mass should be placed below the centroid of the geometry, providing stability to the body. However, and this is one of the biggest intricacies of this design, the component with the highest mass, the Splash-10 F324D, needs to be located on the upper part of the tag, to allow communication with the GPS constellation. Moreover, its wet-dry switches must be in direct contact with the water, which is also the case for the central portion of the acoustic transmitter. This may be assured by either exposing these parts directly to the flow, compromising motion efficiency, or allocating space around these areas for water ducts that will limit the freedom for more compact designs.

Another layout constrain was the VHF transmitter. This device's antenna needs to be exposed, vertically, for maximum signal strength. This results in a massive exposed area that highly decreases the dynamic efficiency of the concept.

Several were the challenges this product prompted such as the aforementioned. Ultimately, all the iterations through this layout definition stage lead to the layout shown in Fig. 5.

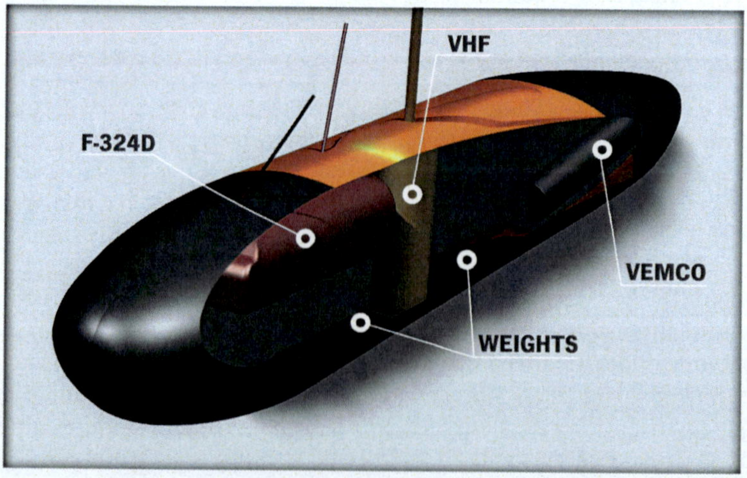

Fig. 5. Components layout for the validated concept.

For the selected shape hydrostatic or hydrodynamic tests were carried out for assessment of its effectiveness in terms of equilibrium and drag. Based on these results other shapes will be created based on strict requirements, such as maximum length and fineness ratio, dictated by the good practices of underwater device's design, the enclosure of all these components respecting their requirements precludes the use of merely scaled versions of these optimized designs. As such, many of the resulting forms were inspired by other existing, either man-made or natural, streamlined shapes.

4 Conclusions

Addressing this work, an initial stage comprehended the analysis of the current state of the art on topics related to the design of shark tagging devices. It was verified that blue sharks are an ideal species to monitor due to their vast ocean coverage. Also, their physical robustness, assures tolerance to the required additional effort of carrying such devices. However, it was concluded that intrusive attachment methods promote physical and behavioral consequences that, ultimately, may lead to the animal's death. This reinforces the importance of implementing a non-intrusive attachment method.

The implementation of a product development plan, supported by the previous research, made possible the definition of engineering requirements for the tag. These, once compared against each other, allowed the identification of those most relevant. Results of this process suggested that great importance should be addressed to the precision and amount of geolocation data collected, as well as to the added resistance offered to the shark's motion by carrying the tag.

This guidance, equipment selection process, and subsequent market research lead to the implementation of three main equipment: a fast lock GPS receiver (Splash 10-P324), an acoustic pinger (VEMCO V13P-1H), and a VHF radio transmitter (ATS F1835B+). This configuration assures both surfaced and submerged refined tracking, as well as high recovery capacity.

In conclusion, the approach followed in this work will lead to the conceptualization of an innovative and non-intrusive tagging device, optimized for high-quality data collection and low impact on the tagged animal. The hydrodynamic study of this concept is necessary to understand the tag's dynamic behavior. Also, the motion resistance at the surface should be studied.

Finally, in what concerns the optimization of the tag as a product, importance should be given to the development, or at least customization, of in-house equipment. This would allow an exponential improvement in the allocation of mass and volumes and, ultimately, open a path for the inclusion of other equipment, such as video cameras. The latter shows enormous potential to significantly enhance the outcome of every mission.

References

Arai, N., Okuyama, J.: Biologging. In: Ishida, T. (ed.) Field Informatics, pp. 21–38. Springer, Heidelberg (2012). https://doi.org/10.1007/978-3-642-29006-0_2

Baeyaert, J., Graça, G., Sobral, A., Prieto, R., Afonso, P., Fontes, J.: EcoDiveAz: summer season technical report (2016)

Chapple, T., Gleiss, A., Jewell, O., Wikelski, M., Block, B.: Tracking sharks without teeth: a non-invasive rigid tag attachment for large predatory sharks. Anim. Biotelemetry **3**(1) (2015). Article number: 14. https://doi.org/10.1186/s40317-015-0044-9

Colmenero, A., Barría, C., Broglio, E., García-Barcelona, S.: Plastic debris straps on threatened blue shark Prionace glauca. Mar. Pollut. Bull. **115**(1–2), 436–438 (2017)

Gleiss, A., Norman, B., Liebsch, N., Francis, C., Wilson, R.: A new prospect for tagging large free-swimming sharks with motion-sensitive data-loggers. Fish. Res. **97**, 1–2 (2009)

Grusha, D., Patterson, M.: Quantification of drag and lift imposed by pop-up satellite archival tags and estimation of the metabolic cost to cownose rays (Rhinoptera bonasus). Fish. Bull. **103**(1), 63–70 (2005)

Hammerschlag, N., Gallagher, A., Lazarre, D.: A review of shark satellite tagging studies. J. Exp. Mar. Biol. Ecol. **398**(1–2), 1–8 (2011)

Hazekamp, A.A.H., Mayer, R., Osinga, N.: Flow simulation along a seal: the impact of an external device. Eur. J. Wildl. Res. **56**(2), 131–140 (2010). https://doi.org/10.1007/s10344-009-0293-0

McIntyre, T.: Trends in tagging of marine mammals: a review of marine mammal biologging studies. Afr. J. Mar. Sci. **36**(4), 409–422 (2014)

Pavlov, V., Rashad, A.: A non-invasive dolphin telemetry tag: computer design and numerical flow simulation. Mar. Mammal Sci. **28**(1), E16–E27 (2012)

Queiroz, N., Humphries, N., Noble, L., Santos, A., Sims, D.: Spatial dynamics and expanded vertical niche of blue sharks in oceanographic fronts reveal habitat targets for conservation. PLoS One **7**(2), e32374 (2012)

Rigby, C., et al.: Prionace glauca, blue shark. The IUCN Red List of Threatened Species (2019). https://www.iucnredlist.org/species/39381/2915850. Accessed 26 Jan 2022

Szpilman, M.: Tubarões no Brasil: guia prático de identificação. Maual Editora Ltda (2004)

Ullman, D.: The Mechanical Design Process, 4th edn. McGraw-Hill (2010)

Witt, M., Doherty, P., Godley, B., Graham, R., Hawkes, L., Henderson, S.: Basking shark satellite tagging project: insights into basking shark (Cetorhinus maximus) movement, distribution, and behavior using satellite telemetry. Final report. Scottish Natural Heritage Commissioned Report (908), 72 (2016)

Zhang, D., van der Hoop, J., Petrov, V., Rocho-Levine, J., Moore, M., Shorter, K.: Simulated and experimental estimates of hydrodynamic drag from bio-logging tags. Mar. Mammal Sci. **36**, 136–157 (2019)

Aspects Regarding Subtractive and Additive CAD-CAM Processes Applied in Prosthetic Dentistry

Edgar Moraru[1], Grigore Octavian Dontu[1], Tudor Catalin Apostolescu[2], Laurentiu Adrian Cartal[1(✉)], Bogdan Grămescu[1], and Nazem Dawod[1]

[1] Politehnica University of Bucharest, 313, Splaiul Independentei, 060042 Bucharest, Romania
adrian.cartal@upb.ro

[2] "Titu Maiorescu" University, Calea Vacaresti 189, Bucharest, Romania

Abstract. The paper presents some technical features regarding the realization of dental prostheses and models by modern CAD-CAM technologies that operate on both the subtractive principle and the additive principle highlighting the advantages and disadvantages for each manufacturing approach compared to traditional methods applied in dental prosthetics. It also describes the processing steps for making dental prostheses by CAD-CAM methods, the equipment used, as well as the characteristics of the resulting restorations.

Keywords: Additive manufacturing · Dental prostheses · CAD-CAM dental milling

1 Introduction

Teeth are anatomical structures with a vital role for the proper functioning of the organism from several points of view, both digestively and psychologically, contributing decisively to a person's overall health. Throughout life, everyone may experience various forms of dental disease that can lead to tooth loss and the need to install dental prostheses, including carious pathological processes, periodontal disease, pathological abrasion, or various dental abnormalities. In addition to the above-mentioned dental pathological conditions that lead to the need for dental prosthetics, injuries of a mechanical nature due to trauma or the aesthetic improvement of existing teeth are other possible indications for the installation of dental prostheses and for restoration of dental and body health [1–4].

The variety of dental prosthetic structures has led the scientific world to classify these orthopedic restorations from several perspectives: in terms of the degree of mobility, depending on the material from which the prosthesis is made, depending on the mode of aggregation with the patient's dental tissues or depending on the aesthetic appearance (Fig. 1 and Fig. 2) [1, 5–7]. However, probably the most important classification criterion is the processing technology for obtaining dental prostheses. Nowadays, there are several classic and digital technological solutions for dental tissues substitution, each of which

J. Machado et al. (Eds.): icieng 2022, LNME, pp. 91–102, 2023.
https://doi.org/10.1007/978-3-031-09382-1_9

having its advantages and drawbacks, and the success over time of dental prosthetic restorations from a technical-medical point of view depends on several factors, such as: mechanical characteristics, degree of biocompatibility or precision of execution of dental prosthesis. Computerized CAD-CAM technologies have become the number one technology for the realizing of dental prosthetic restorations or dental models, due to its outstanding performance and precision, using both the subtractive approach - CAD-CAM milling [8] and the additive approach - selective laser melting SLM (Fig. 2) [1], which lately is becoming more and more attractive [9, 10].

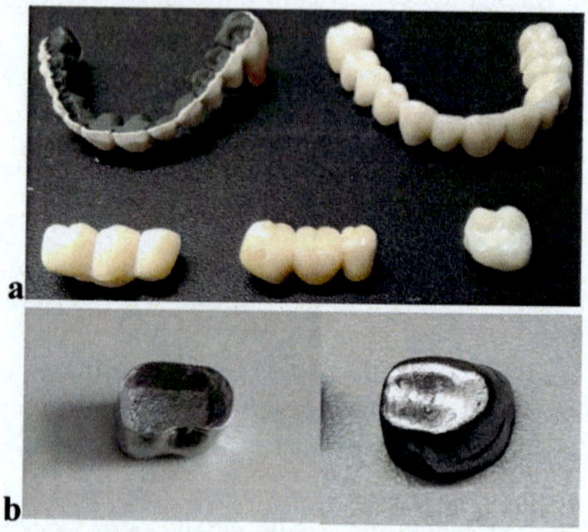

Fig. 1. Various types of CAD-CAM obtained dental prostheses differentiated by aesthetic aspect: a - semi-physiognomic (top left) and physiognomic; b - non-physiognomic

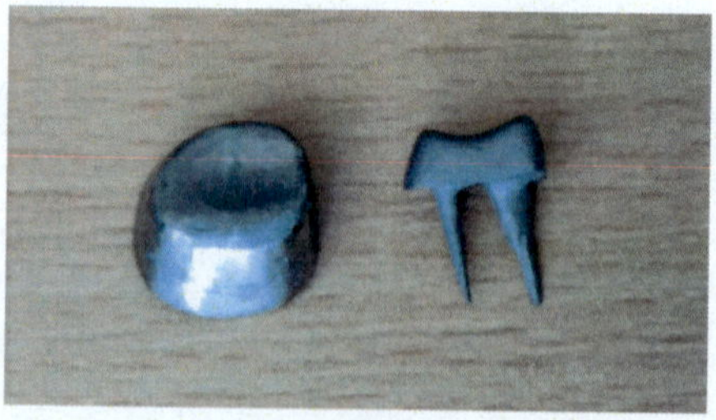

Fig. 2. Dental prosthetic structures with coronary and radicular aggregation made by selective laser deposition

The paper aims to present opposite approaches of subtractive and additive CAD-CAM manufacturing in the field of dental prosthetics by describing the main technological features for the realization of dental restorations or models using various methods and modern manufacturing equipment. Section 2 describes comparatively the classical and the unconventional computerized technologies for the realization of dental prostheses with the presentation of the advantages and disadvantages for each one. It also discusses the presentation of additive manufacturing methods that can be used for various purposes in the prosthetic field. Section 3 presents two different modern subtractive manufacturing systems with highlighting the all important processing stages and technological aspects in order to perform dental restorations, presentation of dental prostheses and their characteristics realized by SLM process and models made by different CAD-CAM additive technologies, as well as some opinions on the development and future trends regarding processing technologies in the field of dental prosthetics.

2 Materials and Methods

The main conventional technologies for obtaining dental restorations include casting, plastic deformation technologies of metallic materials, polymerization, galvanoforming and others [1, 10–12]. Cast metal prostheses have been used in dentistry for many decades, over the years they have repeatedly improved, maintaining, and increasing their positive qualities, but the process of making prostheses by casting is long and time consuming [13]. The polymerization process is used for the manufacture of dental prostheses from acrylic polymers, widely used especially for obtaining mobilizable restorations [14].

Manufacturing using computerized techniques is done by two basic principles: the subtraction method, when the excess is removed from a whole block of material, and the additive method, when an object is built layer by layer [1]. CAD-CAM technologies are a modern method of making dentures with the help of computer modelling and milling on numerically controlled installations. With the computer simulated model, the computer system can generate an algorithm for the trajectory of the milling cutter and the blank, after which the automatic milling process begins [15]. Various materials can be processed to make dentures using CAD-CAM milling technologies: metals, polymers, ceramics, or composite materials, but one essential thing must be considered - the compatibility of the cutter tool with the material to be processed. A diamond tool can be used for processing zirconium oxide, for metal materials - metal carbide cutters, and for polymers a special type of carbide that will prevent the adhesion of the machined polymer. Most dental milling systems use dry milling of materials due to possible contamination of semi-finished materials and the working chamber, except for titanium alloys which require a cooling agent to process due to the resulting high temperatures [1, 16]. It is a technology with significant advances in recent years, but which still has several disadvantages, including limited precision for milling of some ceramic materials, the probability of microcracks, the cost of material or high costs for equipment and their maintenance [17]. Other possible technologies for obtaining dental prostheses by unconventional subtractive methods include EDM [18], which is used less frequently due to high costs, or ultrasonic erosion, which is still in its research phase [1].

Methods that use technologies by adding material for the reconstruction of 3D models have progressed rapidly in various fields of activity. 3D printing is becoming a topic of great interest in dentistry. However, many dental prosthetic works are still performed by conventional manual and often ineffective methods. Selective laser melting (SLM process) is a layered additive manufacturing process in which components with a complex geometry can be made by processing and consolidating consecutive layers of powdered materials. The laser beam is controlled by the computer, which communicates the area and the trajectory to be processed, draws the cross section of the object on the powder according to the respective cross section from the digital file. Selective laser deposition has several important advantages over conventional dental prosthesis technologies and is increasingly being developed in oral medicine applications [19, 20]. Among the main advantages it can be highlighted: the wide range of materials available - metals, ceramics, and polymers; obtaining structures with extremely complex geometries; material saving, material powder can be reused in future processes; efficient and fast method; is a technology in continuous development and improvement. Another important benefit of these additive technologies is the possibility to use other accessible manufacturing methods in this category to obtain dental models/prototypes for the future prosthetic model [21–23]. Table 1 presents the main advantages and disadvantages that the 3 main groups of technologies in the field of dental prosthesis share: the conventional casting technique and the subtractive and additive CAD-CAM manufacturing methods [1].

Table 1. Comparison between main methods of dental prostheses manufacturing

Method	Advantages	Disadvantages
Casting	Reasonable cost Mechanical properties Durability	Fabrication time Human factor Aesthetics
CAD-CAM milling	Wide range of materials Fabrication time Precision	High costs Material overspending Possibility of microcracks
Selective laser deposition	Complex geometries Material economy Wide range of materials	High costs Necessity of post-processing Part density

3 CAD-CAM Approaches in Prosthetic Dentistry

The research on the realization of dental prostheses by modern subtractive technologies was carried out at the Trident Dental Clinic from Bucharest [24] and aimed at obtaining dental prostheses with the help of modern milling installations specialized in dental prosthetics and two different working approaches were researched.

The first researched equipment Ceramill Motion 2 [25] is a 5-axis milling machine specially designed for the dental field and this system can be used both in wet milling mode, which is the optimal regime for polymeric materials, and in dry milling mode for zirconium oxide, without coolant agent, being with most materials in the field of dental prosthetics - metal alloys, polymers, and ceramics and at the same time the installation is quite compact (Fig. 3 – a) [1]. The main advantages of this machine can be listed: high performance milling and grinding due to different processing strategies, wide range of materials that can be processed, intelligent system design that guarantees optimal protection for all electronic components during wet milling. The system is also equipped with the Ceramill Map 400 optical scanner (Fig. 3 – b), specially designed for dental applications.

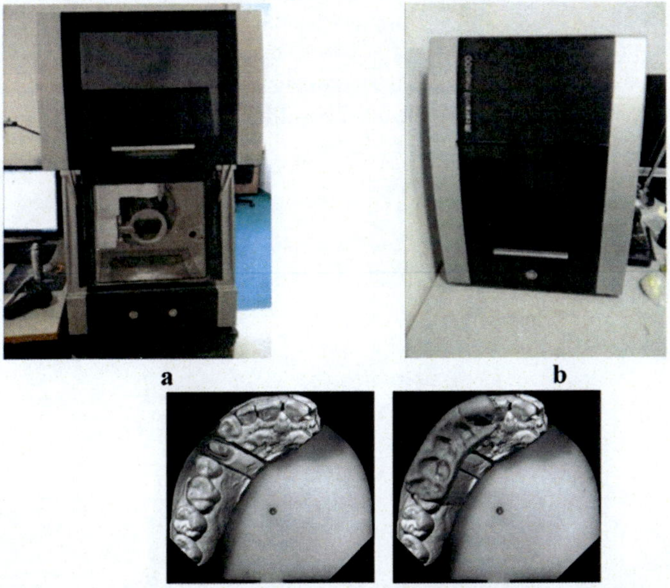

a b

c

Fig. 3. Dental milling CAD-CAM system: a - Ceramill Motion 2 milling installation; b - Ceramill Map400 scanner; c - example of the dental model scan and the antagonistic impression on the Ceramill Map400 scanner

The first step in making dentures is the same as with other conventional technologies - preparing the tooth for prosthesis and obtaining the impression from the patient's oral cavity. After obtaining the impressions, the gypsum dental model is made based on which the scan will be done, and the data will be passed on for processing in the special program of the milling system (scans can also be obtained directly from the dental impression) – Fig. 3 – c. After obtaining digital scans based on dental models, they are loaded into the special design software Ceramill Mind (Fig. 4 – a), developed by Amman Girbach [25]. From the program are set the teeth of interest (Fig. 4 – b), the tooth on which the dental crown will be installed (green), the adjacent teeth (yellow), as well as the antagonistic teeth (purple). The actual design stage of the dental prosthesis in the program follows, according to the results of the scan, and the dental technician models the work before sending it to the milling installation program.

After the design and modelling of the future denture is completed, the digital model is uploaded to the Ceramill Match software. This program is installed on the same computing unit as the Ceramill Mind program which is an integrated system. The respective program is used for the correct positioning of the prosthesis model projected on the virtual material block (Fig. 5 – a), the resulting data being subsequently transmitted to the milling machine. Also, the program will select the required block of material available from the system library and the designed denture can be placed in any free area of the selected block of material. The program also specifies the magnification factor for zirconium oxide, knowing that this material contracts after sintering, so ceramic prostheses must be designed with an optimal scaling factor established for each material [25]. After positioning the prosthesis on the material block, the milling regime is chosen, and the tool trajectories are calculated. The file then opens in another software of the Ceramill Motion system, which effectively controls the milling process itself.

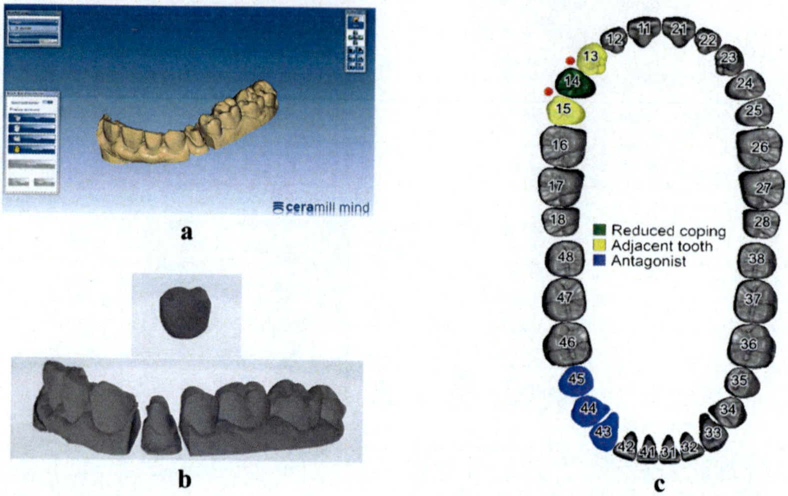

Fig. 4. Modelling of digital dental structures: a - Interface of the Ceramill Mind program with the loaded dental model; b - The future modelled dental crown; c - Representation of elements of interest in the program

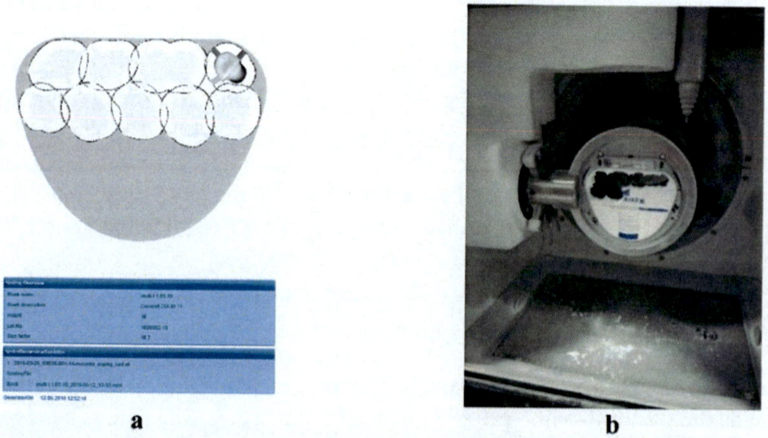

Fig. 5. Virtual material block in the program (a) and material block fixed in the work area (b)

From the program the block of virtual material with the tool trajectories calculated from the previous program is chosen. The material block is mounted and fixed on the milling installation (Fig. 5 – b), after which the actual milling process begins, which on average takes 15 min for a single unitary dental crown. Care must be taken that the

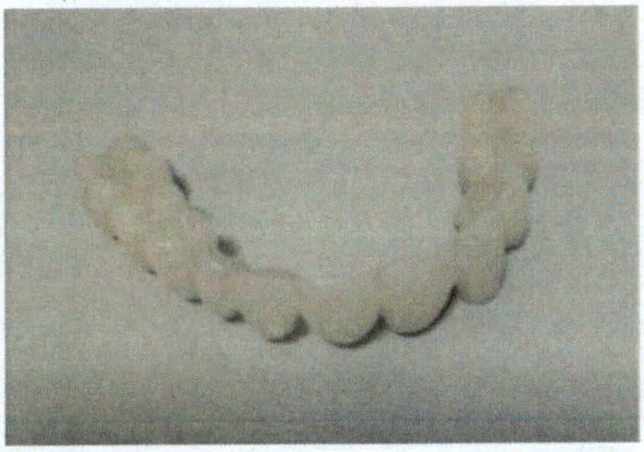

Fig. 6. PMMA dental prosthesis made on the Ceramill Motion 2 installation

assembled and fixed material block corresponds to the modeled virtual material block. The program can also detect tool failure and strictly control milling parameters. Figure 6 shows a PMMA prosthetic restoration performed on the Ceramill Motion 2 dental milling system [1, 24, 25].

Another studied installation that uses the subtractive method to make high quality dental prostheses was Sirona CEREC MC XL (Fig. 7 – a) [26], and the research took place at the same dental clinic [24]. Unlike the Ceramill Motion 2 scanning system, in this case the desired three-dimensional images are obtained directly from the patient's oral cavity. After inserting the scanner into the area of interest, the imaging process will start automatically. With the help of this scanner, precise three-dimensional images in natural color are obtained, which simplifies the orientation in the dental cavity (Fig. 7 – b). All previous prosthetic structures or materials of metal, polymer or composite material can be easily distinguished.

In this case, realization of dentures takes much less time compared to the other manufacturing system described above and generally consists of three steps. The first step is to prepare the tooth for prosthesis, i.e., grinding or removing cavities under anaesthesia. Then, with the help of the modern scanner coupled to the CEREC milling equipment, three-dimensional images of the prepared tooth and the area of interest are obtained (Fig. 7 – b). In this case, the patient no longer feels the discomfort created by the conventional impression, the whole process being practically imperceptible and fast [1].

After obtaining the intraoral scan, the computing unit searches in a special database and automatically determines the optimal prosthetic work proposed. The prosthetic work proposed and generated by the system will be analyzed by the dentist, after which small changes will be accepted or made before the virtual prosthesis is sent to the milling installation.

After the dentist accepts the final version of the future prosthetic restoration, the data are transmitted wirelessly to the milling equipment using diamond tools (Fig. 7 – c). This system processes the future prosthesis from a ceramic block with a high precision (a few micrometers) by wet milling and takes an average of up to 15 min for a unitary dental prosthesis. The ceramic block is fixed in a special fixing device and can be machined simultaneously with two tools. Considering the precision and the execution time, the CEREC system can be considered the most efficient system existing for the realization of fully ceramic dental prostheses at present, but at the same time it is extremely expensive [1, 24, 26].

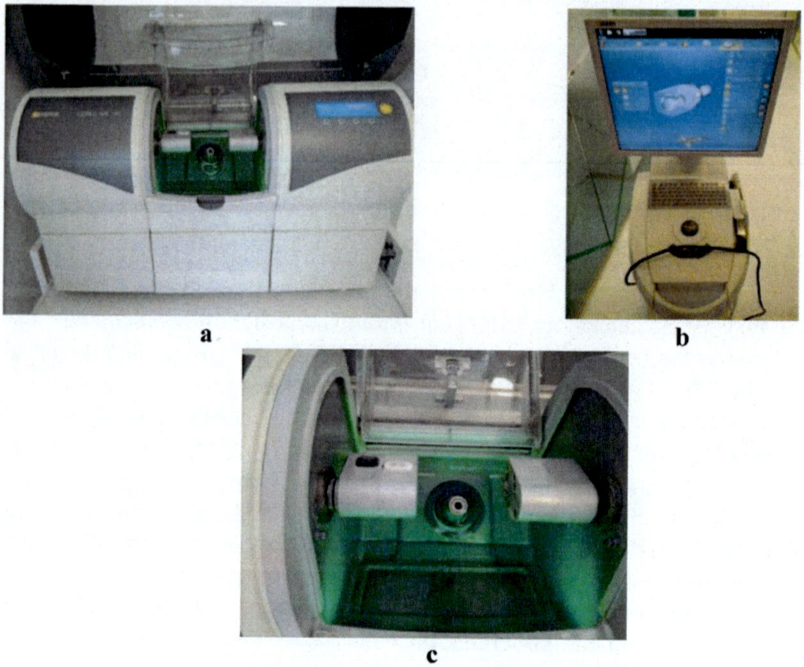

Fig. 7. Dental milling CAD-CAM system: a - CEREC MC XL milling installation; b - The calculation unit with a dental structure proposed by the system after the intraoral scan; c - Working area of the CEREC MC XL dental milling installation

Additive manufacturing approaches in the field of dental prosthetics have recently come into direct competition with subtractive methods for making metal structures for metal-ceramic prostheses or for other purposes in this area, due to exponential advances and the undeniable benefits of these technologies that include geometric complexity of the anatomical structures made or the economy of the raw material, as shown by the previous studies [9, 10, 21–23]. Some dental prostheses made by the technology of selective laser deposition are represented in Fig. 8, where the important disadvantage of this technology is highlighted - the surface quality immediately after processing needs several post-processing steps to acquire the compliant characteristics. Figure 9 shows a comparison of the surface quality of a prosthesis made by selective laser melting in the raw state and the surface condition resulting from the mechanical finishing operation. As can be seen from the microscopic image, the finishing operation used did not completely solve the problem of surface condition, which has a major influence on the physical and mechanical properties of the future prosthesis (wear resistance, hardness, fatigue etc.), knowing the fact that mechanical defects occur in prostheses with a rougher surface, and surface micro-defects lead to a decrease in some important mechanical properties, implicitly and their durability [1].

Also, the roughness of the surfaces of metallic prostheses plays a decisive role in the adhesion of microorganisms that can have negative effects on the body (rougher surfaces facilitate the adhesion of microbial species) and in terms of aesthetics (adhesion of the ceramic layer to the metal surface in case of metal-ceramic restoration) [1].

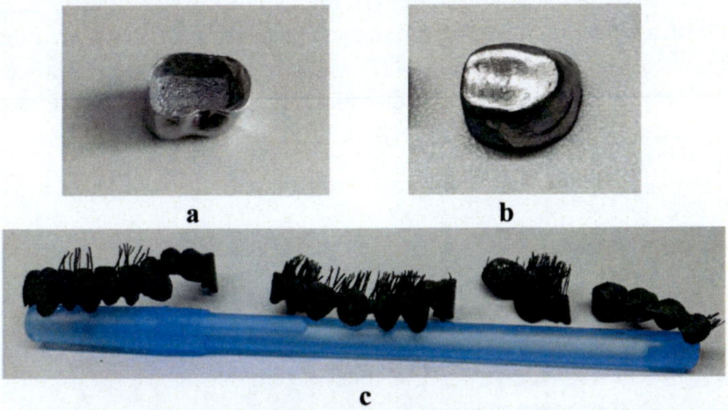

Fig. 8. Dental restorations performed by selective laser deposition: a, b - Ni-Cr dental crowns; c - Co-Cr-W alloy raw dental prostheses (in rough form)

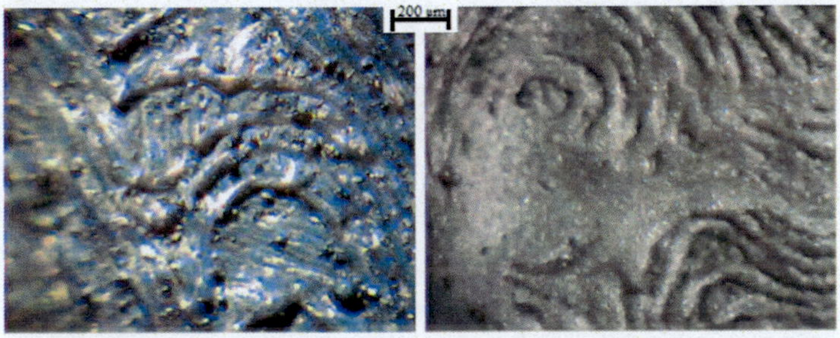

Fig. 9. Microscopic image with internal surface of the dental prosthesis obtained by the SLM process in the rough state (left) and after the finishing (right) [27]

In the dental CAD-CAM industry, it is worth paying attention to the additive technologies that are applied to obtain study models or dental prototypes, so the FDM technology or the additive methods that use liquid photopolymers as raw material that find multiple important applications in dental prosthetic field and not only, and Fig. 10 presents some examples of these, researched in some previous works [10, 21–23] namely dental models made by FDM extrusion technique, prototype of maxillary arch made by vat photopolymerisation procedure and nylon dental models made by selective laser sintering [1]. Experimental research regarding some mechanical characteristics of these dental models realized by various additive technologies was described in previous study [28].

Considering the above mentioned, the main tendencies in the dental prosthetic field of additive technologies refer mainly to the improvement of working parameters and post-processing technologies for obtaining viable and efficient prostheses from mechanical, biological and aesthetical points of view, finding new materials with improved properties that can be optimally processed by AM technologies (including ceramic materials), but also the combination in an integrated system of the two processing approaches, in order to eliminate the disadvantages encountered in each technology. It can be assumed that in the future these two different approaches will be used together for the realization of high quality dentures integrated in a hybrid additive-subtractive manufacturing system, due to the continuous development of macrotechnologies and microtechnologies applied in the medical field [29].

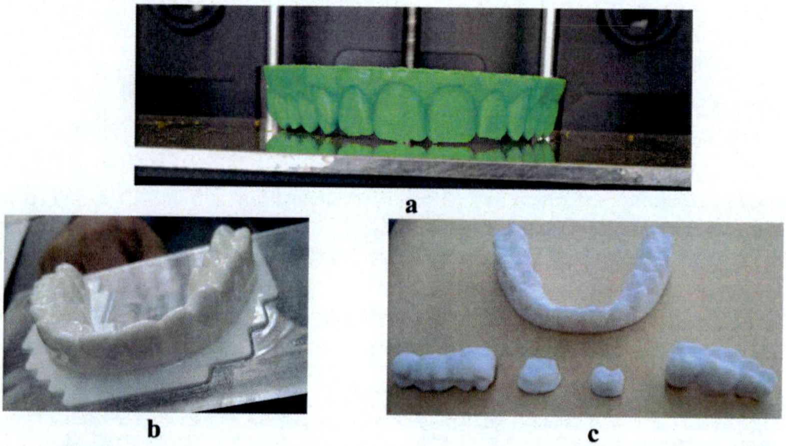

Fig. 10. Additive manufactured dental structures obtained by the FDM method (a), by DLP method (b) and SLS method (c).

4 Conclusions

The paper discussed some innovative aspects regarding manufacturing of dental prostheses by computerized technologies - using subtractive and additive methods. Treatment methods that use prosthetic structures in oral medicine are continuously being improved, and the complications or inconveniences that may occur following prosthetics by conventional methods are successfully replaced using computerized CAD-CAM methods. There is currently various numerical controlled milling equipment for dental restorations made of metallic, polymeric or ceramic materials, and their uses in dental prosthetics and implantology have been constantly developed and improved. Although computerized subtractive methods have special advantages such as the high quality of the prostheses made or the speed of manufacture, the costs, the material overspending, and the difficult use of these systems are still an important drawback for dental CAD-CAM milling. With a diametrically opposed approach of manufacturing, additive CAD-CAM technologies, namely selective laser melting, are constantly evolving in the field of dental prosthetics, due to its exclusive benefits. In addition, in the sector of dental prosthetics, other accessible additive technologies can be used to realize dental models or for other purposes in the field. However, these additive technologies have some limitations, which still stop the wider use of these methods in dentistry. Most likely, in the future, the symbiosis of subtractive and additive approaches will be required to make progress in the field of dental prosthetics and to combine the advantages and eliminate the disadvantages of each manufacturing method, and the two opposite CAD-CAM methods will be used together to get the highest quality dental restorations. Future research aims to expand studies on the properties and characteristics of dental prostheses made by modern additive and subtractive technologies compared to dentures made by conventional methods.

Acknowledgements. Research has been partially supported by the European Social Fund from the Sectoral Operational Programme Human Capital 2014–2020, through the Financial Agreement with the title "Scholarships for entrepreneurial education among doctoral students and postdoctoral researchers (Be Antreprenor!)", Contract no. 51680/09.07.2019 - SMIS code: 124539.

References

1. Moraru, E.: Cercetări privind realizarea protezelor dentare prin depuneri selective cu laser și prin alte tehnologii aditive (Research on the realization of dental prostheses by selective laser deposition and other additive technologies). Ph.D. thesis, Politehnica University of Bucharest, Romania (2021)
2. Berkowitz, B., Holland, G., Moxham, B.: Oral Anatomy, Histology and Embriology. Edited by Mosby (2009)
3. Sabharwal, A., Stellrecht, E., Scannapieco, F.A.: Associations between dental caries and systemic diseases: a scoping review. BMC Oral Health **21** (2021). Article number: 472. https://doi.org/10.1186/s12903-021-01803-w
4. Wostmann, B., et al.: Indications for removable partial dentures: a literature review. Int. J. Prosthodont. **18**(2), 139–145 (2005)
5. Al Jehani, Y.A., Baskaradoss, J.K., Geevarghese, A., AlShehry, M.A.: Current trends in aesthetic dentistry. Health **6**, 1941–1949 (2014)
6. Wassell, R., Walls, A.W.G., Steele, J.G.: Crowns and extra-coronal restorations: materials selection. Br. Dent. J. **192**(4), 199–202 (2002)
7. Christensen, G.J.: Fixed prosthodontics–state of the art. Aust. Dent. J. **40**(3), 164–166 (1995)
8. Alghazzawi, T.F.: Advancements in CAD/CAM technology: options for practical implementation. J Prosthodont. Res. **60**(2), 72–84 (2016)
9. Moraru, E., Dontu, O., Petre, A., Vaireanu, D., Constantinescu, F., Besnea, D.: Some technological particularities on the execution of dental prostheses realized by selective laser deposition. J. Optoelectron. Adv. Mater. **20**(3–4), 208–213 (2018)
10. Moraru, E., Dontu, O., Rizescu, C., Spanu, A., Ciobanu, R., Draghici, C.: Laser-assisted additive technologies for the execution of dental restorative prostheses. IOP Conf. Ser. Mater. Sci. Eng. **997**(1) (2020). Article number: 012050
11. Wendt, S.L., Jr.: Nonprecious cast-metal alloys in dentistry. Curr. Opin. Dent. **1**(2), 222–227 (1991)
12. Chochlidakis, K.M., Papaspyridakos, P., Geminiani, A., Chen, C.J., Feng, I.J., Ercoli, C.: Digital versus conventional impressions for fixed prosthodontics: a systematic review and meta-analysis. J. Prosthet. Dent. **116**(2), 184–190 (2016)

13. Chen, W.-C., Teng, F.-Y., Hung, C.-C.: Characterization of Ni-Cr alloys using different casting techniques and molds. Mater. Sci. Eng. C Mater. Biol. Appl. **35**, 231–238 (2014)
14. Frazer, R.Q., Byron, R.T., Osborne, P.B., West, K.P.: PMMA: an essential material in medicine and dentistry. J. Long Term Eff. Med. Implants **15**(6), 629–639 (2005)
15. Miyazaki, T., Hotta, Y.: CAD/CAM systems available for the fabrication of crown and bridge restorations. Aust. Dent. J. **56**, 97–106 (2011)
16. Sulaiman, T.: Materials in digital dentistry-a review. J. Esthet. Restor. Dent. **32**(2), 171–181 (2020)
17. Coandă, P., Avram, M., Constantin, V.: A state of the art of predictive maintenance techniques. IOP Conf. Ser. Mater. Sci. Eng. **997**(1) (2020). Article number: 012039
18. Bratu, E., Bratu, D., Borsanu, I.: Telescoping implant prosthese with galvano mesostructures. Rom. J. Oral Rehabil. **1**(4), 29–34 (2009)
19. Kruth, J.P., Wang, X., Laoui, T., Froyen, L.: Lasers and materials in selective laser sintering. Assembly Autom. **23**(4), 357–371 (2003)
20. Kruth, J.P., Mercelis, P., Van Vaerenbergh, J., Froyen, L., Rombouts, M.: Binding mechanisms in selective laser sintering and selective laser melting. Rapid Prototyping J. **11**(1), 26–36 (2005)
21. Moraru, E., Besnea, D., Dontu, O., Gheorghe, G.I., Constantin, V.: Applications of additive technologies in realization of customized dental prostheses. In: Gheorghe, G.I. (ed.) ICOME-CYME 2018. LNNS, vol. 48, pp. 8–17. Springer, Cham (2019). https://doi.org/10.1007/978-3-319-96358-7_2
22. Moraru, E., Dontu, O., Besnea, D., Constantin, V.: Study and realization of prosthetic dental models by additive technologies. IOP Conf. Ser. Mater. Sci. Eng. **444** (2018). Article number: 042017
23. Moraru, E., Dontu, O., Spanu, A., Prisecaru, D., Draghici, C.: Some considerations regarding the use of vat photopolymerisation technique in prosthodontics. Int. J. Mechatron. Appl. Mech. **6**(I), 87–90 (2019)
24. Trident Dental Clinic Bucharest. https://clinicatrident.ro/
25. Ceramill Motion 2 CAD-CAM milling machine. https://www.amanngirrbach.com/
26. CEREC MC XL Sirona CAD-CAM installation. https://www.dentsplysirona.com/
27. Moraru, E., Dontu, O., Besnea, D., Rizescu, C., Rizescu, D., Constantin, V.: Theoretical and numerical researches on the electrochemical polishing of dental prostheses realized by selective laser melting. IOP Conf. Ser. Mater. Sci. Eng. **564** (2019). Article number: 012029
28. Moraru, E., Dontu, O., Rizescu, C., Rizescu, D.: Comparative study and mechanical analysis of some additive manufactured dental models. Acta Tech. Napocensis-Ser. Appl. Math. Mech. Eng. **64**(3), 503–508 (2021)
29. Sandu, A., Bogatu, L., Ionascu, G., Manea, E., Gheorghe, V.: Modeling, simulation and experimental research for MEMS cantilevers of complex geometry. In: MATEC Web of Conferences, vol. 290, p. 08002 (2019)

Design and Performance Assessment of a Double-Screen Trommel for MSW Separation

Luís D. Forte[1], Ricardo Faria[1], and Ana C. Ferreira[2,3,4](✉)

[1] João Matos e Ribeiro 2 (JMR2) - Obras Públicas e Engenharia Lda, Braga, Portugal
[2] Mechanical Engineering and Resource Sustainability Center (MEtRICs), University of Minho, Guimarães, Portugal
acferreira@dps.uminho.pt
[3] ALGORITMI Centre, University of Minho, Guimarães, Portugal
[4] Centro de Investigação em Organizações, Mercados e Gestão Industrial (COMEGI), Universidade Lusíada Norte, Vila Nova de Famalicão, Portugal

Abstract. The design of mechanical systems for Municipal Solid Waste (MSW) sorting requires a technical analysis of the structural components and the study of their functional integration. Such analysis should take into account the dimensioning adjusted to the waste quantities to be processed. The objective of this work is to model the geometry and the components of a rotating mechanical device for MSW sorting, also known as trommel. It consists of a cylindrical drum placed on top of drive rollers with a certain inclination, allowing the advancement of the MSW with the rotation movement. The technical design of each structural element was projected, considering the geometric constraints of the sorting facility. It was proposed a mesh with two different orifice sizes to minimize the wear rate that mainly occurs in the MSW feed zone. Given the geometry and the considered operating conditions, it was possible to obtain the maximum mass feed rate of 154.2 ton/h. The trommel efficiency was also estimated, considering that the MSW has an average particle diameter of 152 mm and that each particle impacts, at least, 20 times with the trommel surface before passing through the mesh open area. Thus, for a mesh with an open area fraction of 0.51, the trommel efficiency reaches the value of 82%.

Keywords: Trommel design · Industrial project · Sorting efficiency · MSW

1 Introduction

The currently increase in waste production is related to the intensification of socio-economic activities, deeply linked with urbanization and industrialization. The amount of MSW results therefore from the population growth, the massive use of resources and the recycling system inefficiency [1, 2]. MSW can be divided into different categories, such as the recyclables (e.g., metals, plastic, paper or glass), compostable organic materials (e.g., food waste, forestry and gardening residues, etc.), toxic substances (e.g., used

J. Machado et al. (Eds.): icieng 2022, LNME, pp. 103–115, 2023.
https://doi.org/10.1007/978-3-031-09382-1_10

batteries, all types of chemicals, medicines, and hazardous solid waste). In fact, high quantities of waste are daily produced and its management represents a complex task, which includes several strategies, including efficient waste collection, separation and disposal systems [3, 4]. MSW is characteristically distinguished from other wastes, mainly in terms of its production volume, composition, and management models [2, 5]. Thus, each country defines the best fit solution to manage their MSW, mostly because non-managed waste has a great impact on pollution levels, diseases transmission, increased respiratory problems, wildlife and general environmental impacts [6].

According to the Portuguese Environment Agency (APA), since 2013, the total production of MSW has increased over the years. In the year of 2020, approximately 5.01 million tonnes were produced, which corresponds to the daily production of 1.40 kg per inhabitant [7].

Waste management in Portugal is based on: (1) packaging, paper and glass separation through recycling containers; (2) separation of biodegradable municipal waste for organic recovery through composting and/or anaerobic digestion processes; (3) undifferentiated collection of municipal waste for mechanical and biological treatment and later shipment to recovery processes; (4) undifferentiated collection for energy recovery (waste incineration); and (5) landfill waste deposition of non-recoverable fractions.

Statistically and regarding the collection process, despite the positive trend of the selective collection over the years, the majority of waste is still unsorted collected. Figure 1 presents the relative share of different MSW destinations and management operations in Portugal, for 2020. Despite the increasing need for a drastic reduction in the disposal of waste in landfills, it still represents 40.8% of MSW destinations. The forwarding of waste for mechanical and biological treatment is the second management operation, with a relative share of 20.9%. The shipment of waste for energy recovery and the mechanical treatment represents 18.7% and 5.4%, respectively.

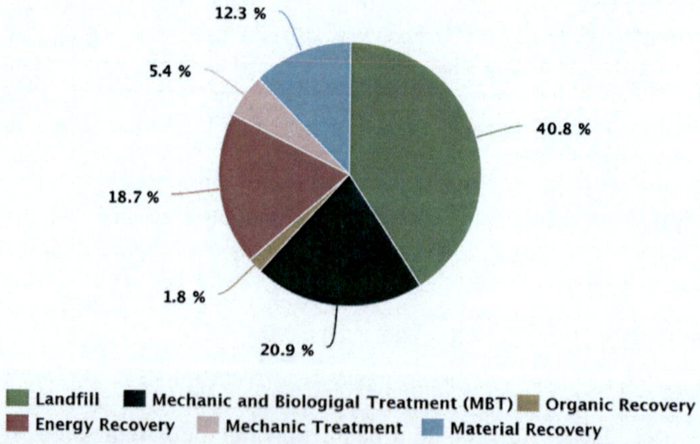

Fig. 1. Different MSW destinations and management operations in Portugal, for 2020 [7].

Automated sorting techniques can be divided into direct and indirect sorting. Direct sorting uses material properties such as the density, electrical conductivity or magnetic susceptibility for media separation by gravity or magnetic external fields, respectively. When the system uses sensors or robots, it is considered as indirect sorting [8].

Some attention is being paid to developments that have taken place in the field of automated sorting of MSW over the last decades [9, 10]. A review published by Dodbiba and Fujita [11] is focused on separating/sorting techniques based on non-sensor systems. Cimpan *et al.* [12] reviewed the physical processing of waste to segregate recyclables from MSW. Gundupalli *et al.* [8] summarize advances in physical processes, sensors and automated systems for sorting and recycling source-separated MSW. Jank *et al.* [13] presented a work regarding a mechanical pre-treatment separation, prior to biowaste shredding. The residues are pressed through a perforated cylinder and thereby divided into an organic fraction ready for further processing. The inclusion of this physical separation process allows a reduction of the plastic film fraction in the organic waste, which is redirected to biological digestion, mostly because the film is pulled through the cylinder orifices.

It is urgent to implement a more sustainable solid waste management, which requires new management systems and the optimization of waste management facilities. Yet, upgrading MSW facilities and their systems require considerable capital investment [10], which often limits the technological options to be selected. In Portugal, mechanical sorting is one of the main processes used, so the optimization and improvement of such systems improve the efficiency of MSW management. Two main principles are used in mechanical sorting: separation by screening and size reduction [14], and few technologies can be identified:

- Shredding – Reduces the size of the waste to be sorted;
- Screening separation – Generally separates waste by size;
- Separation by density – Separates the lighter waste from the heaviest;
- Ballistic separation – Separates fractions based on their mass.

Screening is one of the techniques commonly used for size separation, where a perforated rotating cylinder (e.g., trommel) allow smaller materials to drop out during the rotation process. It is a mechanical separation technique used in different industrial applications [11, 12, 15, 16]. Several studies are focused on the analysis of the factors that influence screening efficiency [17, 18]. This mechanical sorting device provides a lifting and cascading action, which is particularly useful when materials have a very wide size range [19–21]. According to Chen *et al.* [21], trommels are used in MSW sorting because they are relatively inexpensive when compared with other automated systems, requiring lower operating and maintenance costs.

When considering the cost-effectiveness of waste separation and recovery, the performance optimization of the equipment is of utmost importance. Thus, the design, size and operating conditions need to be adjusted to the quantities of MSW that have to be sorted. This requires a geometric and dynamic analysis, including the correct design of the components, their production and the functional integration [22].

The main objective of this paper is the modelling and the performance assessment of a trommel to be installed in a separation and recovery center of MSW. It is intended to model the geometry and the design of structural components, considering the specifications and requirements of the process and the application. It is also intended to evaluate the performance of the equipment by estimating its efficiency.

2 Separation Process and System Modelling

This project regards the design and performance assessment of a double-screen trommel for MSW separation, to be integrated into a functional waste sorting line. The trommel is the first element of the sorting line, whose objective is to separate large-scale waste in a separate stream. Being the first device in a line, it receives all kinds of undifferentiated wastes of all compositions and dimensions, which gives it a high level of wear. Larger residues collide with the trommel mesh, resulting in some elements' rupture. In this section, it is described the sorting line facility and it's presented the trommel operating principles.

2.1 Facility Description

The equipment under study is a 400 mm trommel, which is projected for the primary mechanical sorting process of a waste sorting and recovery facility (Fig. 2).

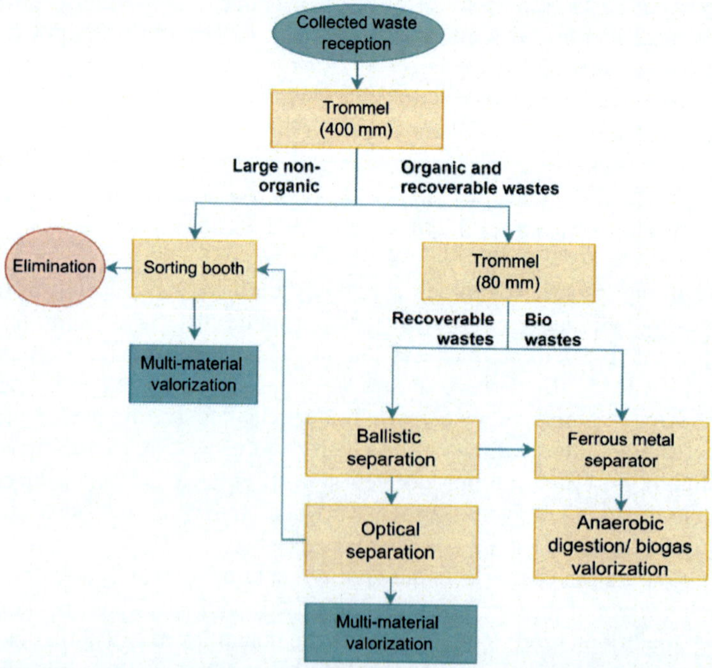

Fig. 2. Schematic of the sorting line operation and valorization chain for the facility under study.

The collected wastes are unloaded on the first conveyor belt, which feeds the 400 mm trommel, responsible for separating the bulky fraction (non-organic) from the medium fraction (organic and recoverable wastes).

Basically, at the trommel exit, the bulky fraction is sent to the first sorting booth where recycling materials are separated for multi-material valorization (e.g., paper, plastic and glass) and the remaining waste is then eliminated.

On the other hand, the organic and recoverable wastes are forwarded to another trommel (80 mm), similar to the first, to separate the fines (organic) from another fraction of recoverable. The fine or organic fraction is transported to the ferrous metal separator. This separator removes all ferrous materials present in the organic fraction, whereas, the remaining waste goes to anaerobic digestion/biogas valorization. The recoverable wastes separated by the 80 mm trommel is forwarded to the ballistic separator, following to the optical separator. The sorted materials, either automatically or manually, are pressed and placed in the compaction station.

2.2 Trommel Operating Principals

Trommel, a rotary screen, is a positive separation equipment used to reduce the size and also to separate waste into different size fractions. The trommel screen consists of a cylindrical metallic surface, which rotates a complete 360° around a longitudinal axis. The cylindrical screen has several apertures through which smaller waste elements pass through the mesh, creating two or more waste streams differentiated by size. Due to the rotating movement, and due to cutting blades, the MSW agglomerates are separated, increasing the effectiveness of the sorting process [21]. Screening efficiency results from the trommel rotational speed, where the fine fraction of the waste falls through the mesh apertures onto a conveyor belt, while the larger particles move on due to the gravitational force imposed by the inclination. Depending on the rotational speed, the waste particles behave differently, which contributes to size segregation. The most common motion mechanisms are lumping, centrifuging and cataracting (Fig. 3).

When the trommel rotation speed is low, the slumping motion mechanism (Fig. 3(a)) takes place and most particles lays at the bottom of the cylinder, increasing the agglomerations on the free cylinder surface. Normally, this mechanism does not occur given the rotation speeds of trommels applied to MSW sorting. As the rotation speed increases, the transition to a centrifuging mechanism occurs. In this case, the screening efficiency decreases, because the particles stick to the surface walls of the rotating cylinder (Fig. 3(b)). Thus, there is an optimal rotation speed that increases the trommel efficiency, avoiding waste agglomeration and simultaneously promoting the particle detaching from the rotating surface (Fig. 3(c)).

The trommel sorting surface consists of a perforated metal mesh. Non-return rings can be mounted on these meshes to avoid material back flow, since the inclination of trommels normally does not exceed 6°. According to the literature, the mesh apertures sizes vary from 50 to 500 mm, in square or circular shapes, depending on the particles size and trommel diameter [23].

2.3 Mathematical Modelling

The efficiency mathematical modelling was based on the work developed by Alter [24], a model in which the design depends on the trommel dimensions, rotation speed and the number of particles' impacts. Thus, the sorting efficiency can be estimated considering

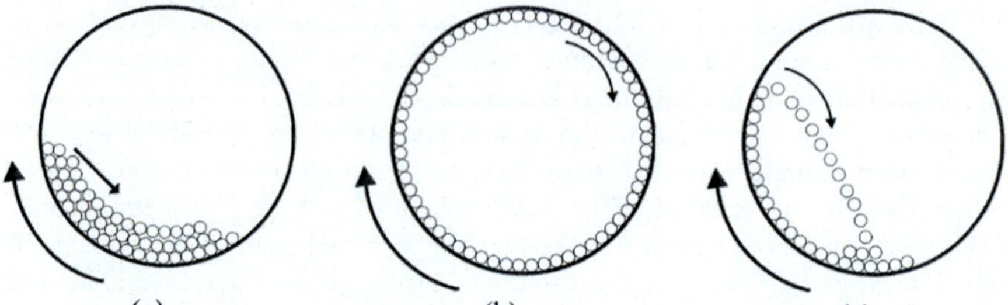

Fig. 3. Particles motion in rotating drums: (a) slumping; (b) centrifuging; (c) cataracting [21].

the number of impacts between the MSW particles and the screen surface, for fixed mesh sizes.

This estimation is based on a probability problem, where the number of impacts is a function of the cumulative probability that particles, with a certain diameter, pass through the mesh open area. Thus, it is assumed that when a spherical particle of diameter d impacts on a square orifice of side a (assuming $a > d$), the probability that it will pass through the orifice is given by Eq. (1):

$$p_i = \left[1 - \frac{d}{a}\right]^2 Q \tag{1}$$

where Q is the ratio of the orifice area to the total screen surface area (mesh open area fraction). Thus, the cumulative probability after n impacts is obtained by Eq. (2):

$$p_i = \sum_{i=1}^{n} p(1-p)^{(i-1)} = 1 - (1-p)^n \tag{2}$$

Considering that MSW is not constituted by uniformly sized particles, it is necessary to consider the fraction number of particles with different sizes for a distribution function $f(\mathrm{x})$. This function corresponds to equivalent spherical size distribution, including shape factors for particles in each size range. Thus, the probability density for all particles is equal to or greater than a minimum size x_0 and equal to or less than a maximum size x_m, determined by the integral from Eq. (3).

$$P(x_0, x_m) = \int_{x_0}^{x_m} f(x)[1 - (1-p)^n]\,dx \tag{3}$$

Therefore, the cumulative probability ranges in size between x_0 and x_m, when colliding n times on the screen surface. The total fraction of particles in the MSW size range can be calculated by Eq. (4).

$$F(x_0, x_m) = \int_{x_0}^{x_m} f(x)\,dx \tag{4}$$

The particles size distribution was assumed as the log-normal function presented by Eq. (5).

$$f(x) = \frac{1}{\sqrt{2\pi}x\sigma_{\ln}} e^{\left(-\ln\frac{x/\bar{x}}{\sqrt{2}\,\sigma_{\ln}}\right)^2} \tag{5}$$

Consequently, the trommel efficiency can be expressed as the ratio between the number of screened particles and the number of particles feed to trommel inlet (Eq. (6)).

$$E(x_0, x_m) = \frac{P(x_0, x_m)}{F(x_0, x_m)} \tag{6}$$

Please note that, when the distribution is known, the number of impacts required to achieve a certain separation fraction can be calculated. However, it is important to mention some model limitations, namely, the difficulty of modelling the particle-to-particle interaction, i.e., collision adhesion or deflection effects. These interactions also increase with the mass feed rate, thus requiring a greater number of impacts as the materials flow increases. The rotational movement, as well as the inclination in relation to the longitudinal axis, are the main factors that create impacts between the screen and the MSW particles. The number of impacts during particles residence time in the trommel is, therefore, a function of the trommel dimensions, rotation speed and inclination angle (β). To fully model the dynamics of the trommel, it is necessary to calculate the flight trajectory and the riding angles of MSW particles, in order to calculate the distance travelled by the particles and their residence time in the trommel drum. These calculations were determined by the mathematical model published in [24]. Those calculations allows, for specific conditions, to determine the angular velocity and the mass feed rate that optimizes the trommel efficiency. The velocity can be calculated by (Eq. (7)):

$$\omega = \sqrt{\frac{\sin\alpha \cdot g}{R}} \tag{7}$$

where α is the angle for which the particle will ride on the surface without falling, typically considered as $\pi/2$, g is the gravitational acceleration and R corresponds to the radius of the trommel. The mass feed rate, M, depends on the material thickness that occupies a cross-sectional area (bR) and that moves longitudinally along the barrel with an inclination β. The product between the cross-sectional area occupied by the particles and its velocity along the barrel corresponds to the volumetric feed rate. When multiplied by the density of the material gives the mass feed rate (Eq. (8)).

$$M = \psi \cdot g^{1/2} b \rho_b \beta R^{3/2} \tag{8}$$

where ρ_b is the density of material and ψ a derived constant factor which was assumed to be 2.31 for the problem under study [24].

3 Proposed Design for the Double-Screened Trommel

The MSW sorting equipment had as requirement the development of a design that allowed its efficiency maximization for a desired sorting capacity of 150 ton/h.

The trommel was designed with an internal radius of 1 475 mm and a sieving length of 8 088 mm, settled on a structure with inclination of $\beta = 5°$. The trommel geometry corresponds to a total length of 10 000 mm and an external diameter of 3 000 mm. The external diameter corresponds to the shields structures that prevent the waste projection.

Each structural element was designed, including the ferrules, the connecting and reinforcing rings, T-bars, calendered bars, screen meshes and cutting blades. Table 1 summarizes some of the characteristics of the components and the assembly steps.

Table 1. Summary of trommel components and assembly steps.

Component	Characteristics	Illustration
Ferrules and connecting rings	• Inlet and an outlet ferrule with similar geometry and same external diameter (2 950 mm). • Ferrules are rectified to be supported on top of the drive rollers that provide the rotation movement. • Connecting rings make the assembling between the ferrules and calendered bars.	
T-bars and calandered bar	• 6 T-bars are soldered to the ferrules. • Calendered bars are added to reinforce the structural stability. • Locking rings are assembled to T-bars and calendered bars to allow the mesh screwing.	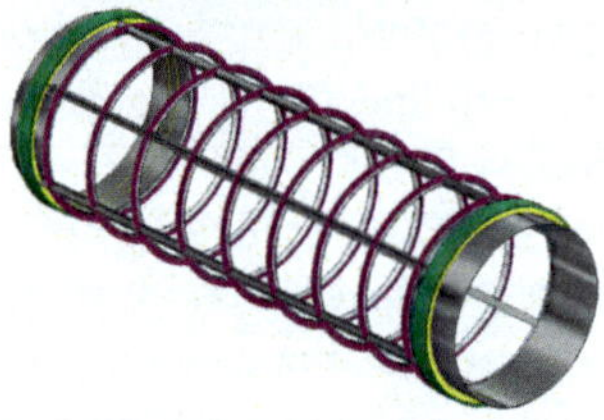
Sorting Meshes	• Perforated components responsible for the sorting process. • Oxycut process is used to produce the mesh orifices. • Meshes are also submitted to calendaring process. • Given its level of wear, it was considered its production in Hardox.	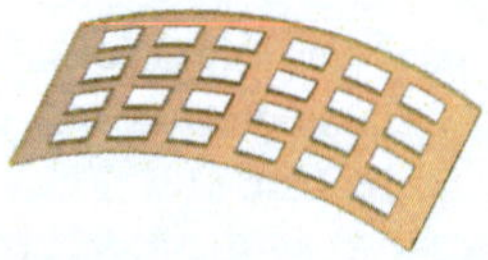
Cutting blades	• Cutting elements added to the equipment to reduce the waste particle size. • The blades are soldered to the meshes and are also manufactured in Hardox.	

Either inlet and outlet ferrules are reinforced with a rectified ferrule assembled with connecting rings. The ferrules are then screwed to 6 T-bars. Since these components are subjected to shear stress, the structure is strengthened by calendered bars and locking rings. To prevent MSW backflow, a non-return ring must be placed on the trommel inlet ferrule. The sorting meshes are soldered to T-bars and calendered bars. The cutting blades are also screwed to meshes along the trommel length, with an inclination of 60°. Figure 4 presents the proposed design for the double-screened trommel and a photo of the manufactured equipment.

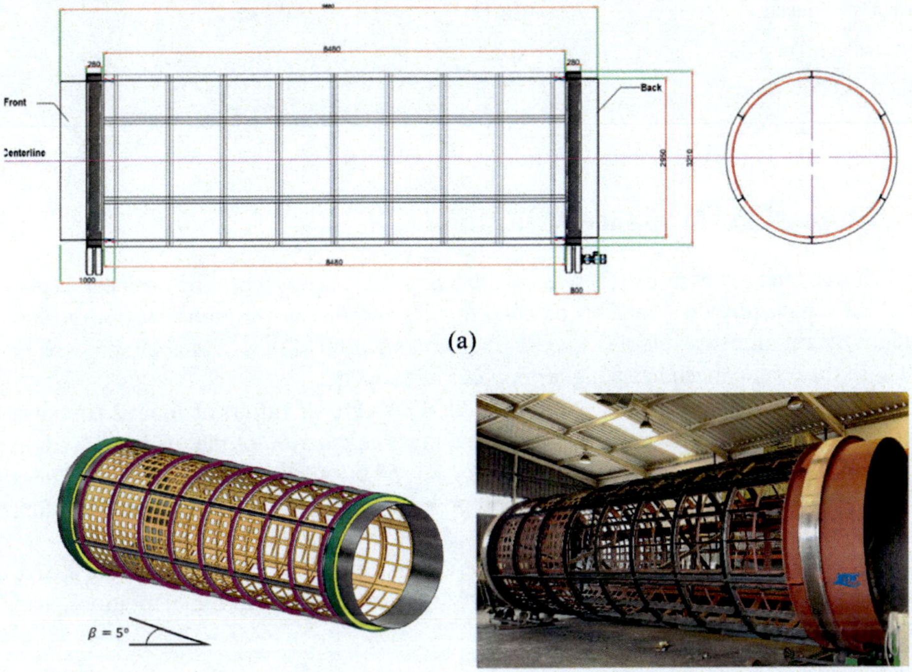

(a)

(b) (c)

Fig. 4. Proposed design for the double-screened trommel: (a) 2D technical drawing; (b) 3D model from CAD software; (c) manufactured trommel.

Generally, commercial trommels are produced with a uniform mesh open area. In this way, there wear rate is higher on the trommel's feeding zone. Thus, it is proposed a design with two different mesh sizes. The mesh characteristics are shown in Table 2.

The first part of the mesh (trommel inlet) is defined with a 150 mm orifice size, allowing a smaller fraction to pass through the mesh and forcing the MSW particles to travel a greater distance along the trommel.

This design option guarantees a higher uniformity of the wear rate throughout the trommel length. The second half-mesh is defined with a 400 mm orifice size, the maximum size of the MSW particles to be sorted downstream. The mesh design parameters were calculated as a function of the total trommel area and the respective orifice areas for both meshes.

Table 2. Design parameters for each size mesh.

Mesh orifice size	150 mm	400 mm
N° of orifices	24	6
Total n° of meshes	18	30
Orifice area (m^2)	0.02	0.16
Total orifice area (m^2)	9.72	20.80
Total mesh area (m^2)	28.1	46.8

4 Trommel Performance Assessment

The mesh open area fraction (Q) was calculated as the ratio between the orifice area and the total screen surface area. This calculation was carried out for both types of meshes and for the trommel as a whole. A mesh open area fraction of 0.51 was obtained, mostly because the coarse mesh has a higher number of elements.

It was also estimated that, on average, each MSW particle suffers of about 20 impacts during its residence in the trommel. For the correct calculation of the trommel feeding rate, it was considered a MSW density of $\rho_b = 252.5$ kg/m^3 and a material thickness of 150 mm. Given the geometry of the proposed design and the considered operating conditions, it was obtained a mass feed rate of 154.2 ton/h.

Yet, the MSW feed rate can be adjusted as a function of the trommel rotational speed. An analysis of the relationship between the mass feed rate and the trommel rotation speed was performed, as shown in Fig. 5.

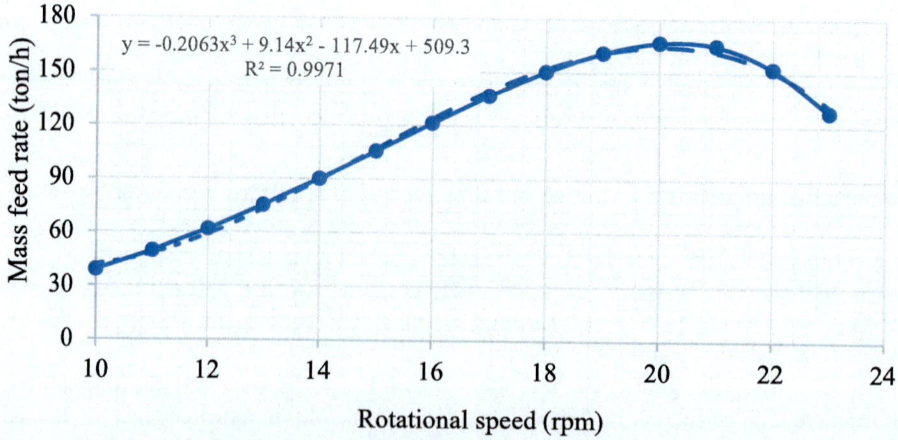

Fig. 5. Mass feed rate variation as a function of trommel rotational speed.

The proposed trommel design reaches its maximum feed rate capacity at a rotational speed of 20 rpm, corresponding to a capacity of about 167.0 ton/h. For higher rotational speed, the sorting capacity of this equipment decreases.

Assuming that the MSW particle size distribution approaches a log-normal, the efficiency of this equipment was estimated. An average particle diameter of de $\bar{x} = 152$ mm and a standard deviation of $\sigma_{ln} = 14.2$ mm was considered. The integrals of Eqs. (3) and (4) were calculated through approximation by the Trapezoidal numeric integration rule, whereas the trommel efficiency value was obtained as a function of the number of the impacts. The calculations were performed in Excel using Visual Basic for Applications (VBA). Figure 6 presents the trommel efficiency as a function of MSW particle size and the number of impacts.

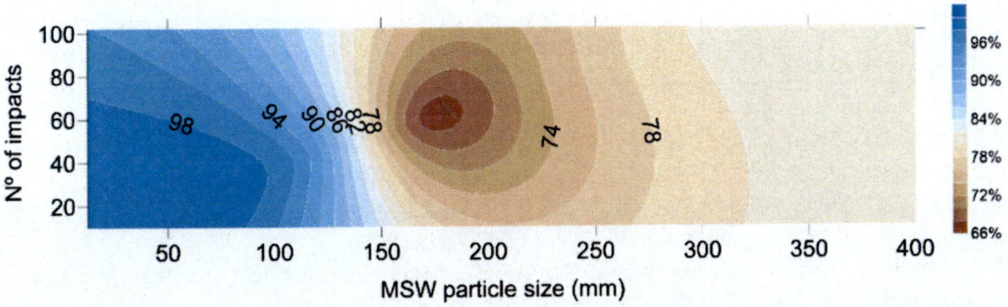

Fig. 6. Trommel efficiency as a function of MSW particle size and number of impacts.

The results were obtained through a parametric analysis by applying the Kriging method, a numerical technique that provides a minimum error-variance estimate of the data. The proposed trommel design has an estimated efficiency of 82% if considering MSW particles with an average size diameter of 152 mm, a mesh open area fraction of 0.51 and 20 impacts with the screen surface.

5 Final Considerations

Processing MSW has become an increasingly common and successful method for minimizing the environmental impact of waste disposal. Pre-processing of MSW makes possible the recovery and reuse of valuable resources such as paper, plastics and metals that may have otherwise been deposited in landfills. Screening is one of the most cost-effective and widely employed physical size separation methods.

This paper proposes the design and performance assessment of a double-screen trommel for MSW separation. The trommel was designed with an internal radius of 1 475 mm and a sieving length of 8 088 mm, settled on a structure with inclination allowing the advancement of the material as it rotates. In order to increase the trommel life span, it was proposed a mesh with two different orifice sizes (150 mm and 400 mm). The finer mesh makes the larger particles travel a greater distance inside the trommel, allowing not only a greater waste sorting efficiency but also a greater uniformity of equipment wear. For the proposed design, it was possible to assess the trommel performance. Given the geometry and the considered operating conditions, it was possible to obtain the maximum feed rate of 154.2 ton/h and a trommel efficiency of about 82%. The rotation speed for which the amount of sorted MSW is maximized corresponds to 20 rpm (mass feed rate of 167.0 ton/h).

Acknowledgement. This work has been supported by FCT within the R&D Units Project Scope UIDP/04077/2020 (METRICS) and R&D Units Project Scope UIDB/00319/2020 (ALGORITMI).

References

1. Sharma, K.D., Jain, S.: Municipal solid waste generation, composition, and management: the global scenario. Soc. Responsib. J. **16**, 917–948 (2020). https://doi.org/10.1108/SRJ-06-2019-0210
2. Palanivel, T.M., Sulaiman, H.: Generation and composition of municipal solid waste (MSW) in Muscat, Sultanate of Oman. APCBEE Proc. **10**, 96–102 (2014). https://doi.org/10.1016/j.apcbee.2014.10.024
3. Beigl, P., Lebersorger, S., Salhofer, S.: Modelling municipal solid waste generation: a review. Waste Manag. **28**, 200–214 (2008). https://doi.org/10.1016/j.wasman.2006.12.011
4. Diaz, L.F., Savage, G.M., Eggerth, L.L., Golueke, C.G.: Integrated waste management. In: Composting and Recycling, pp. 267–286. CRC Press (2020)
5. Möllnitz, S., Khodier, K., Pomberger, R., Sarc, R.: Grain size dependent distribution of different plastic types in coarse shredded mixed commercial and municipal waste. Waste Manag. **103**, 388–398 (2020). https://doi.org/10.1016/j.wasman.2019.12.037
6. Kumar, S., Smith, S.R., Fowler, G., et al.: Challenges and opportunities associated with waste management in India. R. Soc. Open Sci. **4**, 160764 (2017). https://doi.org/10.1098/rsos.160764
7. APA PEA: Municipal waste collection in mainland Portugal. In: Municipal Waste Production and Management (2021)
8. Gundupalli, S.P., Hait, S., Thakur, A.: A review on automated sorting of source-separated municipal solid waste for recycling. Waste Manag. **60**, 56–74 (2017). https://doi.org/10.1016/j.wasman.2016.09.015
9. Shapiro, M., Galperin, V.: Air classification of solid particles: a review. Chem. Eng. Process. Process Intensif. **44**, 279–285 (2005). https://doi.org/10.1016/j.cep.2004.02.022
10. Wagland, S.: Waste collection, sorting, and pretreatment. In: Substitute Natural Gas from Waste, pp. 37–48. Elsevier (2019)
11. Dodbiba, G., Fujita, T.: Progress in separating plastic materials for recycling. Phys. Sep. Sci. Eng. **13**, 165–182 (2004). https://doi.org/10.1080/14786470412331326350
12. Cimpan, C., Maul, A., Jansen, M., et al.: Central sorting and recovery of MSW recyclable materials: a review of technological state-of-the-art, cases, practice and implications for materials recycling. J. Environ. Manag. **156**, 181–199 (2015). https://doi.org/10.1016/j.jenvman.2015.03.025
13. Jank, A., Müller, W., Schneider, I., et al.: Waste separation press (WSP): a mechanical pretreatment option for organic waste from source separation. Waste Manag. **39**, 71–77 (2015). https://doi.org/10.1016/j.wasman.2015.02.024
14. Fitzgerald, G.C.: Pre-processing and treatment of municipal solid waste (MSW) prior to incineration. In: Waste to Energy Conversion Technology, pp. 55–71. Elsevier (2013)
15. Moon, B.Y., Kim, K.H., Kwak, K.H., et al.: Dynamic analysis and design of separation screen mechanism in a plant of moisturized wastes. Int. J. Mod. Phys. B **22**, 1449–1454 (2008). https://doi.org/10.1142/S0217979208046918
16. Liu, K.: Some factors affecting sieving performance and efficiency. Powder Technol. **193**, 208–213 (2009). https://doi.org/10.1016/j.powtec.2009.03.027

17. Gikas, P., Zhu, B., Batistatos, N.I., Zhang, R.: Evaluation of the rotary drum reactor process as pretreatment technology of municipal solid waste for thermophilic anaerobic digestion and biogas production. J. Environ. Manag. **216**, 96–104 (2018). https://doi.org/10.1016/j.jenvman.2017.07.050
18. Möllnitz, S., Küppers, B., Curtis, A., et al.: Influence of pre-screening on down-stream processing for the production of plastic enriched fractions for recycling from mixed commercial and municipal waste. Waste Manag. **119**, 365–373 (2021). https://doi.org/10.1016/j.wasman.2020.10.007
19. Wolfe, A.L., Liu, R., Stewart, B.W., et al.: A method for generating uniform size-segregated pyrite particle fractions. Geochem. Trans. **8** (2007). Article number: 9 https://doi.org/10.1186/1467-4866-8-9
20. Sanchez, S.M., Wan, H., Lucas, T.W.: Two-phase screening procedure for simulation experiments. ACM Trans. Model. Comput. Simul. **19**, 1–24 (2009). https://doi.org/10.1145/1502787.1502790
21. Chen, Y.-S., Hsiau, S.-S., Lee, H.-Y., et al.: Size separation of particulates in a trommel screen system. Chem. Eng. Process. Process Intensif. **49**, 1214–1221 (2010). https://doi.org/10.1016/j.cep.2010.09.003
22. Rousseaux, P., de Castilhos, A.B., Vermande, P., Veron, J.: Determination of trommel meshes for optimal separation of compostible fraction of municipal solid wastes from Florianópolis, SC, Brazil. Resour. Conserv. Recycl. **6**, 217–229 (1992). https://doi.org/10.1016/0921-3449(92)90032-W
23. Wheeler, P.A., Barton, J.R., New, R.: An empirical approach to the design of trommel screens for fine screening of domestic refuse. Resour. Conserv. Recycl. **2**, 261–273 (1989). https://doi.org/10.1016/0921-3449(89)90003-7
24. Alter, H., Gavis, J., Renard, M.L.: Design models of trommels for resource recovery processing. Resour. Conserv. **6**, 223–240 (1981). https://doi.org/10.1016/0166-3097(81)90051-1

A Comparative Study of Weld Bead Geometry of AISI 1023 with Varying Plate Width Using SAW and ASAW

Tejendra Singh Singhal[1], Jinesh Kumar Jain[1(✉)], Manoj Kumar[2], and Rajeev Agrawal[1]

[1] Department of Mechanical Engineering, Malaviya National Institute of Technology, Jaipur, Rajasthan, India
jineshjain.mech@mnit.ac.in
[2] Department of Mechanical-Mechatronics Engineering, The LNM Institute of Information Technology, Jaipur, Rajasthan, India

Abstract. Submerged arc welding (SAW) is a high heat input process which favours fine weld bead with high metal deposition. Although higher metal deposition is an assuring output of this process but optimization of welding parameters is still crucial for proper responses. Most of the work till now has been done by using various alloys and also by varying plate thickness but no work has been reported on impact of varying plate width on weld bead geometry. In this experiment, impact of varying plate width is investigated through Taguchi's L9 orthogonal array and a total of 18 experiments (9 for SAW and 9 for Advance submerged arc welding (ASAW) process) were conducted on 12 mm thick plate of plain carbon steel (AISI 1023). In this method, three levels of four process parameters are selected and their effect on weld bead geometry was observed. The S/N ratios was calculated to determine the optimum parameters. A novel water-cooled SAW process also used and comparison study between SAW and ASAW was done. optimum values of weld penetration, bead width, height of reinforcement, % D, HAZ area can be achieved have shown in the article.

Keywords: Penetration · Plate width · Bead width · Reinforcement height · HAZ

1 Introduction

Steel is the most commonly used material in the structural components of bridges, buildings, and many other types of machine parts. It is commonly used in turbine discs, shaft couplings, and ship gear blanks. Plain carbon steel (PCS) is the most common type of steel because it is inexpensive, strong despite having a low carbon content, and easily formable to mold into any desired shape. Because of its low carbon content, it has excellent weldability. Low carbon steel has great capability to get weld when worked with correct parameters.

Welding is the process of producing a localized joint for material joining. Welding has had an impact on many businesses by increasing productivity and product life. SAW

J. Machado et al. (Eds.): icieng 2022, LNME, pp. 116–135, 2023.
https://doi.org/10.1007/978-3-031-09382-1_11

is a type of fusion welding process that is preferred over other processes because it produces high heat, which allows metal to be deposited at a faster rate on the workpiece plate, resulting in deeper penetration [1]. SAW is characterized by the formation of an electric arc between the workpiece plate and the wire tip. A granular flux was used to submerge the weld bead and wire tip region to improve weld bead quality; this flux is also used to improve weld pool region properties by shifting the alloying element to the weld pool region [2]. Thus, to produce a quality bead profile, selection of these chemical compounds also plays a key role [3]. The SAW process has some advantages, such as a more stable arc and higher productivity, as well as no contact with environmental contaminants. SAW provides a higher welding current, approximately four times that of other processes such as GMAW, and SMAW results in a higher heat input [4].

The study on the effect of various input variables on weld bead profile have been reported by many investigators. Gunaraj and Murugun [5] studied that increasing the arc length results in more melting at the workpiece surface gives shallow penetration and when the welding speed increases, the heat input decreases and the welding torch travels at a higher speed over the workpiece surface results in a lower rate of metal deposition on the weld joint. So, because of a lower metal deposition rate and less heat input, penetration decreases. And with increasing wire feed rate welding arc current increases and similarly heat input, and the rate of metal deposition also increases. Hence, for a higher rate of metal deposition and heat input, Penetration increases. In other words, heat input, welding current and welding feed rate having a positive effect on penetration whereas welding voltage and welding speed having a negative effect on penetration.

Dae-Won Cho et al. [6] studied that with direct current electrode negative polarity without any change in the Open circuit voltage (OCV) and wire feed rate, a quick reduction in dilution takes place. This is because of a reduction in welding current and therefore, arc force which finally affects the penetration of the arc. In single DC welding, a different type of weld bead geometries measured by varying the torch angle result in a positive torch angle produced shallow penetration and a negative torch angle produced deeper penetration. In single AC welding, shallow penetration is produced because of frequent changes in the molten pool flow pattern. W H Song et al. [7] studied that by increasing the wire feed rate, welding current increase results in improved line power per unit length of the weld bead and with an increase in welding current, the heat content of the droplets increases and molten droplet quickly striking the weld pool causes more heating of the base plate, results in deeper penetration.

S Shen et al. [8] studied that higher heat input produces a large molten pool in the fusion zone and as the cooling rate changes in reverse with heat input, the molten metal remains in the liquid state for a long period at high heat input, results in deeper penetration. an increase in a nozzle to plate distance (N), the heat input decreases results in reinforcement of weld bead decreases and with the increase in heat input the reinforcement of weld bead also increases. M.E. Khallaf [9] studied that the crack length decreases with an increase in travel speed, results in low heat input, and as the heat input decreases, penetration decreases, and percentage dilution decreases.

Junior et al. [10] studied that the weld bead dimension increases with an increase in heat input without significantly changing the weld dilution. Hence, to control, the corrosion characteristics of weld metal, use coupling variations of heat input with cold wire addition. Serdar Karaoglu et al. [11] studied that the speed sensitivity of weld bead width and current sensitivity of penetration depth reach their maximum values at higher heat input value (which means higher the welding voltage and welding current and lower the welding speed).

P V S S Sridhar et al. [12] observed that with an increase in current bead width increases and after a specific point it decreases with further increase in current but it is higher than initial bead width and with an increase in voltage up to 26 V, and keeping other parameters as constant, results in arc length increases which provide more arc spread. Hence, the bead width increases. and as a result, for SS 304 austenitic stainless steel, maximum bead width of 15.62 mm was obtained at 26 V. An increase in welding voltage gives a flatter bead profile result in reduces reinforcement height. An increase in welding current penetration increases and shows the linear relationship and as a result for SS 304 austenitic stainless-steel maximum penetration of 5.72 mm can be obtained at 390 A current.

L J Yang et al. [13] studied that a constant current power source with acidic flux having no impact on bead width but when the basic flux is used, bead width increases. Gunaraj and Murugun [14] studied that by increasing the nozzle to plate distance (N), the arc length increases resulting in the expansion of the arc cone. Similarly, because of Joule's heating effect, at higher values of N, the metal fusion rate increase, Hence, bead width increases. By increasing the welding feed rate, the welding current increases, and hence, the heat input, wire melting and deposition rate increase. Therefore, the bead width increases. By increasing the welding speed, the welding torch travels at a higher speed over the workpiece, results in lower heat input and metal deposition rate. Hence, the bead width decreases. Singhal and Jain [15] has explained the cladding process through welding with different approaches and explained the advantages of various processes. Singhal And Jain [16] have studied various sustainability measures including various form of welding with different process parameters and limits and also about how these processes could become sustainable and can produce benefits to the society.

Although in some applications SAW welding process results in higher dilution, to overcome this an advance submerged arc welding was introduced which has an assisting power source for preheating the welding wire [17]. Shukla and Pandey [18] studied with five process parameters and preheat current to observed that dilution increases with welding speed, WFR and voltage while preheat current decreases dilution. Om and Pandey [19] modelled and optimized the impact of process parameters in cladding operations. OCV, NPD, WFR, preheat current and speed were used with different levels. Study showed that with increase in speed and preheat current, penetration and bead width decreases. Bead width and reinforcement increases with OCV and wire feed rate respectively. Preheat current and speed reduces dilution. Dilution minimum to 30.94% were achieved in this study. Although preheating of welding wire decreases the dilution but it bends the welding wire on higher preheat current values. To overcome this problem, a water-cooled submerged arc welding torch is developed through which we can use preheat current up to 320 A. Ankush C et al. [20] studied on impact of process variables on bead properties and observed that preheat current and arc voltage increases metal deposition rate and flux consumption per unit length while speed decreases metal deposition rate. H K Arya et al. [21] explained the impact of varying plate thickness and observed that increment in plate thickness decreases penetration because heat conduction loss in thickness direction becomes predominate for thick plates while little increment in bead width observed due to heat sinking in thickness direction.

Most of the work was done on impact of varying plate thickness on bead geometry but few works were done on impact of varying plate width on weld bead geometry. In this paper, SAW and ASAW process were used in which welding wire was preheated via auxiliary power source and welding torch was cooled through water. This process produces lower dilution by provides high preheat current. This paper focused on impact of varying plate width on weld bead profile through Taguchi method and multiple regression analysis.

2 Materials and Methods

In this experiment, Plain Carbon Steel (AISI 1023) is used as a workpiece plate. Plain Carbon Steel (AISI 1023) provided good weld ability because of lower carbon contents. The Experiment was conducted with varying the plate width (50, 75 and 100 mm), 200 mm length and 12 mm plate thickness. The experiment was carried out using Cu coated wire, having 4 mm diameter.

The composition of the workpiece plate and wire is shown in Table 1. Bead-on-plate method used to deposit metal on the workpiece surface along the centre line in the longitudinal direction. Experiments were conducted with direct current reverse polarity and the nozzle to plate distance (NTPD) kept as constant 30 mm. Workpiece and wire melting were done with ESAB CPRA 800 S power source in SAW process and preheating of wire was done with ESAB WARRIOR 400 I, a secondary power source with ESAB CPRA 800 S main power source in ASAW process, constant voltage power source was used in this experiment. during first 9 samples welding wire is preheated and welding torch is cooled continuously by an arrangement, in which an electric motor circulate water in the nozzle. The experiment setup shown in Fig. 1 and 2.

The input process variables such as voltage, WFR, speed and plate width were selected to find their impact on weld bead profile. The input variable levels and their respective values are selected based on previous research as shown in Table 2. Experiments was conducting by using Taguchi L_9 orthogonal array in which 3 levels and 4 variables were selected [22–24]. The experimental design using the L9 orthogonal array is shown in Table 3. Total 18 experiments were conducted using various combinations of process parameters (9 with preheat and 9 without preheat) as per design matrix. Plates were cut into desired size by plasma arc cutting and cleaned with hand grinder to remove dust and any small scratches from workpiece plate. Flux is heated up to 200 °C for 1 h to remove moisture contents. Place the workpiece plate on fixture at SAW setup and used two backup plate at both sides to eliminate arc blow, measured the NTPD as 30 mm, mark the end point of plate and submerged the plate by flux. Perform the operations one by one, after welding operation kept the weld plate at dry place to cool. After that with help of abrasive cutter cut the uniform section of sample from the plate, take three samples from each plate, with use of grinding machine clean the samples, remove burrs.

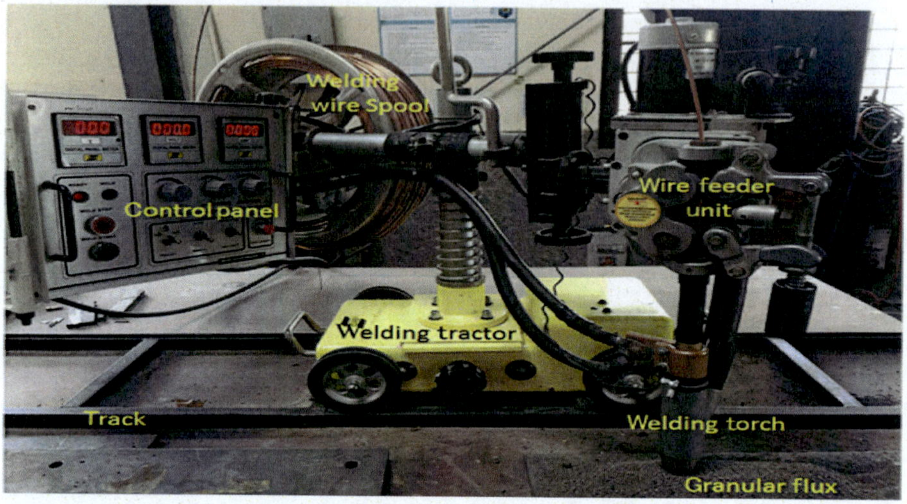

Fig. 1. Experiment setup for SAW [2]

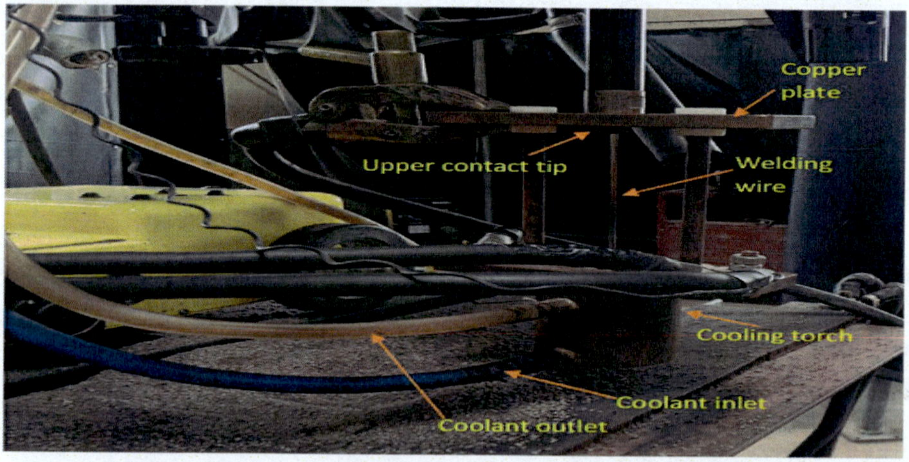

Fig. 2. Setup of water-cooled nozzle in advance SAW process [20]

Table 1. Chemical composition of workpiece plate and welding wire

	Composition (wt%)						
	Carbon	Silicon	Sulphur	Manganese	Phosphorus	Copper	Chromium
Workpiece plate	.22	.30	.043	.45	.042	.15	.082
Wire electrode	.10	.07	.020	.60	.018	.28	–

Then each sample polished with different emery papers starting from 120 up to 2000 through belt grinder and rotary polishing machine. These samples etched with 2% nital solution after etching with help of high-resolution scanner weld bead parameters were measured.

Table 2. SAW input variable values

Process parameter	Units	Symbols	Levels		
			−1	0	+1
Welding voltage	volts	V	36	40	44
Wire feed rate	mm/min	F	21	23	25
Welding speed	mm/s	C	7	8	9
Plate width	mm	D	50	75	100

2.1 Taguchi Method

Taguchi's quality engineering method, which employs design of experiment (DOE), is one of the important statistical tools for developing high-quality systems at a low cost. This method is a systematic and effective method for optimizing designs for performance, cost, and quality. The fundamental stage in Taguchi's strategy for achieving high quality without increasing costs is to optimize input variables [25, 26].

In the SAW process, Taguchi method is used for optimization procedure to determine the optimal input variables. Unal et al. [27] used the Taguchi method to optimize design during the manufacturing process. measured the friction welding variables that influence the output parameters (primarily tensile properties). Singh et. al. [28] has done experiment on shrinkage reduction using polypropylene and optimized the results using taguchi and ANOVA. The paper shows the optimization of the effects of various process parameters in reduction of shrinkage in plastic components manufacturing.

2.2 Multiple Regression Analysis

To determine the relationships between variables, the multiple regression analysis technique is used.

Table 3. Experiment design using Taguchi L_9 orthogonal array

S. No.	Voltage	Wire feed rate	Welding speed	Plate width	Status
1	36	21	7	50	With preheat
2	36	23	8	75	
3	36	25	9	100	
4	40	21	8	100	
5	40	23	9	50	
6	40	25	7	75	
7	44	21	9	75	
8	44	23	7	100	
9	44	25	8	50	
10	36	21	7	50	Without preheat
11	36	23	8	75	
12	36	25	9	100	
13	40	21	8	100	
14	40	23	9	50	
15	40	25	7	75	
16	44	21	9	75	
17	44	23	7	100	
18	44	25	8	50	

3 Results and Discussion

In this study, experiments were conducted accurately and their responses were measured, then critical process parameters identified by analyzing their effects on weld bead geometry. The output parameters considered in this work are penetration, bead width, % dilution, reinforcement height, HAZ. These output parameters were measured through Adobe Acrobat® software (Measuring tool plug in of adobe has been used to measure the responses after scan). The weld bead geometry obtained on AISI 1023 plate are shown in Fig. 3. The results obtained for each response are described in detail below, along with their analysis.

3.1 Effect of Input Variables on Penetration

Weld penetration was measured through the weld joint's cross section. By machining and polishing, good surface finish was obtained on weld joint area. Table 4 shows the experiment design and S/N ratio for ASAW and SAW process. The values of average S/N ratios in ASAW for penetration and optimum input variables value for penetration is A_3 - B_3 - C_1 - D_2, arc voltage 44 V, wire feed rate 25 mm/min, speed 7 mm/s, plate width 75 mm. The optimum input variable value for SAW is A_2 - B_3 - C_1 -

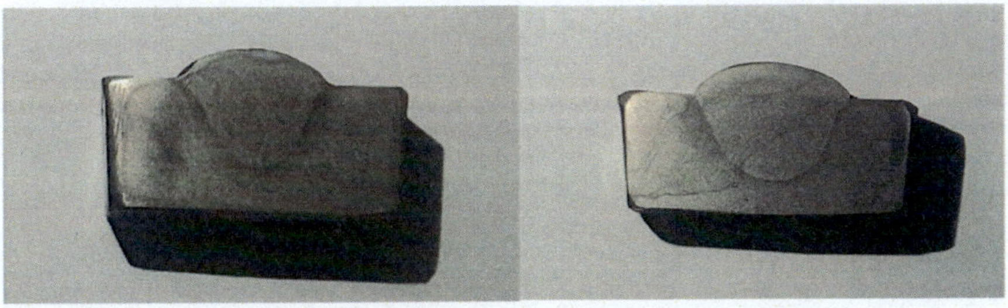

Fig. 3. Scanned images of bead geometry

D_2, arc voltage 40 V, wire feed rate 25 mm/min, speed 7 mm/s, plate width 75 mm (As shown in Table 5).

Table 4. Experiment design and S/N ratio for weld penetration

S. No.	Voltage	Wire feed rate	Welding speed	Plate width	Penetration (mm) ASAW	Penetration (mm) SAW	S/N ratio of penetration ASAW	S/N ratio of penetration SAW
1	36	21	7	50	5.45	6.9	14.7279	16.7770
2	36	23	8	75	6.37	8.1	16.0828	18.1697
3	36	25	9	100	5.97	7.94	15.5195	17.9964
4	40	21	8	100	5.4	6.6	14.6479	16.3909
5	40	23	9	50	6.72	7.21	16.5474	17.1587
6	40	25	7	75	7.01	10.28	18.0727	20.2399
7	44	21	9	75	6.13	7.48	15.7492	17.4780
8	44	23	7	100	8.01	7.33	18.0727	17.3021
9	44	25	8	50	7.75	8.44	17.7860	18.5268

It is shown that for ASAW wire feed rate having maximum impact on penetration followed by voltage, welding speed and plate width respectively. For SAW wire feed rate having maximum impact on penetration followed by plate width, welding speed and welding voltage respectively (Fig. 4).

The relationship between penetration and input variables is determined using multiple regression (MR) analysis. The linear equation of predicted weld penetration for ASAW is observed from regression analysis is shown below:

$$P = -5.49 + 0.1708\ V + 0.3958\ F - 0.442\ S - 0.00360\ D \tag{1}$$

The linear equation of predicted weld penetration for SAW

$$P = -0.75 + 0.0129\ V + 0.473\ F - 0.313\ S - 0.0045\ D \tag{2}$$

The coefficient of determination (R2) obtained for the model is 89.15% (ASAW) and 63.18% (SAW).

Table 5. Mean values of S/N ratio for input variables

Level	Welding voltage (V)		Wire feed rate (F)		Welding speed (c)		Plate width (D)	
	ASAW	SAW	ASAW	SAW	ASAW	SAW	ASAW	SAW
1	15.44	17.65	15.04	16.88	16.96	18.11	16.35	17.49
2	16.42	17.93	16.90	17.54	16.17	17.70	16.63	18.63
3	17.20	17.77	17.13	18.92	15.94	17.54	16.08	17.23
Delta	1.76	0.28	2.08	2.04	1.02	0.56	0.55	1.4
Rank	**2**	**4**	**1**	**1**	**3**	**3**	**4**	**2**

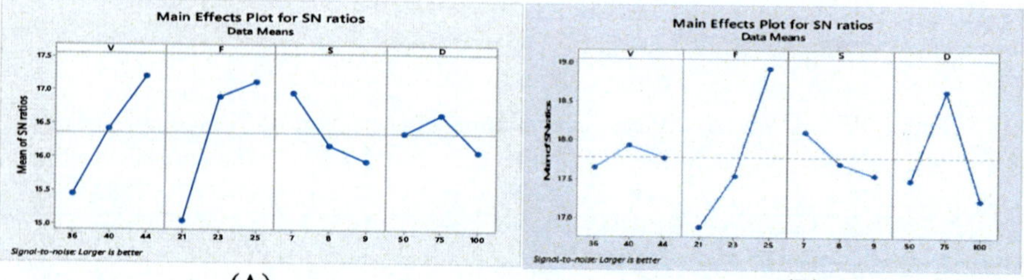

(A) (B)

Fig. 4. Effect of input variables on mean S/N ratio for weld penetration (A) ASAW (B) SAW

3.2 Effect of Input Variables on Weld Bead Width

Measurement of weld bead width was observed from each experiment to analysed the influence of each input variable. Table 6 shows the experiment design and S/N ratio was found with help of above equation. The values of average S/N ratios for bead width and optimum input variables value for bead width in ASAW is A_3 - B_1 - C_2 - D_1, voltage 44 V, wire feed rate 21 mm/min, speed 8 mm/s, plate width 50 mm and for SAW is A_3 - B_1 - C_1 - D_1, arc voltage 44 V, wire feed rate 21 mm/min, speed 7 mm/s, plate width 50 mm as shown in Table 7.

It is shown that for ASAW plate width having maximum impact on weld bead width followed by welding voltage, wire feed rate and speed respectively and for SAW welding voltage having maximum impact on weld bead width followed by plate width, wire feed rate and welding speed.

The relationship between bead width and input variables for ASAW is determined using MR analysis. The linear equation of predicted bead width is observed from regression analysis is shown below:

$$W = 18.37 + 0.0104\ V - 0.1008\ F - 0.097\ S - 0.01467\ D \tag{3}$$

For SAW

$$W = -2.46 + 0.5367\ V - 0.0725\ F - 0.113\ S - 0.01647\ D \tag{4}$$

The coefficient of determination (R2) obtained for the model for ASAW is 55.36% and for SAW it is 96.79% (Fig. 5).

Table 6. Experiment design and S/N ratio for weld bead width

S. No.	Voltage	Wire feed rate	Welding speed	Plate width	Bead width (mm) ASAW	Bead width (mm) SAW	S/N ratio of bead width ASAW	S/N ratio of bead width SAW
1	36	21	7	50	15.35	13.93	−23.722	−22.87
2	36	23	8	75	14.87	13.33	−23.446	−22.49
3	36	25	9	100	14.02	12.59	−22.935	−22.00
4	40	21	8	100	14.28	14.46	−23.094	−23.20
5	40	23	9	50	14.31	15.07	−23.112	−23.56
6	40	25	7	75	14.01	14.69	−22.928	−23.34
7	44	21	9	75	14.85	17.59	−23.434	−24.90
8	44	23	7	100	14.4	17.31	−23.167	−24.76
9	44	25	8	50	15.24	17.83	−23.659	−25.02

Table 7. Mean values of S/N ratio for input variables

Level	Welding voltage (V)		Wire Feed rate (F)		Welding speed (c)		Plate width (D)	
	ASAW	SAW	ASAW	SAW	ASAW	SAW	ASAW	SAW
1	−23.37	−22.46	−23.42	−23.66	−23.27	−23.66	−23.50	−23.82
2	−23.05	−23.37	−23.24	−23.61	−23.40	−23.57	−23.27	−23.58
3	−23.42	−24.90	−23.17	−23.45	−23.16	−23.49	−23.07	−23.32
Delta	0.38	2.44	0.24	0.21	0.24	0.17	0.43	0.50
Rank	**2**	**1**	**3**	**3**	**4**	**4**	**1**	**2**

3.3 Effect of Input Variables on Reinforcement Height

Measurement of reinforcement height was observed from each experiment to analysed the influence of each input variable. Table 8 shows the experiment design and S/N ratio was found with help of above equation. The values of average S/N ratios for reinforcement height and optimum input variables value for reinforcement height IN ASAW process is A_1 - B_3 - C_2 - D_3, arc voltage 36 V, wire feed rate 25 mm/min, speed 8 mm/s, plate width 100 mm and for SAW process is A_2 - B_3 - C_2 - D_3, arc

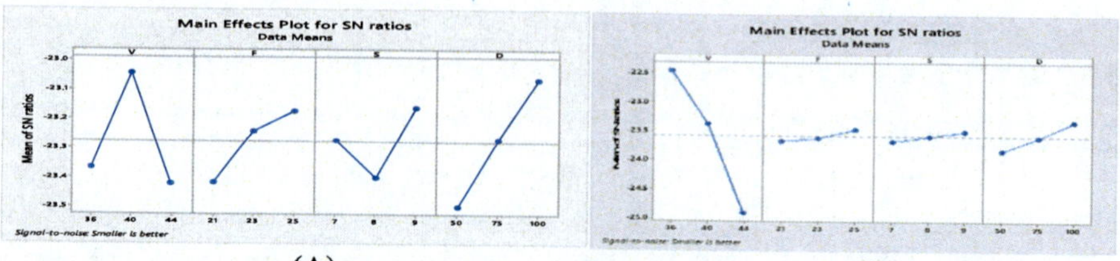

(A) (B)

Fig. 5. Effect of input variables on mean S/N ratio for weld bead width (A) ASAW (B) SAW

voltage 40 V, wire feed rate 25 mm/min, speed 8 mm/s, plate width 100 mm as shown in Table 9.

It is shown that for ASAW plate width having maximum impact on reinforcement height followed by welding speed, welding voltage and wire feed rate while for SAW wire feed rate having maximum impact on reinforcement height followed by welding voltage, plate width and welding speed respectively.

The relationship between reinforcement height and input variables is determined using MR analysis. The linear equation of predicted reinforcement height is observed from regression analysis for ASAW is shown below:

$$R = 4.35 - 0.0287\ V + 0.0150\ F - 0.1183\ S + 0.00753\ D \tag{5}$$

For SAW

$$R = -0.85 + 0.0483\ V + 0.1058\ F - 0.1567\ S + 0.00800\ D \tag{6}$$

The coefficient of determination (R2) obtained for the model is 71.50% for ASAW and 80.36 for SAW (Fig. 6).

Table 8. Experiment design and S/N ratio for reinforcement height

S. No.	Voltage	Wire feed rate	Welding speed	Plate width	Reinforce-ment (mm) ASAW	Reinforce-ment (mm) SAW	S/N ratio of reinforce-ment ASAW	S/N ratio of reinforce-ment SAW
1	36	21	7	50	3.06	2.27	−9.7144	−7.1205
2	36	23	8	75	3.48	2.74	−10.8316	−8.7550
3	36	25	9	100	3.26	2.78	−10.2644	−8.8809
4	40	21	8	100	3.49	3.17	−10.2644	−8.8809
5	40	23	9	50	2.74	2.6	−8.7550	−8.2995
6	40	25	7	75	3.31	3.31	−10.3966	−10.3966
7	44	21	9	75	2.87	2.53	−9.1576	−8.0624
8	44	23	7	100	3.21	3.27	−10.1301	−10.2910
9	44	25	8	50	3.03	3.15	−9.6289	−9.9662

Table 9. Mean values of S/N ratio for input variables

Level	Welding voltage (V)		Wire feed rate (F)		Welding speed (c)		Plate width (D)	
	ASAW	SAW	ASAW	SAW	ASAW	SAW	ASAW	SAW
1	−10.27	−8.252	−9.91	−8.401	−10.08	−9.269	−9.366	−8.462
2	−10.003	−9.572	−9.906	−9.115	−10.439	−9.581	−10.129	−9.071
3	−9.639	−9.440	−10.097	−9.748	−9.392	−8.414	−10.417	−9.731
Delta	0.631	1.320	0.191	1.347	1.047	1.167	1.051	1.269
Rank	**3**	**2**	**4**	**1**	**2**	**4**	**1**	**3**

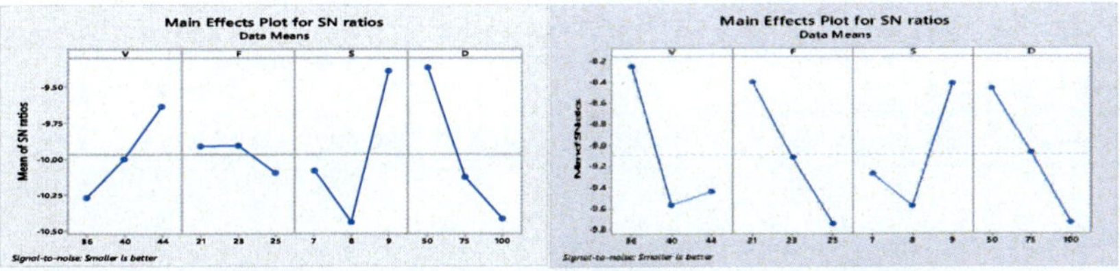

(A) (B)

Fig. 6. Effect of input variables on mean S/N ratio for reinforcement height (A) ASAW (B) SAW

3.4 Effect of Input Variables on Percentage Dilution

Measurement of % dilution was observed from each experiment to analyzed the influence of each input variable. Table 10 shows the experiment design and S/N ratio was found with help of above equation. The values for ASAW of average S/N ratios for % dilution and optimum input variables value for % dilution is A_2 - B_1 - C_2 - D_1, arc voltage 40 V, wire feed rate 21 mm/min, speed 8 mm/s, plate width 50 mm and for SAW is A_2 - B_3 - C_3 - D_1, arc voltage 40 V, wire feed rate 25 mm/min, speed 9 mm/sec, plate width 50 mm as shown in Table 11.

It is shown that for ASAW welding voltage having maximum impact on % dilution and further followed by plate width, wire feed rate and welding speed respectively and for SAW wire feed rate having maximum impact on % dilution and further followed by welding voltage, welding speed and plate width.

The relationship between % D and input variables is determined using MR analysis. The linear equation of predicted % D is observed from regression analysis for ASAW is shown below:

$$\%\,\mathrm{D} = 0.593 + 0.00205\ \mathrm{V} - 0.00054\ \mathrm{F} + 0.00425\ \mathrm{S} - 0.000363\ \mathrm{D} \quad (7)$$

For SAW

$$\%\,\mathrm{D} = 0.565 + 0.00140\ \mathrm{V} + 0.00121\ \mathrm{F} + 0.00597\ \mathrm{S} - 0.000237\ \mathrm{D} \quad (8)$$

The coefficient of determination (R2) obtained for ASAW and SAW for the model is 53.43 and 33.12% respectively (Fig. 7).

Table 10. Experiment design and S/N ratio for % dilution

S. No.	Voltage	Wire feed rate	Welding speed	Plate width	% Dilution (mm) ASAW	% Dilution (mm) SAW	S/N ratio of % Dilution ASAW	S/N ratio of % dilution SAW
1	36	21	7	50	66.42538	67.10704	3.55332	3.46464
2	36	23	8	75	65.11864	65.7173	3.72589	3.64641
3	36	25	9	100	65.24532	67.59646	3.70901	3.40152
4	40	21	8	100	67.89780	68.82648	3.36289	3.24489
5	40	23	9	50	68.14047	68.4153	3.33190	3.29694
6	40	25	7	75	67.74635	69.10098	3.38228	3.21032
7	44	21	9	75	68.12559	68.95744	3.33379	3.22838
8	44	23	7	100	64.78925	65.18138	3.76994	3.71753
9	44	25	8	50	68.80413	69.64202	3.24771	3.14257

Table 11. Mean values of S/N ratio for input variables

Level	Welding voltage (V)		Wire feed rate (F)		Welding speed (c)		Plate width (D)	
	ASAW	SAW	ASAW	SAW	ASAW	SAW	ASAW	SAW
1	3.663	3.504	3.417	3.3132	3.569	3.464	3.378	3.301
2	3.359	3.251	3.609	3.554	3.445	3.345	3.481	3.362
3	3.450	3.363	3.446	3.251	3.458	3.309	3.614	3.455
Delta	0.304	0.253	0.193	0.302	0.123	0.155	0.236	0.153
Rank	**1**	**2**	**3**	**1**	**4**	**3**	**2**	**4**

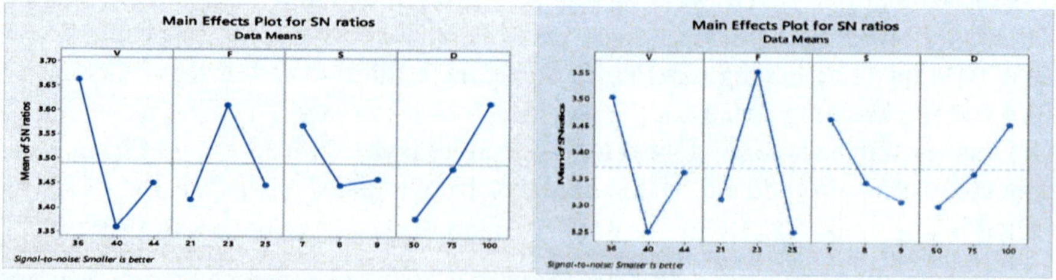

(A) (B)

Fig. 7. Effect of input variables on mean S/N ratio for % dilution (A) ASAW (B) SAW

3.5 Effect of Input Variables on HAZ

Measurement of HAZ was observed from each experiment to analyzed the influence of each input variable. Table 12 shows the experiment design and S/N ratio was found with help of above equation. For the ASAW process the values of average S/N ratios for HAZ and optimum input variables value are is A_1 - B_3 - C_3 - D_1, arc voltage 36 V, wire feed rate 25 mm/min, speed 9 mm/s, plate width 50 mm whereas for SAW these are A_2 - B_1 - C_1 - D_1, arc voltage 40 V, wire feed rate 21 mm/min, speed 7 mm/s, plate width 50 mm as shown in Table 13.

It is shown that for ASAW and SAW both plate width will have maximum impact on HAZ followed by welding voltage, speed and wire feed rate respectively.

The relationship between HAZ and input variables is determined using MR analysis. The linear equation of predicted HAZ is observed from regression analysis for ASAW is shown below:

$$\mathrm{HAZ} = 51.25 - 0.3867\ \mathrm{V} + 0.160\ \mathrm{F} + 0.713\ \mathrm{S} - 0.11613\ \mathrm{D} \tag{9}$$

For SAW

$$\mathrm{HAZ} = 70.2 + 0.049\ \mathrm{V} - 0.518\ \mathrm{F} - 1.14\ \mathrm{S} - 0.1632\ \mathrm{D} \tag{10}$$

The coefficient of determination (R2) obtained for the model for ASAW and SAW is 98.26 and 54.41% respectively (Fig. 8).

Table 12. Experiment design and S/N ratio for HAZ

S. No.	Voltage	Wire feed rate	Welding speed	Plate width	HAZ (mm) ASAW	HAZ (mm) SAW	S/N ratio of HAZ ASAW	S/N ratio of HAZ SAW
1	36	21	7	50	39.98	42.65	−32.036	−32.598
2	36	23	8	75	38.01	36.46	−31.598	−31.236
3	36	25	9	100	36.24	30.14	−31.183	−29.582
4	40	21	8	100	33.36	39.9	−30.464	−32.019
5	40	23	9	50	39.32	47.4	−31.892	−33.515
6	40	25	7	75	35.71	36.13	−31.058	−31.157
7	44	21	9	75	35.71	36.13	−31.055	−31.157
8	44	23	7	100	30.76	34.91	−29.759	−30.859
9	44	25	8	50	38.48	39.38	−31.704	−31.905

Table 13. Mean values of S/N ratio for input variables

Level	Welding voltage (V)		Wire feed rate (F)		Welding speed (c)		Plate width (D)	
	ASAW	SAW	ASAW	SAW	ASAW	SAW	ASAW	SAW
1	−31.61	−31.14	−31.19	−31.93	−30.99	−32.04	−31.88	−32.67
2	−31.18	−32.73	−31.08	−31.87	−31.26	−31.72	−31.28	−31.68
3	−30.84	−31.31	−31.36	−31.38	−31.38	−31.42	−30.47	−30.82
Delta	0.77	1.59	0.27	0.54	0.38	0.62	1.41	1.85
Rank	**2**	**2**	**4**	**4**	**3**	**3**	**1**	**1**

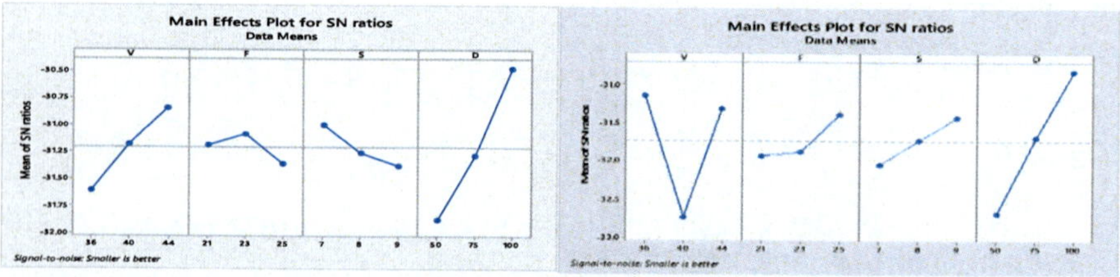

(A) (B)

Fig. 8. Effect of input variables on mean S/N ratio for HAZ (A) ASAW (B) SAW

4 Comparative Study of Bead Geometry on Varying Plate Width

In this experiment, variation in plate width was considered as welding process parameter and effect of this parameter on weld bead geometry was observed for submerged arc welding as well as advance submerged arc welding process. Mean value of penetration, weld bead width, reinforcement, % dilution and HAZ area were calculated and graphically represented.

4.1 Effect of Plate Width on Penetration

As shown in Table 14, a constant increment in the penetration can be seen with decreasing plate width. As small width of plate will result in more heat accumulation per unit area for a longer duration. Hence more metal will melt.

Also, penetration in ASAW is less than conventional SAW process, as we are using low heat input in ASAW (Fig. 9).

4.2 Influence of Width of Plate on Bead Width

As shown in Table 15, a constant increment in the bead width can be seen with decreasing plate width. As small width of plate will result in more heat accumulation per unit area for a longer duration. Hence more metal will melt.

Also, weld bead width in ASAW is less than conventional SAW process, as we are using low heat input in ASAW (Fig. 10).

Table 14. Mean values of penetration for plate width

Plate width (mm)	Penetration in SAW (mm)	Penetration in ASAW (mm)
100	7.29	6.46
75	8.62	6.50
50	8.65	6.64

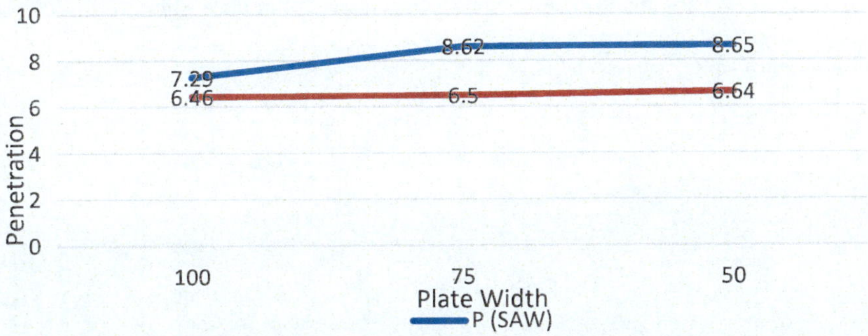

Fig. 9. Effect of plate width on weld penetration

Table 15. Mean values of bead width for plate width

Plate width (mm)	Bead width in SAW (mm)	Bead width in ASAW (mm)
100	14.78	14.23
75	15.20	14.57
50	15.61	14.96

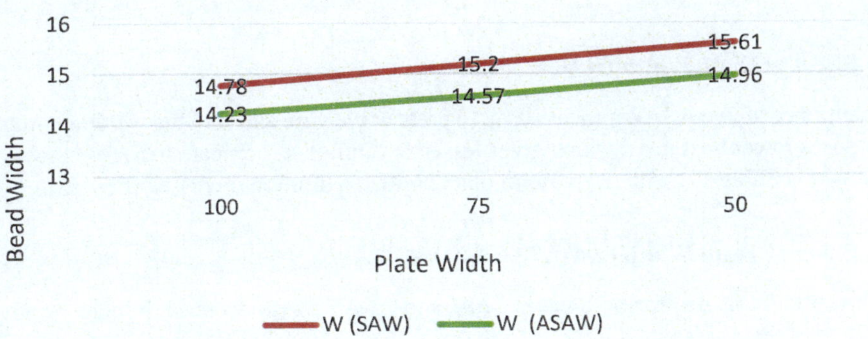

Fig. 10. Effect of plate width on weld bead width

4.3 Effect of Plate Width on Reinforcement Height

As shown in Table 16, a constant decrement in the reinforcement height can be seen with decreasing plate width. As small width of plate will result in more heat accumulation per unit area for a longer duration. Hence more metal will melt.

Also, reinforcement height in ASAW is more than conventional SAW process, as we are using low heat input in ASAW which having higher cooling rate and lesser fluidity of molten metal (Fig. 11).

Table 16. Mean values of reinforcement height for plate width

Plate width (mm)	Reinforcement height in SAW (mm)	Reinforcement height in ASAW (mm)
100	3.07	3.32
75	2.86	3.22
50	2.67	2.94

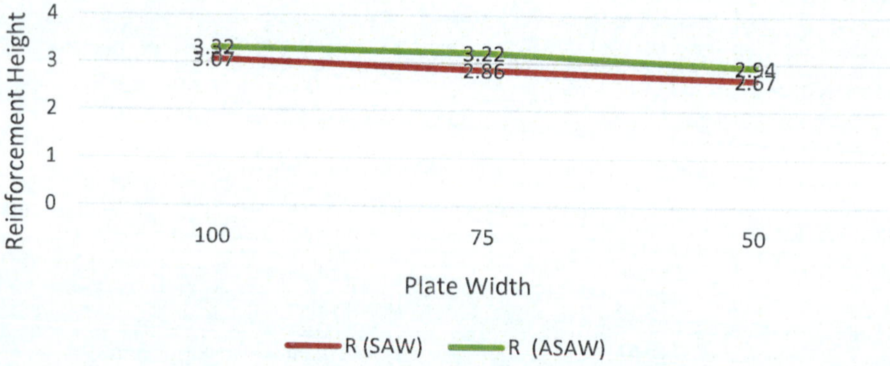

Fig. 11. Effect of plate width on reinforcement height

4.4 Effect of Plate Width on % Dilution

As mentioned above ASAW process used to preheat the wire and cool the welding torch, it produces lower heat input which gives lesser % dilution as compare to SAW process. As shown in Table 17, with decrease in plate width, % dilution increases (Fig. 12).

4.5 Effect of Plate Width on HAZ

As shown in Table 18, for both cases SAW and ASAW, with decrease in plate width, HAZ linearly increases.

Also, HAZ in ASAW is less than conventional SAW process, as we are using low heat input in ASAW (Fig. 13).

Table 17. Mean values of % dilution for plate width

Plate width (mm)	% dilution in SAW	% dilution in ASAW
100	67.19	65.97
75	67.92	66.99
50	68.38	67.78

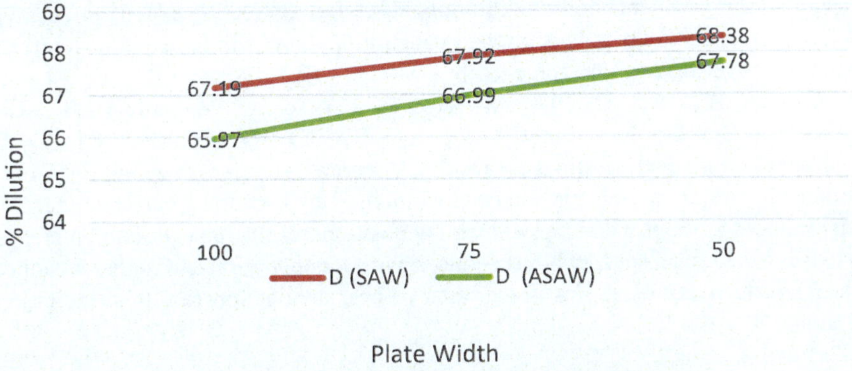

Fig. 12. Effect of plate width on % dilution

Table 18. Mean values of HAZ for plate width

Plate width (mm)	HAZ in SAW (mm^2)	HAZ in ASAW (mm^2)
100	34.98	33.45
75	38.51	36.66
50	43.14	39.26

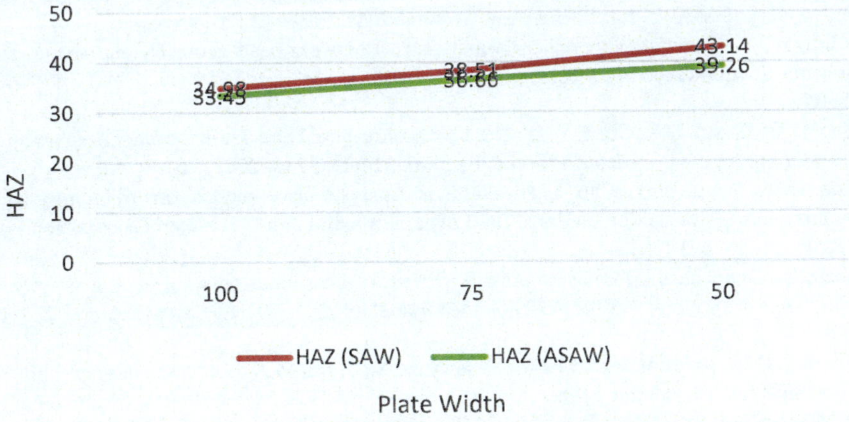

Fig. 13. Effect of plate width on HAZ

5 Conclusions

In this experiment, Taguchi technique and MR analysis were used to predict the optimum values of input variable for penetration, weld bead width, reinforcement height, % dilution and HAZ on 12 mm thickness plain carbon steel AISI 1023 material in SAW and ASAW. The experiment involves Taguchi L_9 orthogonal array to carry out on total 18 samples (9 for ASAW and 9 for SAW process) and average value of S/N ratios of the output parameters using larger is better for penetration and smaller is better for weld bead width, reinforcement height, % dilution, HAZ. By conducting experiment through Taguchi method, optimum values of weld penetration, bead width, height of reinforcement, % D, HAZ area can be achieved by A_3 - B_3 - C_1 - D_2, A_3 - B_1 - C_2 - D_1, A_1 - B_3 - C_2 - D_3, A_2 - B_1 - C_2 - D_1, A_1 - B_3 - C_3 - D_1 for ASAW and A_2 - B_3 - C_1 - D_2, A_3 - B_1 - C_1 - D_1, A_2 - B_3 - C_2 - D_3, A_2 - B_3 - C_3 - D_1, A_2 - B_1 - C_1 - D_1 for SAW process respectively. From these results, V increases arc length which spread over more area of workpiece. Hence, weld bead width increases. Welding current increases F which produces more heat and rate of metal deposition, results in penetration increases. C increases troch travel speed which produces narrow penetration and higher weld bead width. As plate width decreases, lesser area of heat conduction results in increase in HAZ area.

References

1. Aswal, V.K., Jain, J.K., Sonia, P.: Review on the behavior of various parameters on heat distribution in the SAW process. Mater. Today Proc. **47**(1), 6734–6739 (2021)
2. Choudhary, A., Kumar, M., Unane, D.: Experimental investigation and optimization of weld bead characteristics during submerged arc welding of AISI 1023 steel. Defence Technol. **15**(1), 72–82 (2019)
3. Viano, D.M., Ahmed, N.U., Schumann, G.O.: Influence of heat input and travel speed on microstructure and mechanical properties of double tandem submerged arc high strength low alloy steel weldments. Sci. Technol. Weld. Joining **5**(1), 26–34 (2000)
4. Sonia, P.: Review on heat and mass transfer in submerged arc welding (SAW) and gas metal arc welding (GMAW). Int. J. Eng. Adv. Technol. **8**(1), 503–512 (2019)
5. Murugan, N., Gunaraj, V.: Prediction and control of weld bead geometry and shape relationship in submerged arc welding of pipes. J. Mater. Process. Technol. **168**(3), 478–487 (2005)
6. Cho, D.W., Kiran, D.V., Song, W.H., Na, S.J.: Molten pool behavior in the tandem submerged arc welding process. J. Mater. Process. Technol. **214**(1), 2233–2247 (2014)
7. Cho, D.W., Song, W.H., Cho, M.H., Na, S.J.: Analysis of submerged arc welding process by three-dimensional computational fluid dynamics simulations. J. Mater. Process. Technol. **213**(1), 2278–2291 (2013)
8. Shen, S., Oguocha, I., Yannacopoulos, S.: Effect of heat input on weld bead geometry of submerged arc welding ASTM A709 Grade 50 steel joints. J. Mater. Process Technol. **212**(1), 286–294 (2012)
9. Khallaf, M.E., Ibrahim, M.A., El-Mahallawy, N.A., Taha, M.A.: On crack susceptibility in the submerged arc welding of medium- carbon steel plates. J. Mater. Process Technol. **68**(1), 43–49 (1997)

10. Junior, R.C., et al.: Influence of heat input and cold wire feeding rate on pitting corrosion resistance of submerged arc welding duplex stainless steel welds. J. Mater. Eng. Perform. **28**(1), 967–974 (2019). https://doi.org/10.1007/s11665-019-03967-7
11. Karaoglu, S., Secgin, A.: Sensitivity analysis of submerged arc welding process parameters. J. Mater. Process. Technol. **201**(1), 500–507 (2008)
12. Sridhar, P.V.S.S., Biswas, P., Mahanta, P.: Effect of process parameters on bead geometry, tensile and microstructural properties of double-sided butt submerged arc welding of SS 304 austenitic stainless steel. J. Braz. Soc. Mech. Sci. Eng. **42**(10), 1–15 (2020). https://doi.org/10.1007/s40430-020-02636-4
13. Yang, L.J., Chandel, R.S., Bibby, M.J.: The effects of process variables on the weld deposit area of submerged arc welds. Weld. J. **72**, 11–16 (1993)
14. Gunuraj, V., Murugan, N.: Prediction of heat affected zone characteristics in submerged arc welding of structural steel pipes. Weld. J. **81**(3), 45–53 (2002)
15. Singhal, T.S., Jain, J.K.: GMAW cladding on metals to impart anti-corrosiveness: machine, processes and materials. Mater. Today Proc. **26**(2), 2432–2441 (2020)
16. Singh, T., Jain, J.K.: A quantitative and qualitative review of sustainable manufacturing. In: Pandey, K.M., Misra, R.D., Patowari, P.K., Dixit, U.S. (eds.) Recent Advances in Mechanical Engineering. LNME, pp. 941–960. Springer, Singapore (2021). https://doi.org/10.1007/978-981-15-7711-6_93
17. Jain, A., Sonia, P., Kumari, S., Pushp, P.: Study of intermetallic compound (IMC) formed in welding of steel with magnesium: a review. Mater. Today Proc. **26**(2), 1159–1166 (2020)
18. Shukla, D.K.; Pandey, S.: Dilution control by advanced submerged arc welding. In: Advanced Materials Research, vol. 488, no. 1, pp. 1737–1741 (2012)
19. Om, H., Pandey, S.: Effect of process parameters on weld penetration shape factor in ASAW based surfacing. IJIRSET **3**(6), 13590–13596 (2014)
20. Choudhary, A., Kumar, M., Unune, D.R.: Investigation effects of resistance wire heating on AISI 1023 weldment characteristics during ASAW. Mater. Manuf. Process. **33**(7), 759–769 (2018)
21. Arya, H.K.: Effects of welding speed, welding current and plate thickness on temperature variation and angular distortion in butt joint welding. In: International Conference on Materials Science, pp. 131–136 (2011)
22. Chandel, R.S., Seow, H.P., Cheong, F.L.: Effect of increasing deposition rate on the bead geometry of submerged arc welds. J. Mater. Process. Technol. **72**(1), 124–128 (1997)
23. Kumanan, S., Dhas, R., Edwin, J., Gothman, K.: Determination of submerged arc welding process parameters using Taguchi method and regression analysis. Indian J. Eng. Mater. Sci. 177–183 (2007)
24. Yang, L.J., Chandel, R.S., Bibby, M.J.: The effects of process variables on the weld deposit area of submerged arc welds. Weld. J. **72**(11) (1993)
25. Sharma, A., Bhojak, V., Vikas Kukshal, S.K., Biswas, A.P., Patnaik, T.K.: Mechanical and Erosion characteristics of natural fiber reinforced polymer composite: effect of filler size. In: Katiyar, J.K., Bhattacharya, S., Patel, V.K., Kumar, V. (eds.) Automotive Tribology. EES, pp. 101–116. Springer, Singapore (2019). https://doi.org/10.1007/978-981-15-0434-1_6
26. Tarng, Y.S., Yang, W.H., Juang, S.C.: Int. J. Adv. Manuf. Technol. **16**(1), 688–694 (2000)
27. Resit, U., Edwin, B.D.: Taguchi approach to design optimization for quality and cost: an overview. In: Proceedings of the 13th Annual Conference of the International Society of Parametric Analysts (1991)
28. Singh, T., Singh, M.P., Alam, M.M.: Taguchi and ANOVA analysis of shrinkage of injection molded polypropylene Component. Int. J. Sci. Eng. Res. **5**(7), 750–756 (2014)

Ecofriendly Technologies of Fish Oils and Collagen Extraction from Fishery Waste – A Review

Nádia Valério[1,2](✉), Margarida Soares[1], Maria Cândida Vilarinho[2], and Joana Carvalho[1,2]

[1] CVR – Centre for Waste Valorisation, Guimarães, Portugal
jcarvalho@cvresiduos.com

[2] MEtRICs – Mechanical Engineering and Resource Sustainability Center, University of Minho, Guimarães, Portugal

Abstract. Following the exponential growth of the human population, a remarkable increase in the amount of fish waste has been produced worldwide. Fish processing industry generates a considerable number of by-products which represents a considerable environmental problem. Accordingly, reuse and valorisation of these by-products is a key process for marine resources preservation. This literature review focuses on eco-friendly extraction technologies of collagen and fish oils obtained by fish by-products. Furthermore, we will also carry out a literature review of ecotoxicological studies to evaluate the ecological impact of the extraction of the high-commercial products.

Keywords: Fish waste · Marine collagen · Fish oil · Marine sustainable sources · Circular economy · Green extraction technology

1 Introduction

Currently, Humanity faces the challenge of providing food and livelihoods for a world population of over 9 billion people. At the same time, Humankind is concerned about climate changes and environmental degradation, where natural resources are taken to extreme levels. According to the Food and Agriculture Organization (FAO) [1], fish represents about 17% of protein for animal consumption and 6.5% of protein for human consumption. Also in accordance to this organization, products from the fisheries and aquaculture sector represent about 20% of the intake of animal protein [1]. Fishery product consumption is related to its high nutritional value, since fish is a source of essential amino acids, and Omega-3 polyunsaturated fatty acids, which intake has been associated with beneficial health effects, including cardiovascular protection. The increase in fishery production resulted in an increase in the production of waste. FAO data, referring to 2018, indicate waste generation of more than 20 million tons of fishery by-products (head, skin, fins, bones, red meat, offal, shavings, and scales) [2]. Regarding the management of this type of waste, these are according to the Commission decision 2014/955/EU - European

J. Machado et al. (Eds.): icieng 2022, LNME, pp. 136–147, 2023.
https://doi.org/10.1007/978-3-031-09382-1_12

Waste List, chapter 02: "Wastes from agriculture, horticulture, aquaculture, forestry, hunting, food preparation and processing [3]. Thus, considering the waste management hierarchy and the potential of this type of product, its valorization must be privileged.

The Blue Economy concept is used to define all the wealth potential contained in the oceans. It is based on the efficient and full use of natural resources, without harming ecosystems, combining their current and future uses with regions, industry, and people, aligning interests, and reconciling expectations, contributing to the sustainable development of communities. Thus, the sustainable development advocated by the Blue Economy implies that eco-nomic development is inclusive and environmentally sound and must be carried out in a way which does not deplete the natural resources that societies depend on in the long term.

Therefore, this literature review will focus on the analysis of the green extraction methods available for the extraction of fish oil and collagen from fish waste. Furthermore, we will also review ecotoxicological studies of the fish oil and collagen extracted from fish waste. However, the literature shows a lack of information about this subject, which shows the urgency to enhance this field research.

1.1 Fishery in Portugal

Portugal is in the westernmost part of Europe on the west coast, with a coastline of around 1 214 km. Since the Portuguese Age of Discovery, fishing is a major economic activity in the country. Portugal is one of the countries in the world with the highest fish consumption per capita [4]. The Direction of Sea Policy (Direção-Geral da Política do Mar - DGPM) develops, evaluates, and updates the National Strategy for the Sea. The organization is also the responsible for the promotion of the national and international cooperation in maritime affairs [5]. As reported by Statistics Portugal and the Directorate-General for Natural Resources, Safety and Maritime Services (DGRM) in the 2020 Fishery Statistics compendium, the overall capture of fish caught in Portugal was 163 837 tons. In turn, 110 454 tons corresponded to fresh or chilled fish, traded in auction, which profits around 262 M euros [6]. However, the fisheries volume in 2020 shows a 13,1% reduction compared to the previous year. The main species caught showing a decrease, compared to 2019, at a national level, were mackerel, with a catch of 23 666 tons, anchovy with 5 575 tons, tuna with 6 822 tons and 14 609 tons of horse mackerel [6]. On the other hand, some species of fish showed an increase of the volume caught, such as sardines, with a volume of 14 526 tons in 2020, which represented an increase of 49.8% compared to 2019 [6]. Regarding fish catches in the Azores, these represented 7 682 tonnes. Although this number was lower in volume of fish caught in 2019, tuna fishing represented an increase of about 3.5%, contrary to what was seen in the rest of the country [6].

1.2 Fish Oil and Collagen Extracted from Fish Waste

In the primary processing of fish, all edible or non-edible surpluses can be considered by-products [7, 8]. Several operations are involved in fish processing, which generate by-products such as heads, offal, roe, shells [8]. In recent years, several studies have been carried out on the use of compounds produced from by-products of the fishery industry. Those biomaterials have applications in several sectors, such as nutraceuticals, emitted foods, pharmaceuticals, biomedicals, livestock, aquaculture feed, agriculture, biodiesel and fine chemicals, among others [8]. However, the literature highlights that the greatest bet has been on the production of products for human consumption, with the greatest profitability in the production (extraction and purification) of enzymes, bioactive peptides and biopolymers for biotechnological or pharmaceutical applications [9]. This approach has several benefits within the circular economy paradigm.

1.2.1 Collagen

Collagen is the predominant protein in the extracellular matrix of tissues, with fibrils formed by the self-assembling of triple helices [8, 10].

Marine organisms are the safest and most convenient source for obtaining a high quality of collagen, particularly in comparison with collagen obtained from terrestrial animals [8, 11]. Crustaceans, mollusks, annelids, the skin, scales, bones and fins of fish, squid, among others, have been widely studied as an alternative source of collagen [8, 12].

The use of collagen has been expanding in several areas, such as food, pharmaceutical, biomedical and cosmetic [8, 13, 14].

1.2.2 Fish Oil

Fish oils are extremely popular in the nutritional sector due to their high content of long chain polyunsaturated Omega-3 fatty acids and their beneficial impact on human health, recognized for four decades, continuing to be a target of commercial interest and in several studies [14].

The most relevant omega-3 fatty acids are docosahexaenoic acid and eicosapentaenoic acid, originally synthesized from microalgae and then accumulated in phytoplankton, which is part of the dietary intake of fish [14].

In general, marine oils have several applications that include food, feed, aquaculture and nutraceuticals [15].

In the food field, fish oil can be used for edible purposes and has been used in the bakery industry. However, a wide variety of foods, including bread, baby food, maternal drinks, margarine and salad sauces containing marine oils have been prepared [16]. Fish oils are extremely popular in the nutritional sector, due to their beneficial impact on human health, being a target of commercial interest and in several studies, since they help in the fight against numerous diseases [15].

2 Green Extraction Methods

Different extraction methods can be applied, based on the marine resource used. However, it is important to consider the environmental concerns. Green extraction methods are less harmful than chemical methods. They can consist of lower energy consuming methods, using lower temperatures, and more environmentally friendly processes with lower environmental impacts. However, they must be able to generate high extraction efficiency at the same time [17].

Among several innovative techniques, ultrasounds-assisted, supercritical fluid extraction, enzyme extraction and microwave-assisted extraction have demonstrated a great interest in this area [17, 18]. Table 1 presents some experimental studies of fish oils and collagen extraction using eco-friendly technologies.

Table 1. Biomaterials extracted from fishery waste.

Waste	Biomaterial	Extraction method	Reference
Tuna by-products	Fish oil	Enzymatic extraction	[19]
Salmon by-products	Fish oil	Enzymatic extraction	[20]
Baltic herring whole fish and by-products	Fish oil	Enzymatic extraction	[21]
Sábalo skins and viscera	Collagen and Proteases	Enzymatic extraction	[22]
Elasmobranch byproducts	Collagen	Enzymatic extraction	[23]
Atlantic cod swim bladders	Collagen	Enzymatic extraction	[24]
Tuna by-products	Fish oil	Supercritical fluids (CO_2)	[25]
Common carp tissues	Fish oil	Supercritical fluids (CO_2)	[26]
Atlantic cod skin	Collagen	Supercritical fluids	[27]
Clown featherback (Chitala ornata) skin	Collagen	Ultrasound extraction	[28]
Monopterusalbus fillets	Fish oil	Ultrasound extraction	[29]
Bighead carp viscera	Fish oil	Ultrasound extraction	[30]
Sardine fish waste	Lipids	Microwave-assisted extraction	[31]
Chub mackerel, sardine, and horse mackerel	Lipids	Microwave-assisted extraction	[32]
Common carp	Gelatin	Microwave-assisted extraction/ultrasound extraction	[33]

2.1 Ultrasound Extraction

The use of ultrasound has increased in the last years. Among the several advantages, this extraction technique allows to minimize processing, maximize the quality and productivity with economic benefits at the same time [18, 34]. Many studies have demonstrated that ultrasounds technique increases the extraction yield [35].

This extraction process is based on waves sound mechanicals, oscillating in a propagation medium. Frequencies are such that they are above human ear levels (20 kHz to

10 MHz). There are two types of ultrasounds: high intensity (low frequency and high power) and low intensity (high frequency and low power) [18]. The effect of ultrasounds technique is associated to cavitation (rapid occurrence and formation of the bubbles that grow and collapse when the ultrasound waves propagate across the medium) to disrupt cell walls, and mechanical impact (heating, dynamic agitation, shear stresses, and turbulence) to increase the surface area of contact [18, 35, 36].

Several studies have been carried out to show the efficiency of ultrasounds process to extraction of collagen and fish oil from fish waste.

Petcharat [28] showed that ultrasound-assisted process applied to clown featherback (*Chitala ornata*) skin, a by-product from the filleting process, under suitable conditions, could be used to enhance the extraction yield with reduced effects on the molecular integrity of the resulting collagen.

Abdullah [29] they used Ultrasound technique in ethanol to extract the oil from the *Monopterusalbus* fillets. Results obtained demonstrated that the highest yields of oil extraction were obtained through the followed parameters: ultrasonic power of 200 W, ethanol volume of 500 mL and 60 min of sonication time.

Xiao [30] studied the ultrasound extraction of bighead carp viscera oil, used cyclohexane as solvent. The results showed an oil extraction rate of 94.82% in the followed conditions: 4:1 liquid-to-solid ratio at 50 °C with 57 min and 400 W extraction power.

2.2 Supercritical Fluid – Assisted Extraction

Supercritical fluid extraction is an approach that use solvents at temperatures and pressures above their critical points. The phenomena were discovered in 1822 by Charles Cagniard de la Tour, but it was only in 1960 that this method that this method had success as a extraction method [37].

The supercritical fluid extraction is a green extraction method with ecological advantages over the classical separation processes [26]. This method reduces the use of potentially toxic solvents, it has a fast extraction and separation phases with no heavy metal contaminants or inorganic salts [17].

Several compounds have been studied supercritical fluid extraction such as carbon dioxide, hexane, pentane, butane, nitrous oxide, sulfur-hexafluoride and fluorinated hydrocarbons. Yet, the most usual is the supercritical carbon dioxide (CO_2) since is a non-toxic, cheap solvent with mild critical values for the operating parameters, critical temperature of 31,1 ºC and critical pressure of 7.4 MPa. This features assure the extraction success of thermally sensitive biomaterials, such as Omega-3 Fatty acids and collagen [17, 26].

At supercritical conditions, the gas and liquid phases properties of the substances converge, enhancing their solvation and flow properties [37].

This technique might be divided into three phases: (1) In the first phase, the solvent diffuses into the matrix and decreases the mass transfer resistance; (2) In second phase, soluble compounds are dissolved, and they are further transferred by diffusion first into the solid surface and then into the bulk of the fluid; (3) The third phase comprises the end of the extraction process, with the solute and bulk fluid phase transportation out of the extraction equipment [37].

Several experimental studies have been carried out to show the efficiency of supercritical fluids extraction process to obtain collagen and fish oil from fish waste. Kuvendziev and his colleagues [26] used common carp tissues, namely viscera, as a source to obtain unsaturated fatty acids through supercritical fluid CO_2 extraction with profitable results. In the experimental research performed by Ahmed and his colleagues [25], they extracted fish oil from skins, scales and bones of bigeye tuna also using supercritical carbon dioxide with high fatty acids extraction. In both studies, the authors also extracted fish oils using the Soxlet method to compare the yield of extraction by supercritical fluids with the yield of extraction by Soxlet [25, 26].

Although the literature review demonstrates the use of supercritical fluids to extract other biomaterials, such as fish oils, in the study led by Sousa [27], the authors evaluated collagen extraction using supercritical fluids technology, in particular with water acidified by pressurization with CO_2. That method, previously employed for collagen extraction from marine sponges, was applied to extract collagen from Atlantic cod skins. The authors had an efficient extraction yield, comparatively to the conventional methodologies based in acetic acid solutions [27].

2.3 Microwave-Assisted Extraction

As previously mentioned, the microwave-assisted extraction has showed a big interest in this area. This technique is not frequently used for fish oil extraction but several studies show that the efficiency is similar or greater than traditional techniques [17, 18]. This method has numerous advantages, such as greater extraction rates, lower temperatures, and automatization. However, the heat generation may cause an oxidative degradation [17, 38].

Microwave-assisted extractions is based on the utilization of microwaves to heat solvents in contact with matrix. Therefore, the moisture is decreased, and it causes a significant pressure generation. This pressure breaks the cell membranes cell, allowing to access to biomaterials [17, 38].

There are several considerations that can affect the extraction yield, such as: time, frequency of microwaves, capacity, solvent, and particle size [17].

Several studies have been performed in this area. Rahimi and his colleagues [31] studied the extraction of lipid from sardine fish waste. They compared with Soxhlet and Hara and Radin methods. The results showed that the highest extraction yield was achieved for microwave-assisted technique, using distilled water as an extraction solvent at 10 min extraction time. So, according to this study, this method can improve the lipid extraction from fish waste, being an attractive alternative for the traditional methods.

Same occurs in the study of Ramalhosa [32], where a microwave-assisted extraction method was compared with traditional techniques for lipid extraction from Chub mackerel, sardine, and Horse mackerel. Microwave-assisted technique was the method with best extraction efficiency and repeatability, with several advantages (requiring less laboratory material, solvent volume, time), although other traditional methods also worked well. Therefore, this study showed that this method is suitable for fish fat extraction achieving total lipid content like or higher than conventional methods with higher repeatability.

Mirzapour-Kouhdasht [33] studied the effects of ultrasound and microwaves-assisted extraction methods on molecular and physicochemical characteristics of the gelatin produced from Common carp. Ultrasound technique increased the ex-traction yield of gelatin. However, contrasted with ultrasound method, microwave-assisted led to a considerably lower extraction yield. The time of extraction demonstrated a reverse correlation with the gel strength and viscosity of gelatin.

2.4 Enzymatic - Assisted Extraction

Enzymatic-assisted extraction is a technology with effective peptides generation obtained from several sources. This technology uses proteolytic enzymes that under specific and controlled conditions promote collagen and fish oil extraction [39].

This methodology can be carried out using bacterial proteases or enzymes extracted from organic sources, including fish waste, which promotes a circular economy. Further, the addition of multiple specific or nonspecific proteases extracted from fish viscera can enhance protein hydrolysis [40].

The extraction yield of the biomaterials is mostly influenced by enzyme concentration and time. Generally, an increase in the enzyme concentration corresponds to an increase in yield extraction [41]. Yet, the temperature level also affects the extraction yield [20].

Marine collagen extraction from fish waste has been studied with successful results. Gomez and her colleagues [22] used sábalo viscera to extract digestive aspartic proteases and then used the enzyme to extract collagen from the same fish skin. The authors tested collagen extraction from three different methods, acid extraction using acetic acid, pepsin from porcine extraction, and sábalo enzyme extraction. The collagen recovery yields were 14.92%, 17.7%, and 25.29%, respectively, demonstrating that the sábalo enzyme has the highest collagen extraction yield. Similar occurs in the experimental study of Sousa and her colleagues [24]. In their study, the authors extracted collagen from Atlantic cod swim bladders using the acid extraction method and pepsin extraction, which resulted in 5.72% and 11.14% extraction yields, respectively. Those results confirmed that enzymatic-assisted collagen extraction enhances the extraction yield comparatively to the acid extraction method. Furthermore, in the study led by Seixas [23], the authors tested collagen pepsin extraction from high cartilaginous tissues, Elasmobranch by-products. The authors also confirmed that the enzymatic collagen extraction yield was higher than the acid extraction yield.

Regarding the enzymatic fish oils extraction, several researchers have been studied fatty acids extraction from a variety of fish wastes. Aitta and colleagues [21] successfully extracted fish oil from Baltic Herring using three commercial proteolytic enzymes, Alcalase, Neutrase, and Protamex. The authors also compared enzyme extraction times of 35 and 70 min, however, the results were not conclusive. Mohammad [19] also tested fish oil extraction from proteolytic enzyme using tuna fish by-products. Moreover, the researchers evaluated the fish oil extraction yields through enzymatic extraction and wet pressing and they showed that the fish oil extraction yield was higher with the enzymatic extraction method than the wet pressing. The efficiency of enzymatic fish oil extraction was also assessed in the research of Routray [20]. In the experimental study, the researchers extracted fish oils from salmon by-products through enzyme treatment. In that study, where the authors tested different enzyme concentrations and temperatures, they concluded that the highest oil extraction was 19% with a temperature level of 50°C and enzyme concentration of 2.5%.

3 Ecotoxicology

Primarily, ecotoxicology was defined by Ernst Haeckel, in 1866, as "an economy of nature" and "science between the interactions of organisms and the surrounding environment" [42]. Subsequently, the definition of the term ecotoxicology progressed to a study, with methodologies used in the area of Toxicology, being defined by Truhaut, in 1977, as a specialized area of environmental toxicology, in which the toxic effects of substances are studied, natural and artificial, in living beings, such as animals or plants, aquatic or terrestrial [43].

Currently, ecotoxicological studies are relevant in the assessment of ecological and environmental risks. That conduced to the development of several ecotoxicity assessment policies.

The European Union (EU), through the commitments adopted in the European Ecological Pact, places the environment at the forefront during the elaboration of new policy measures for the EU. Thus, in this agreement, the European Commission recognizes the need to prevent and reverse pollution, also committing itself to new environmental regulations aimed at reducing air, water, and soil pollution. With this plan, the European Commission intends to make the European economy as sustainable as possible, with the goal that in 2050, the impact on the EU's climate will be neutral [44].

When compared with chemical analyses, an ecotoxicological evaluation study, present the advantage of detecting the toxicity of a sample, evaluating the combined effects of the different constituents of the sample. On the other hand, chemical analysis only allows quantifying the isolated substances in the sample, and the substance is toxic if its concentration exceeds the permitted limit, it is considered toxic. Consequently, ecotoxicological studies to assess the hazardous characteristics of a compound must be carried out simultaneously with chemical analysis, to obtain all possible information about the toxicity of a product [45, 46].

It was performed a literature review about the ecotoxicological studies on fish oil and collagen extracted from the fishing waste, using the platforms "Scopus", "Google Scholar" and "Science Direct" using the keywords "ecotoxicology", "fish oil", "collagen" and "marine waste". The results obtained in the literature review showed a lack of relevant information about the toxicity of those extracted products in the last five years.

Regarding the ecotoxicological studies of marine collagen, only two experimental study was considered, since most of the studies found evaluate changes in the collagen of marine organisms induced by the presence of exogenous substances.

The research performed by Sousa and her colleagues [27] showed that the collagen extracted from Atlantic cod skins through water acidified with CO_2 has no cytotoxicity when in concentrations up to 5 mg/mL. The *in vitro* experimental study was performed using Human fibroblasts. However, to evaluate ecotoxicity, it is important to use different organisms from different trophic levels.

In the study performed by Allouche [47], the authors demonstrate that water contamination with collagen-rich products has high harmful effects on populations of marine nematodes. However, in this study, the source of collagen extraction was not identified.

Regarding the "fish oil" biomaterial, as mentioned before, the literature review shows a lack of information about the ecotoxicity associated to fish oils extracted from fish waste. Most of the research reviewed focus more on intrinsic fish oil profile variation in different fish species than in the extracted fish oils. The reviewed information also focus on the impact of fish oil on diet of different organisms [48, 49].

Since the results obtained in this literature review showed a lack of information about ecotoxicity studies performed on the fish oils and collagen extracted from fish waste, it is urgent to evaluate the ecotoxicological features of the biomaterials extracted to evaluate promote the ecological sustainability of the extraction processes.

4 Conclusions

Marine resources are limited, so its sustainable management is essential. Fish waste represents both a major environmental problem and a massive economic loss, so a better fish-waste management is needed to overcome these important issues, also playing a key role in the Waste Framework Directive, which aims to prevent the generation of waste as much as possible, and to use the waste generated as a resource for reuse, recycling, and recovery. In this way, the use of fish by-products could contribute to the development of products with high-commercial value, and consequently, to economic growth. As mentioned before, most green extraction methods are less harmful and lower energy consumer than conventional ones. They also reveal great biomaterials extraction efficiency. So, the literature review carried out highlights that eco-friendly extraction have great efficiency regarding collagen and fish oils extraction. To analyze the process efficiency, most researchers compared the biomaterial green extraction method with the conventional extraction method. In some cases, the results obtained showed that, ecofriendly technologies applied were more efficient than the conventional ones.

Nevertheless, some extraction technologies are more suitable to some biomaterials than others. The scientific papers reviewed showed that the enzymatic and ultrasound-assisted technologies are the most promising for collagen extraction. In relation to fish oils extraction, the reviewed articles showed that supercritical fluids and ultrasound technologies are the most suitable. However, this literature review revels that it is necessary to optimize the extraction parameters to promote a more efficient use of the local marine resources.

Moreover, the literature review carried out showed a lack of information about ecotoxicological studies of fish oils and collagen extraction using the green technologies described. To make the extraction technologies ecologically sustainable, it is imperative to carry out an assessment of the ecotoxicity caused by these biomaterials.

Acknowledgements. This work has been co-financed by FITEC - Fund for Innovation, Technology and Circular Economy, Interface Program, within the scope of the project CVR – Base Funding and CVR.TechRe4C - Reinforcing CVR in Ecotoxicity, Sustainability and 4.0 Industry Technologies. Co-financed by the European Union through the European Regional Development Fund (ERDF), included in the Norte 2020 – Norte's Regional Operational Programme, Portugal 2020 (NORTE-01-0246-FEDER-000056).

References

1. FAO: The state of world fisheries and aquaculture. In: Meeting the Sustainable Goals (2018)
2. FAO: The state of world fisheries and aquaculture 2020. Sustainability in action (2020)
3. European Commission: Commission decision 2014/955/EU - list of waste. Off. J. Eur. Union. **7**, 43 (2014)
4. Almeida, C., Karadzic, V., Vaz, S.: The seafood market in Portugal: Driving forces and consequences. Mar. Policy **61**, 87–94 (2015)
5. European Commission: Blue economy report (2020)
6. INE/DGRM: Estatísticas da Pesca 2020 (2021)
7. Le Gouic, A.V., Harnedy, P.A., FitzGerald, R.J.: Bioactive peptides from fish protein by-products (2018)
8. Shahidi, F., Varatharajan, V., Peng, H., Senadheera, R.: Utilization of marine by-products for the recovery of value-added products. J. Food Bioact. **6** (2019)
9. Ambigaipalan, P., Shahidi, F.: Bioactive peptides from shrimp shell processing discards: antioxidant and biological activities. J. Funct. Foods. **34**, 7–17 (2017)
10. Coppola, D., Lauritano, C., Palma Esposito, F., Riccio, G., Rizzo, C., de Pascale, D.: Fish waste: from problem to valuable resource. Mar. Drugs **19**, 1–39 (2021)
11. Bernhardt, A., Paul, B., Gelinsky, M.: Biphasic scaffolds from marine collagens for regeneration of osteochondral defects. Mar. Drugs **16**, 91 (2018)
12. Ehrlich, H., Wysokowski, M., Żółtowska-Aksamitowska, S., Petrenko, I., Jesionowski, T.: Collagens of poriferan origin. Mar. Drugs **16**, 79 (2018)
13. Atay, E., Fabra, M.J., Martínez-Sanz, M., Gomez-Mascaraque, L.G., Altan, A., Lopez-Rubio, A.: Development and characterization of chitosan/gelatin electrosprayed microparticles as food grade delivery vehicles for anthocyanin extracts. Food Hydrocoll. **77**, 699–710 (2018)
14. Maschmeyer, T., Luque, R., Selva, M.: Upgrading of marine (fish and crustaceans) biowaste for high added-value molecules and bio(nano)-materials. Chem. Soc. Rev. **49**, 4527–4563 (2020)
15. Shahidi, F., Ambigaipalan, P.: Novel functional food ingredients from marine sources. Curr. Opin. Food Sci. **2**, 123–129 (2015)
16. Suleria, H., Osborne, S., Masci, P., Gobe, G.: Marine-based nutraceuticals: an innovative trend in the food and supplement industries. Mar. Drugs **13**, 6336–6351 (2015)
17. Ivanovs, K., Blumberga, D.: Extraction of fish oil using green extraction methods: a short review. Energy Procedia **128**, 477–483 (2017)
18. Al Khawli, F., et al.: Innovative green technologies of intensification for valorization of seafood and their by-products. Mar. Drugs **17**, 1–20 (2019)
19. Taati, M.M., Shabanpour, B., Ojagh, M.: Investigation on fish oil extraction by enzyme extraction and wet reduction methods and quality analysis. AACL Bioflux. **11**, 83–90 (2018)
20. Routray, W., Dave, D., Ramakrishnan, V.V., Murphy, W.: Production of high quality fish oil by enzymatic protein hydrolysis from cultured atlantic salmon by-products: investigation on effect of various extraction parameters using central composite rotatable design. Waste Biomass. Valor. **9**(11), 2003–2014 (2017). https://doi.org/10.1007/s12649-017-9998-6
21. Aitta, E., Marsol-Vall, A., Damerau, A., Yang, B.: Enzyme-assisted extraction of fish oil from whole fish and by-products of baltic herring (Clupea harengus membras). Foods **10**, 1811 (2021)
22. Acevedo Gomez, A.V., Gomez, G., Chamorro, E., Bustillo, S., Leiva, L.C.: Digestive aspartic proteases from sábalo (Prochilodus lineatus): characterization and application for collagen extraction. Food Chem. **269**, 610–617 (2018)
23. Seixas, M.J., Martins, E., Reis, R.L., Silva, T.H.: Extraction and characterization of collagen from elasmobranch byproducts for potential biomaterial use. Mar. Drugs **18**, 1–18 (2020)

24. Sousa, R.O., et al.: Acid and enzymatic extraction of collagen from Atlantic cod (Gadus Morhua) swim bladders envisaging health-related applications. J. Biomater. Sci. Polym. Ed. **31**, 20–37 (2019)
25. Ahmed, R., Haq, M., Cho, Y.-J., Chun, B.-S.: Quality evaluation of oil recovered from by-products of bigeye tuna using supercritical carbon dioxide extraction. Turk. J. Fish. Aquat. Sci. **17**, 663–672 (2017)
26. Kuvendziev, S., Lisichkov, K., Zeković, Z., Marinkovski, M., Musliu, Z.H.: Supercritical fluid extraction of fish oil from common carp (Cyprinus carpio L.) tissues. J. Supercrit. Fluids. **133**, 528–534 (2018)
27. Sousa, R.O., et al.: Collagen from Atlantic cod (Gadus morhua) skins extracted using CO_2 acidified water with potential application in healthcare. J. Polym. Res. **27** (2020). Article number: 73. https://doi.org/10.1007/s10965-020-02048-x
28. Petcharat, T., Benjakul, S., Karnjanapratum, S., Nalinanon, S.: Ultrasound-assisted extraction of collagen from clown featherback (Chitala ornata) skin: yield and molecular characteristics. J. Sci. Food Agric. **101**, 648–658 (2021)
29. Abdullah, S., Mudalip, S.K.A., Shaarani, S.M., Pi, N.A.C.: Ultrasonic extraction of oil from monopterus albus: effects of different ultrasonic power, solvent volume and sonication time. J. Appl. Sci. **10**, 2713–2716 (2010)
30. Xiao, L., Ji, Y., Wenjiao, L., Xing, Z., Chengjin, M.: Ultrasound-assisted extraction of bighead carp viscera oil and its physiochemical properties. J. Jishou Univ. Sci. Ed. (2017)
31. Rahimi, M.A., Omar, R., Ethaib, S., Siti Mazlina, M.K., Awang Biak, D.R., Nor Aisyah, R.: Microwave-assisted extraction of lipid from fish waste. In: IOP Conference Series: Materials Science and Engineering, vol. 206 (2017)
32. Ramalhosa, M.J., Paíga, P., Morais, S., Rui Alves, M., Delerue-Matos, C., Oliveira, M.B.P.P.: Lipid content of frozen fish: comparison of different extraction methods and variability during freezing storage. Food Chem. **131**, 328–336 (2012)
33. Mirzapour-Kouhdasht, A., Sabzipour, F., Taghizadeh, M.S., Moosavi-Nasab, M.: Physicochemical, rheological, and molecular characterization of colloidal gelatin produced from Common carp by-products using microwave and ultrasound-assisted extraction. J. Texture Stud. **50**, 416–425 (2019)
34. Gallo, M., Ferrara, L., Naviglio, D.: Application of ultrasound in food science and technology: a perspective. Foods **7**, 164 (2018)
35. Arzeni, C., Martínez, K., Zema, P., Arias, A., Pérez, O.E., Pilosof, A.M.R.: Comparative study of high intensity ultrasound effects on food proteins functionality. J. Food Eng. **108**, 463–472 (2012)
36. Carrillo-Lopez, L.M., Alarcon-Rojo, A.D., Luna-Rodriguez, L., Reyes-Villagrana, R.: Modification of food systems by ultrasound. J. Food Qual. **2017**, 1–12 (2017)
37. Ibáñez, E., Mendiola, J.A., Castro-Puyana, M.: Supercritical fluid extraction. In: Encyclopedia of Food and Health, pp. 227–233. Elsevier (2016)
38. Alfio, V.G., Manzo, C., Micillo, R.: From fish waste to value: an overview of the sustainable recovery of omega-3 for food supplements. Molecules **26**, 1002 (2021)
39. Coppola, D., et al.: Marine collagen from alternative and sustainable sources: extraction, processing and applications. Mar. Drugs **18**, 214 (2020)
40. Nasri, M.: Protein hydrolysates and biopeptides: production, biological activities, and applications in foods and health benefits: a review. In: Advances in Food and Nutrition Research, pp. 109–159. Elsevier (2017)
41. Salvatore, L., et al.: Marine collagen and its derivatives: Versatile and sustainable bio-resources for healthcare. Mater. Sci. Eng. C. **113**, 110963 (2020)
42. Sandermann, H.: Molecular ecotoxicology of plants. Trends Plant Sci. **9**, 406–413 (2004)
43. Truhaut, T.: Ecotoxicology: objectives, principles and perspectives. In: Ecotoxicology and Environmental Safety, pp. 151–173 (1977)

44. Europeia, C.: Pacto Ecológico Europeu
45. Boros, B.-V., Ostafe, V.: Evaluation of ecotoxicology assessment methods of nanomaterials and their effects. Nanomaterials **10**, 610 (2020)
46. Walker, C.H., Greig-Smith, P.W., Crossland, N.O., Brown, R.: Ecotoxicology. In: Animals and Alternatives in Toxicology, pp. 223–251. Macmillan Education, London (1991)
47. Allouche, M., et al.: Laboratory bioassay exploring the effects of anti-aging skincare products on free-living marine nematodes: a case study of collagen. Environ. Sci. Pollut. Res. **27**(10), 11403–11412 (2020). https://doi.org/10.1007/s11356-020-07655-1
48. Ghamkhar, R., Hicks, A.: Comparative environmental impact assessment of aquafeed production: sustainability implications of forage fish meal and oil free diets. Resour. Conserv. Recycl. **161**, 104849 (2020)
49. Gribble, M.O., et al.: Mercury, selenium and fish oils in marine food webs and implications for human health. J. Mar. Biol. Assoc. U. K. **96**, 43–59 (2016)

Study Concerning the Correspondence Between Optical Materials and Optical Coatings Using Matlab

Dana Rizescu and Ciprian Ion Rizescu(✉)

University POLITEHNICA of Bucharest, 313 Splaiul Independentei, Bucharest, Romania
dana.rizescu@upb.ro

Abstract. Optical coatings generally consist of thin films made up of single or multiple layers of either metallic or dielectric materials. When properly designed and fabricated, these coatings can substantially modify the reflection and transmission properties of an optical component. Almost all lenses have AR coatings to maximize transmission factor and minimize ghost images factor. Even we talk about glasses or complex optical systems, AR coatings of one or many layers, depending on what is intended to gain. When it comes about mirrors, AR eliminated old technologies based on complex matrix and super polished, raising reflection factor light weight and low cost. The paper is focused on a guide for selecting optical materials and proper optical coating types depending on the wavelength range and the applications requirements. The software permits selection of optical materials and optical coatings for certain application considering spectral domain, spectral characteristics such as transmittance/reflectance, material, coating technology, number of layers and incidence angle. The selection guide was developed in Matlab environment. The program is a useful tool for knowledge of coating types and selection of proper optical coating for a specific application and for a group of convenient substrates. Selecting the right optical glass for a particular application can significantly impact cost, durability, and performance. In a future development of the interface, some features will be added, such as adding materials catalogues and matching criteria for better combinations between substrates and coatings.

Keywords: Optical materials · Optical coatings · AR coatings

1 Introduction

1.1 How to Choose a Coating

Anti-reflection coatings are designed to work under certain illumination conditions, light direction and environment, because their performances depend on wavelength, polarization, intensity, incidence angle, humidity and light temperature. The simplest AR

J. Machado et al. (Eds.): icieng 2022, LNME, pp. 148–160, 2023.
https://doi.org/10.1007/978-3-031-09382-1_13

coatings, single layer MgF2 or Al, have lower performances in many applications comparing with multi-layer coatings [1]. Selecting the right optical glass for a particular application can significantly impact cost, durability, and performance. That's why it makes sense to work with engineers who know their materials. When selecting materials for optical applications outside the visible spectrum, for example, there are many different factors to consider.

1.2 Coatings Characteristics

a) Choosing a coating
   Choosing a coating is influenced by reflectance or transmittance prescribed for certain wavelength [2], but there are taken into consideration all seven criteria mentioned above:

   - Wavelength;
   - Reflectance or transmittance;
   - Polarization;
   - Incidence angle;
   - Substrate;
   - Light intensity;
   - Environment conditions.

b) Wavelength
   All AR coatings have different reflectance or transmittance values related to wavelength; these can be wide band or narrow band, AR of wide band covering large zones of the spectrum. Based on these criteria, AR coatings can be classified as it follows:

1. Wideband

   - Metallic;
   - Dielectrics single layer or multi-layer [3];
   - Hybrids (metallic and dielectrics).

2. Narrow band (multi-layer, dielectrics)
   Wide band categories include all types of AR coatings [4], meanwhile narrow band category is limited to multilayer dielectrics due to interferential optical effects get between layers.

c) Reflectance and transmittance
   Usually, reflectance defines both reflective and antireflective coatings, for reflective one (i.e. mirrors) is necessary a high reflectance and the opposite for antireflective coating, which means high performance.

d) Polarization
There are two types of polarization and two terms associated: s-polarization and p-polarization. First one, s-polarization has the roots in German word "senkrecht" which means orthogonal, because electric field vector is normal on its incidence plan. The second term, p-polarization is based on German word „parallel" because polarization axis is parallel or included in incidence plan [5].

e) Incidence angle
Incidence angle is measured between light ray and optical surface normal. Coatings are used to enlarge or to narrow this angle in order to improve performances at a specific angle.

f) Substrate
The basis material on which the coating is deposited is generally called the substrate. Same coating will act different if it has a different substrate: this means that over-layer material is chosen, especially in multi-layer coatings, accordingly to under-layer material. Technical explanation is: different substrates have different optical characteristics. Choosing substrate and coating in a proper way is the key solution to obtaining the required optical performances of the coated optical material.

g) Light intensity
Some coatings are soft and other hard. For imaging applications, when light intensity is lower, soft coatings are exposed to incident beam, but for powerful laser applications (molding or surgery) too soft coatings will be destroyed by incident beam.

Composite materials consisting of solid solutions have several advantages over single, "pure" materials. Improved optical and mechanical properties can be obtained with the simplicity and facility of a single evaporation source, rather than multiple sources. The technology developed for the demanding laser applications can also be applied to more common coatings such as ophthalmic and camera lenses.

h) Environment conditions
Optical coatings must be carefully treated. Even if, for example, the hardest coatings, which are resistant to laser radiation, resist even the actions of scratches or abrasives, these coatings are still more fragile than many bottles. Variations in humidity and temperature can seriously affect optical coatings. In principle, both variations in humidity and temperature affect the refractive index and sometimes even the thickness of the layer. In many cases, the sensitivity of the cover to the environment is low enough to be ignored [6].

2 Optical Coating Technology

Optical coatings are used to change the properties of optical components: reflectance, transmittance, absorbance or polarization. The coating is obtained in high vacuum ($<10^{-6}$ Torr) using one of the procedures mentioned below. As cover materials should be mentioned: metals, dielectric, or semiconductor materials. The most used dielectric materials are oxides, fluorides and sulfides. Tellurium and selenium are used in specialized applications.

Evaporation

The coating material must be converted from solid to vapor and then condensed on the optical surface. The most used evaporation methods are: thermal evaporation and ion bombardment with electron beam [7].

Thermal Evaporation

In this process, the coating material is heated resistively in a tungsten, molybdenum, or platinum vessel in the case of metals and high temperature dielectric materials. For low temperature dielectric materials, the indirect heating method with refractory filament can be used.

Electron Ion Beam Bombardment

In this process, the coating material is heated directly by electron bombardment. It uses a beam of electrons of variable voltage and current (5-15kV and a few amps), the beam being concentrated in a mobile spot that acts on a special copper crucible that is cooled with water. The crucible has several pockets that can contain different materials that are evaporated. This method of evaporation is simple and reduces the effect of contamination that would occur on the interaction between the coating material and the crucible. The heating is localized and the temperatures can be so high that refractory materials can be vaporized [8].

Method Based on Bombarding the Cathode with Positive Ions

The method called "sputtering" is used to deposit thin layers. It is particularly useful for wide area substrates. The substrate and the coating material are held together very close to each other in a plasma enclosure. The substrate is cooled. Heavy plasma ions bombard the coating material, releasing very small particles of material [9].

They "condense" on the cooled substrate. The method can ensure a uniform coating on wide surfaces and uses the coating material much more efficiently than the evaporation-based method.

All optical coating technologies provide an adequate layer thickness control which is very important for optical coatings spectral characteristics. There are two main methods of controlling the thickness of the layer: controlling the interference and measuring the resonance of an oscillating crystal [10].

3 Optical Coating Materials

Optical coatings, used by well-known laboratories in the field of optical components, are obtained on substrates materials and could be grouped according to an unanimously accepted criterion [11]. The table below shows the main materials [12, 13].

Examples of materials used for coatings are presented in Table 1.

Table 1. Coatings materials.

Metals	Alloy	Transparent conductors	Dielectrics	Special applications
Aluminum	Constantan	Indium oxide	Silicon dioxide	Silicon
Chrome	Cupil	Special indium oxide	Silicon nitrite	Germanium
Copper	Gold alloys	Oxide: Antimony	Niobium oxide	Sapphire
Gold	Cusil	Others	Aluminum dioxide	Zinc sulfide/Zinc selenide
Platinum	Incusil			Special Diamond (DLC)
Palladium	Aluminum: Silver			
Indium	Inconel			
Molybdenum	Steel			
Tungsten	Silver alloy			
Titanium	Others			
Nickel				
Silver				
Others				

4 Optical Materials and Coatings Selection Guide

The authors designed a dedicated interface to ease the selection of the best match between substrate and coating materials, based on criteria mentioned above. The software behind the interface permits selection of optical materials and optical coatings for certain application considering spectral domain, spectral characteristics such as transmittance / reflectance, material, coating technology, number of layers and incidence angle. Main interface window has two pages and permits choosing the substrate from a proposed list of optical well-known glasses and crystals. As it is presented in Fig. 1, the selection in the first page is made using radio buttons placed on the left side of the window, while the characteristic spectral domain of the material is displayed [13]. The

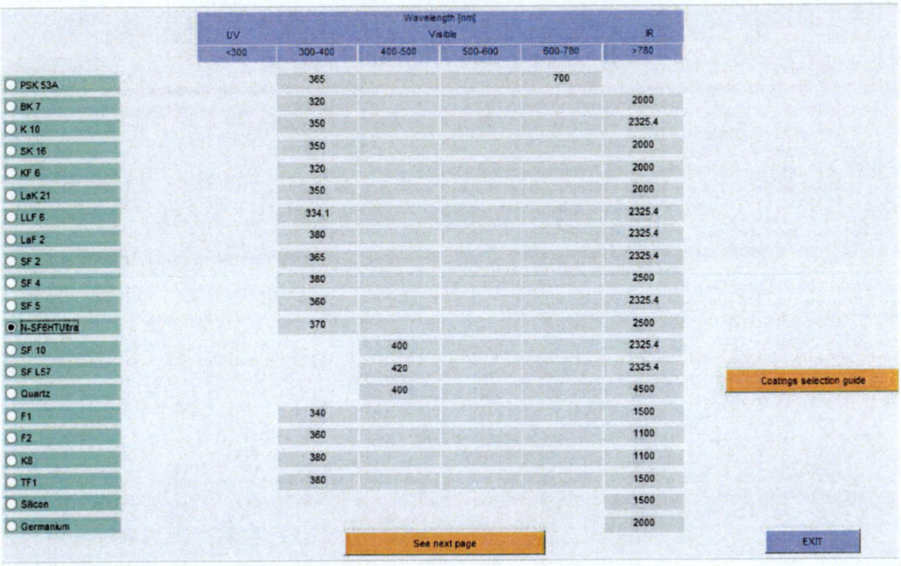

Fig. 1. Main interface page 1.

pushbutton labeled "Coatings selection guide" sends the user to another window where optical coatings could be selected according to their type.

When an optical material is chosen, the software displays in a new window its main and useful characteristics together with dedicated applications, transmittance over wavelength range, other related materials, as it is shown in Fig. 2. Also, the interface is provided with a button which sends the user through a hyperlink to a producer catalog where complete data of the selected material can be found. The catalog could be replaced anytime or, with small modifications in the program, could be chosen from a list of catalogues. Figure 3 represents a detail with the button for the hyperlink.

In order to select the proper optical coating, the user can access the second main page of the interface, Fig. 4, where he can choose from a list of materials or can study the spectral characteristics of the most used metal coatings. Also, from this second page, the interface user can directly access the coatings selection guide.

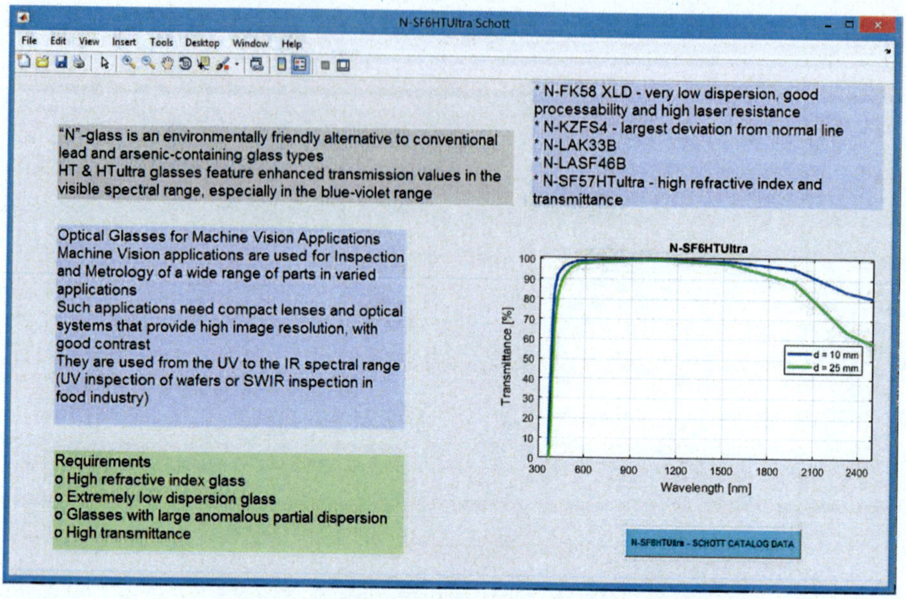

Fig. 2. Optical material selection from first page.

Fig. 3. Link to catalog [12].

Figure 5 presents the resulting page when a material is selected by clicking the corresponding radiobutton. As it was previously shown, the transmittance over the wavelength, the main applications, properties and a link to a producer catalogue are displayed. For a quick look over the metallic coatings, the interface is provided with two pushbuttons. When the pushbutton labeled "Coatings – table" is pressed, the software displays a page with the values of reflectance of the most used metals for coatings and their useful spectral domain, Fig. 6 [14].

The other pushbutton labeled "Coatings – plots" sends to the graphic representations of the reflectance over wavelength, which could be a useful tool to compare and choose a metallic coating considering the reflectance provided in a specific wavelength range, Fig. 7.

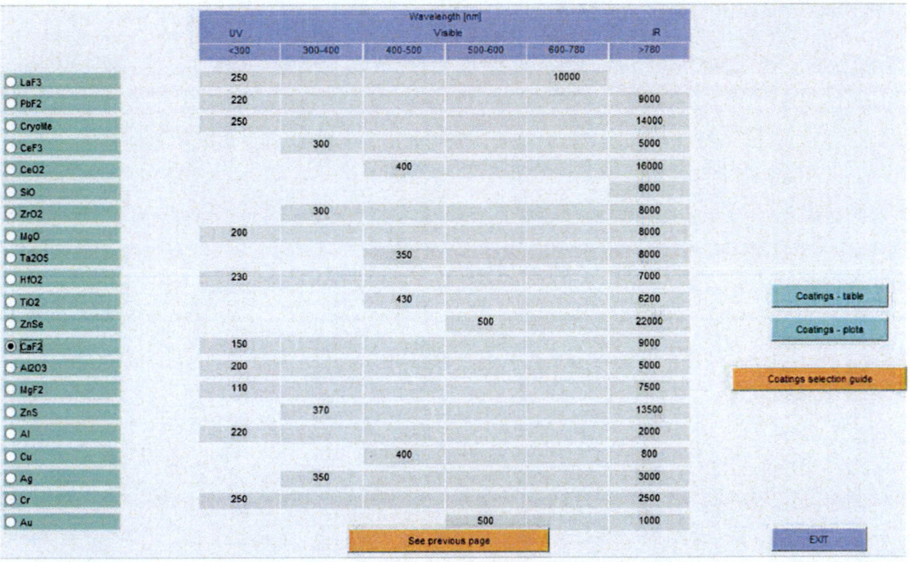

Fig. 4. Main interface page 2.

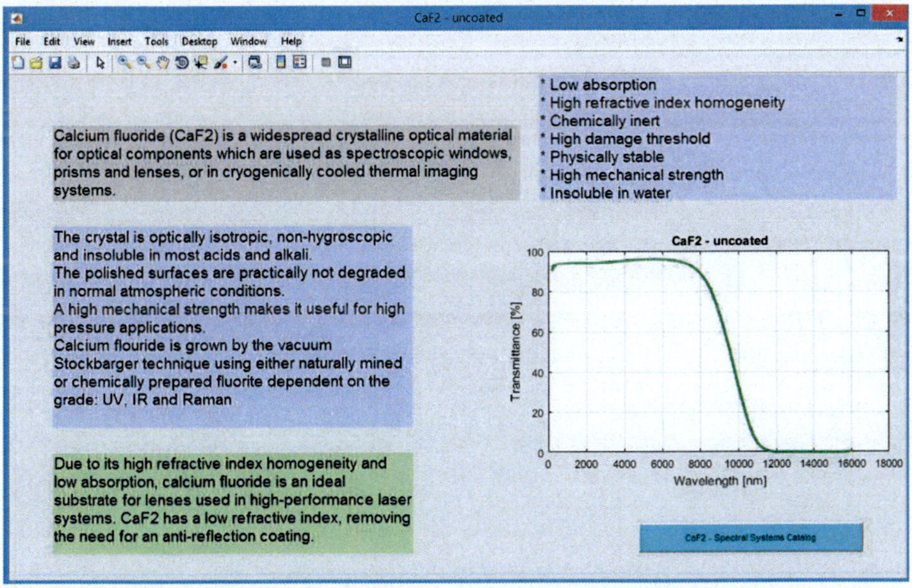

Fig. 5. Optical material selection from second page.

REFLECTANCE AT NORMAL INCIDENCE (%)						
λ (μm)	ALUMINIUM	SILVER	GOLD	COPPER	RHODIUM	PLATINUM
0.220	91.5	28.0	27.5	40.4	57.8	40.5
0.240	91.9	29.5	31.6	39.0	63.2	46.9
0.260	92.2	29.5	35.6	35.5	67.7	51.5
0.280	92.3	25.2	37.8	33.0	70.7	54.9
0.300	92.3	17.6	37.7	33.6	73.4	57.6
0.315	92.4	5.5	37.3	35.5	75.0	59.4
0.320	92.4	8.9	37.1	36.3	75.5	60.0
0.340	92.5	72.9	36.1	38.5	76.9	62.0
0.360	92.5	88.2	36.3	41.5	78.0	63.4
0.380	92.5	92.78	37.8	44.5	78.1	64.9
0.400	92.4	94.8	38.7	47.5	77.4	66.3
0.450	92.2	96.6	38.7	55.2	76.0	69.1
0.500	91.8	97.7	47.7	60.0	76.6	71.4
0.550	91.5	97.9	81.7	66.9	78.2	73.4
0.600	91.9	98.1	91.9	93.3	79.7	75.2
0.650	90.3	98.3	95.5	96.6	81.1	76.4
0.700	89.9	98.5	97.0	97.5	82.0	77.2
0.750	88.0	98.6	97.4	97.9	82.6	77.9
0.800	86.3	98.6	97.7	98.1	83.1	78.5
0.850	86.0	98.7	97.8	98.3	83.4	79.5
0.900	89.0	98.7	98.0	98.4	83.6	80.5
0.950	91.8	98.8	98.1	98.4	83.9	80.6
1.000	94.0	98.9	98.2	98.5	84.2	80.7
1.500	96.8	98.9	98.2	98.5	87.7	81.8
2.000	97.2	98.9	98.3	98.6	91.4	81.8
3.000	98.0	98.9	98.4	98.6	95.0	90.6
4.000	98.2	98.9	98.5	98.7	95.8	93.7
5.000	98.4	98.9	98.6	98.7	96.4	94.9
6.000	98.5	98.9	98.6	98.7	96.8	95.6
7.000	98.6	98.9	98.7	98.7	97.0	95.9
8.000	98.7	98.9	98.8	98.8	97.2	96.0
9.000	98.7	98.98	98.9	98.8	97.4	96.1
10.000	98.7	98.9	98.9	98.9	97.6	96.2
15.000	98.9	99.0	99.0	99.0	98.1	96.7
20.000	99.0	99.2	99.0	99.2	98.3	97.0
30.000	99.2	-	99.0	-	-	-
40.000	-	-	99.1	-	-	-

Metallic coatings are used primarily for mirrors and are not considered thin films coatings in the strict sense of the term. They aren't based on the principle of interference, but more on the optical properties of the coating material. However, metallic coatings are often "over-coated" with thin dielectric layers to increase reflectance in a wavelength range or for certain angles of incidence.

Coating type	Wavelength [nm]	Reflectance %
Aluminium	400-1200	90
Aluminiu - protected	400-800	87
Aluminium - enhanced	450-750	93
UV - Enhanced aluminium	250-400	86
Silver	400-1200	98
Silver - protected	400-20000	95
Gold	700-20000	99
Gold - protected	650-16000	98

Fig. 6. Coatings - table.

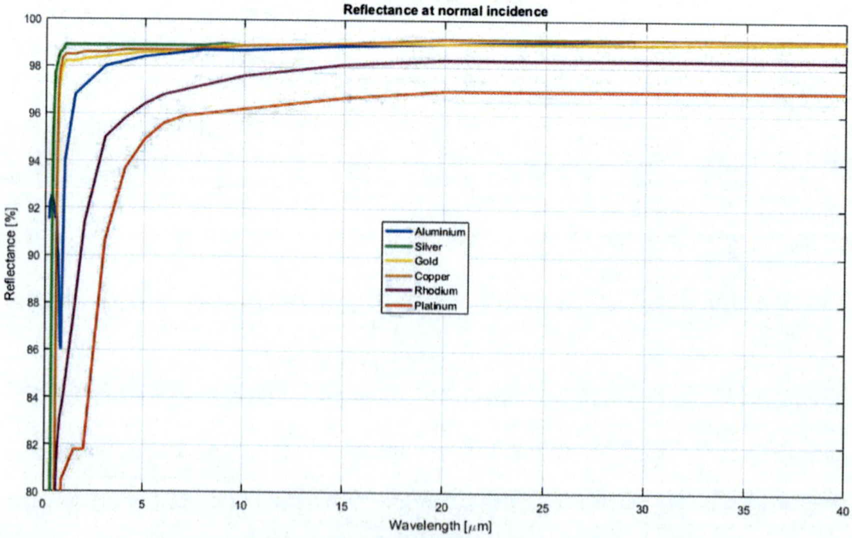

Fig. 7. Coatings - plots.

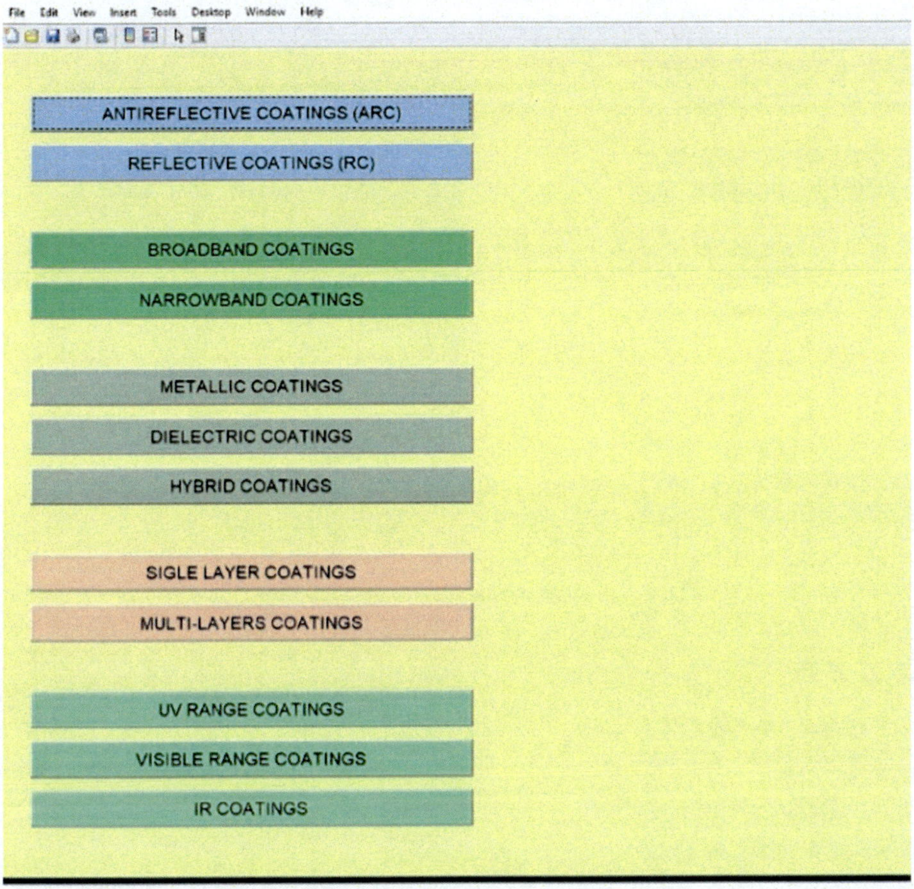

Fig. 8. Coatings selection guide main view.

As it was mentioned above, the two main pages of the interface are provided with a pushbutton labeled "Coatings selection guide". The main view presented in Fig. 8 is generated when this pushbutton is pressed and shows the most general coatings types. Each type can be further developed from the corresponding button and representative materials are displayed, Fig. 9. When a certain coating type [15] is selected, the user can go further and choose from the main options displayed, as it is shown in Fig. 10.

When all choses are made, the properties, applications and spectral characteristics of the coating material are shown in a dedicated page, Fig. 11.

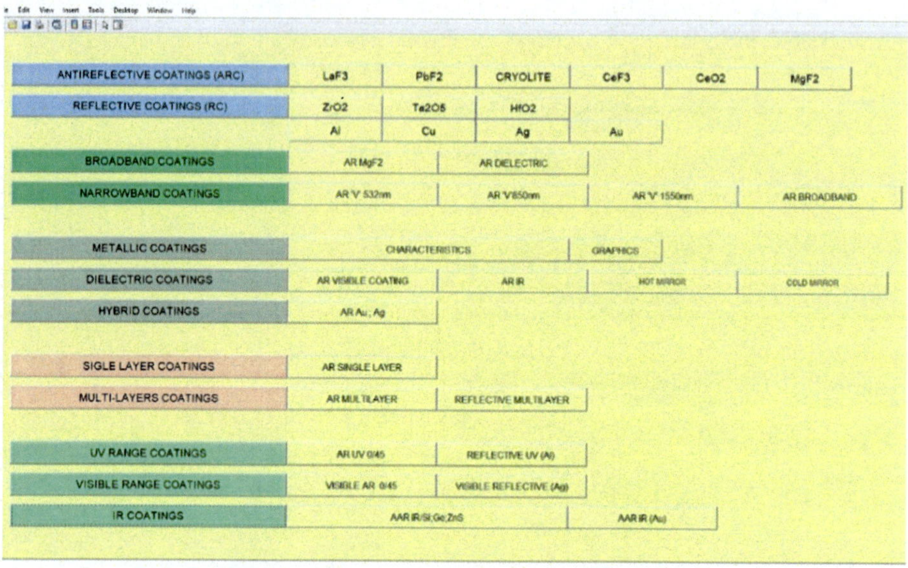

Fig. 9. Coatings selection guide detailed view.

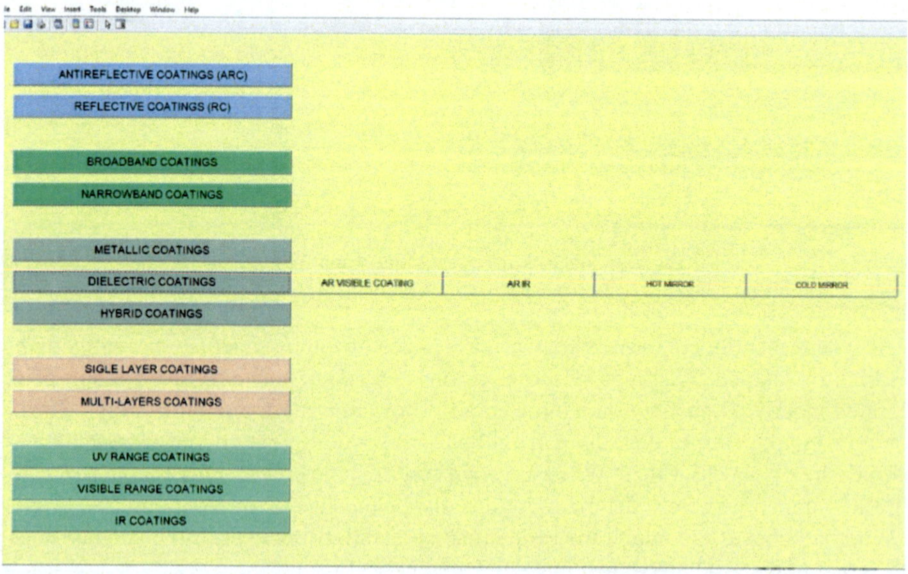

Fig. 10. Coatings selection guide example.

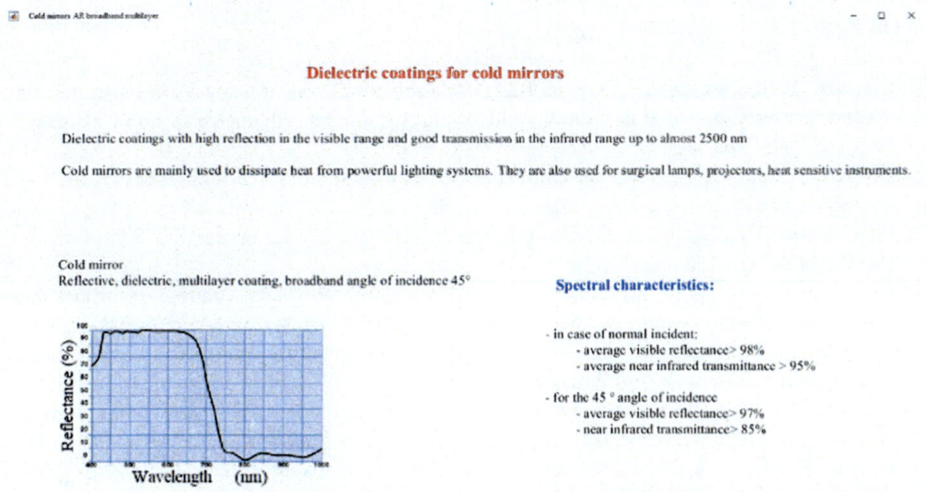

Fig. 11. Selected coating page.

5 Conclusions

The authors developed a software in Matlab environment for an easy selection of optical materials, either substrates or optical coatings, useful for education, engineers.

The software has an interface which is user friendly, provided with radiobuttons and pushbuttons, organized in two main pages where most encountered materials in optics applications are displayed. When one material is chosen, the user can find from a dedicated page the main properties, applications, advantages, plots of spectral characteristics and also links to web pages of material producers, where complete data are available.

The guide interface which was developed by the authors allows to add anytime more pages with more optical materials as substrates and coatings. Raw data of spectral characteristics, as wavelength domain, reflectance / transmittance can also be imported from available excel files, as an example, so the plots can be created and included in the dedicated pages.

The interface could be a useful tool for knowledge of coating types and selection of proper optical coating for a specific application and for a group of convenient substrates. Selecting the right optical glass for a particular application can significantly impact cost, durability, and performance.

The authors developed their program in Matlab environment because Matlab is used both in academic and industrial fields, the students and engineers being familiar with this software. In a future development of the interface, some features will be added, such as adding materials catalogues and matching criteria for better combinations between substrates and coatings.

The purpose of the interface developed by the authors is to give the specialists a guide for choosing the optimal combination between the substrate and the optical coating material, without achieving an automatic pairing. Thus, it is allowed to add new optical materials and choose an optimal combination depending on the application and the availability of materials.

References

1. Bernacki, B.E., Johnson, T.J., Myers, T.L.: Modeling thin layers of analytes on substrates for spectral analysis: use of solid/liquid n and k values to model reflectance spectra. Opt. Eng. **59**(9), 092005 (2020)
2. Gaspari, F.: Optical Property of Thin Film, Energy Materials, in Comprehensive Energy Systems (2018)
3. Li, Y., Heinz, T.F.: Optical models for thin layers, 2D Materials 5, February 2018. https://doi.org/10.1088/2053-1583/aab0cf
4. Krogman, K.C., Druffel, T., Sunkara, M.K.: Anti-reflective optical coatings incorporating nanoparticles. Nanotechnology **16**(7), S338–S343 (2005)
5. Edmund Optics homepage. https://www.edmundoptics.com/knowledge-center/application-notes/optics/introduction-to-polarization/. Accessed 15 Jan 2021
6. Rizescu, C.I., Ghita, I., Rizescu, D.: methods for testing the strength of layers for different optical coatings. In: Cioboată, D.D. (ed.) International Conference on Reliable Systems Engineering (ICoRSE) - 2021. LNNS, vol. 305, pp. 119–128. Springer, Cham (2022). https://doi.org/10.1007/978-3-030-83368-8_12
7. Guglya, A., Marchenko, I.: Ion beam-assisted deposition technology as a method of nanocrystalline coating formation. Probl. Atom. Sci. Technol. **90**, 125–136 (2014)
8. Sultan, Z., Sultana, N.: Analysis of reflectance and transmittance characteristics of optical thin film for various film materials, thicknesses and substrates. J. Elect. Electron. Syst. **4**(3), 2332 (2015)
9. Ranjgar, A., Norouzi, R., Zolanvari, A., Sadeghi, H.: Characterization and optical absorption properties of plasmonic nanostructured thin films. Armenian J. Phys. **6**, 198–203 (2013)
10. Lu, H., Gunther, J., Andersson-Engels, J.: Determination optical properties of tissue-like phantoms using diffuse reflectance and transmittance spectroscopy. In: Proceedings of SPIE, pp.10820-108202X (2018)
11. Czapla, B., Hanssen, L.: Direct method of extracting complex refractive index from routine Fourier transform infrared reflectance/transmittance measurements. In: SPIE Optics Photonics 2020 Digital Forum, San Diego, CA (2020). https://doi.org/10.1117/12.2568502. Accessed 8 Nov 2021
12. SCHOTT Homepage. https://www.schott.com/en-gb/products/optical-glass-p1000267/downloads. Accessed 29 Jan 2022
13. JENOPTIK Homepage. https://www.jenoptik.com/products/optical-systems/optical-precision-components/optical-coatings/standardized-optical-coatings. Accessed 29 Jan 2022
14. Spectral Systems homepage. https://www.spectral-systems.com/wp-content/uploads/2014/04/spectral-systems-catalog-2014.pdf. Accessed 11 Oct 2021
15. THORLABS Homepage. https://www.thorlabs.com/newgrouppage9.cfm?objectgroup_id=6973. Accessed 29 Jan 2022

Bat Algorithm for Discrete Optimization Problems: An Analysis

Bruno Sousa[1], Rita Guerreiro[1], André S. Santos[2(✉)], João A. Bastos[3], Leonilde R. Varela[4], and Marlene F. Brito[5]

[1] Institute of Engineering, Polytechnic Institute of Porto - ISEP/IPP, Porto, Portugal
{1180543,1180753}@isep.ipp.pt

[2] Interdisciplinary Studies Research Center (ISRC), Institute of Engineering, Polytechnic Institute of Porto - ISEP/IPP, Porto, Portugal
abg@isep.ipp.pt

[3] Institute of Engineering, INESC TEC, Polytechnic Institute of Porto - ISEP/IPP, Porto, Portugal
jab@isep.ipp.pt

[4] Algoritmi Research Centre, University of Minho, Guimarães, Portugal
leonilde@dps.uminho.pt

[5] Instituto de Ciência E Inovação Em Engenharia Mecânica E Engenharia Industrial - INEGI, Porto, Portugal
mab@isep.ipp.pt

Abstract. In this article the application of the discrete version of the bat algorithm to flowshop scheduling problems is presented and compared with Simulated Annealing, Local Search, as well as versions of each that start from constructive heuristics (Palmer and CDS). Bat algorithm is a novel metaheuristic, developed for continuous problems that has shown exceptional results. This paper intends to assess its effectiveness and efficiency for discrete problems when compared with other optimization techniques, including Simulated Annealing and Local Search, whose results are already proven. First, it was developed a literature review about those algorithms, then they were implemented in VBA with Microsoft Excel. Once implemented, the parameterization was carried out, ensuring an adequate application of the algorithms before they can be compared. Then, the methods were applied for 30 normally distributed instances, in order to draw broader conclusions. Finally, a statistical evaluation was carried out and concluded the inferiority of the Local Search in relation to the metaheuristics and the superiority of the hybrid version of the Bat Algorithm with CDS in relation to Simulated Annealing, with significantly better solutions, in an equal computation time.

Keywords: Metaheuristics · Flowshop · Bat algorithm · Simulated annealing · Palmer and Campbell · Dudek and smith

J. Machado et al. (Eds.): icieng 2022, LNME, pp. 161–172, 2023.
https://doi.org/10.1007/978-3-031-09382-1_14

1 Introduction

1.1 A Subsection Sample

In the last decades, many algorithms have been used to model and solve manufacturing management problems [1–6], among which, flowshop scheduling problems play a crucial role, and have been solved by using diverse kind of optimization approaches, including mathematical modelling, branch and bound, heuristics, among others [7–10]. These approaches work for small instances but face some difficulties when used for large-scale problems. Compared to exact methods, metaheuristics do not guarantee the achievement of an optimal solution, but they do manage to approach close values for large problems in a significantly short time span, which guarantees the possibility of their use. After 2000, many metaheuristics were developed to solve this type of problems, including ant colony optimization (ACO), genetic algorithm (GA), artificial bee colony (ABC), Tabu Search (TS), Simulated Annealing (SA), among others. This evolution allowed the possibility of solving increasingly larger problems in shorter intervals [11, 12]. Thus, this study intends to help in this evolutionary process, with a hybrid approach between flowshop heuristics and metaheuristics.

Following on from that, Local Search and two metaheuristics, Simulated Annealing (SA) and Discrete Bat Algorithm, with a combination of flowshop heuristics (Palmer and CDS), were implemented on 30 randomly generated flowshop problems with 150 tasks and 5 machines in order to test the efficiency of the Discrete Bat Algorithm. It is important to test and compare the resolution of the same problem by different existing algorithms, since there is still not enough research and experience to ensure that the best technique for one problem is the best for other problems. The tool used to carry out this implementation was the Microsoft Excel using Visual Basic for Applications (VBA). The developed tool is prepared to apply these algorithms to a problem with n tasks and m machines. For a fair comparison of these algorithms, a parameterization study is necessary, ensuring that each one of them is well adapted to the problem at hand.

2 Literature Review

2.1 Heuristics and Metaheuristics

There are many decision problems in business and economics related to production, routing, resource allocation, and planning that can be formulated as optimization problems. However, most of these problems difficult to solve within a reasonable computation time. For this reason, heuristics are often preferred in order to obtain a solution that is not only possible, but also good [13]. Metaheuristics incorporate concepts based on biological evolution, nature, mathematical and physical sciences, and statistical mechanics.

A metaheuristic is an iterative process that guides a subordinate heuristic, exploring the entire space of solutions, structuring information in order to efficiently find approaches to the optimal solution [14]. While a heuristic, such as Local Search, studies a limited part of the solution space, converging to a local optimum, a metaheuristic attempts to converge to a global optimum.

2.2 Local Search (LS)

Local Search is a heuristic in which solutions are always changing through local moves that change the solution [14]. Starting from a solution, usually random, a search is made in the neighbourhood for solutions with better performance, to replace the current solution. The process will repeat itself until it is not possible to find a solution with better performance in the neighbourhood of the current solution. In this work we use the Steepest Descent technique, which explores the entire neighbourhood of the current solution and then it will move to the best solution in that neighbourhood if, and only if, it's better than the current solution [15].

2.3 Simulated Annealing (SA)

Simulated Annealing (SA) is a single solution metaheuristic, based on the principles of statistical mechanics and especially on the annealing process, which extends the concept of local research, allowing movements to inferior solutions. In this algorithm a solution is accepted if its objective function is better than the previous one or with a probability that depends on the deterioration of the objective function value. This probability is given by expression (1).

$$P(T, f(x1), F(x)) = \exp\left(-\frac{f(x1) - f(x)}{T}\right) \tag{1}$$

The temperature T, represented in expression (1), is a control parameter. The value of T is initially high, which allows a high number of lower moves to be accepted. This temperature gradually decreases after a certain value L (equilibrium state) of iterations [13]. The value of T can decrease linearly ($T = T - \beta$), geometrically ($T = \alpha T$) or logarithmically ($T = T0 \log(i)$). The value L of iterations must be defined according to the size of the problem instance and should be proportional to the size of the neighbourhood. As for the stopping criterion, there are several methods that can be used, from reaching an a priori determined final temperature to a predetermined number of iterations without improving the best solution, among others.

2.4 Discrete Bat Algorithm (DBA)

The Bat Algorithm (BA) is a population metaheuristic, based in swarm intelligence, which was developed by Dr. Xin-She Yang in 2010 [16]. It is inspired by the sonar behaviour of micro-bats, namely echolocation, used in prey detection and obstacle avoidance. Indeed, bats emit a loud sound and analyse the echo feedback to understand their surroundings and move to a better location. However, this metaheuristic was not developed for discrete problems and therefore it was necessary to find a discrete approach, so that it could be used in such problems [16, 17]. In [18], the authors, presented a possible approach to the algorithm for flowshop problems, which was used in this paper, with some minor changes as can be seen in Table 1.

Table 1. Discrete bat algorithm.

Begin
Initialize the population *ps*, t = 1 and the other parameters
Evaluate fitness for each individual and found global best and population best
While (the termination condition does not satisfy)
For i = 1: ps
Generate frequency f: $f = \lfloor f_{max} + (f_{min} - f_{max}) * (\frac{t}{t_{max}}) \rfloor$
Obtain velocity $v_{i,f}^{N}$: **divide solution in partitions with f size**
Determine $v_{tmp,f}^{N}$ by NEH method: **sort each partition in descending order of processing time**
Update $x_i^N(t)$ using $v_i^N(t)$;
Evaluate fitness of individual and update population best;
Perform Pulse Emission Rate Local Operation: **swap or insert partitions according to a random value being higher or lower than** $r_i(t) = 1 + exp\left(-\frac{10}{t_{max}} * \left(t - \frac{t_{max}}{2}\right) + r_i(1)\right)^{-1}$
Evaluate fitness of individual and update population best;
Perform Loudness Operation: **replace a partition or an inverse partition of global best according to a random value being higher or lower that** $Ld_i = \frac{fit_i - fit_{min}}{fit_{max} - fit_{min}}$
Make Solution Feasible;
Evaluate fitness of individual and update population best;
End for
Find out current global best position;
For i = 1: ps
Execute swap operation based on global best;
End For
Evaluate fitness for each individual and find out global best;
For i = 1: ps
Execute insert operation based on global best;
End For
Evaluate fitness for each individual and find out global best;
For i = 1: ps
Execute single point move backward operation based on global best;
End For
Evaluate fitness for each individual and find out global best;
t = t + 1
End While
Output result
End

2.5 Palmer and Campbell, Dudek and Smith (CDS)

A constructive heuristic is a type of heuristic method, which starts with an empty solution and repeatedly extends the current solution until a complete solution is obtained.

Some examples are Palmer and Campbell, Dudek and Smith (CDS) heuristics. Palmer's heuristic first calculates a slope order for each job, and then sequences the jobs according to the slope orders [19]. This is based on the idea of giving priority to jobs having the strongest tendency to progress from short times to long times in the sequence of operations [20]. For smaller problems, Palmer's heuristic provides good schedules [21]. Campbell, Dudek and Smith (CDS) presented a heuristic, which is an extension to Johnson's rule [19]. This heuristic gives generally good results and due to its performance record it has also been used as a standard of comparison [21]. Their method splits the original m-machine problem into a set of n-1 two-machine problems and solves each of these by Johnson's rule [19].

2.6 Parameters

The Simulated Annealing meta heuristic parameters are the initial temperature, equilibrium state, L, that is, the number of iterations for which the value of T decreases, value of α, for the geometric cooling criterion and finally the stop criterion which in this work will be up to a limited number of iterations. Thus, starting with the value of alpha, and though the research conducted it was apparent that the values are ideally between 0.8 and 0.99 [22–24]. For the value of L, the majority of the authors considers that the value must be equal to the number of tasks/jobs [24–26], however Ogbu & Smith, in 1990 [27] propose that the value of L is within the range [30n,10n], where n is the number of tasks/works. In [28] it is recommended that the acceptance probability is close to 1 at the initial temperature, the last parameter (initial temperature) must allow for an large initial solution acceptance rate. In [23], as mentioned in [26], the authors propose to use the following Eq. (2) where Pi represents the acceptance probability wanted and Δf(x) corresponds to the dispersion of the sample. Lee & Lee, in 2014 [29] also introduce this equation stating that the values of pi must lie between 0.5 and 0.95.

$$T_i = -\frac{\Delta f(x)}{\ln(Pi)} \tag{2}$$

In the case of the DBA, the parameters are essentially based on the number of bats, stopping criteria, and the range of partition sizes (fmin and fmax). Starting with the number of bats and stopping criterion, one of the main points of this paper was to compare the performance of DBA with Simulated Annealing, so the multiplication of these parameters will initially be equal to the stopping criterion of the latter, in order to generate the same number of solutions for each instance. In the case of the size of the partitions, since this is a specific approach to the Bat Algorithm, there are no studies found that suggest values with better performance, so values within the order used in [18] will be used.

3 Case Study

3.1 Problem

To carry out this work, 30 instances of equal size were used with processing time values being random generated between 1 and 100. So, for each one, it was necessary to find a production schedule that minimized the makespan, that is, total processing time. Each algorithm was applied 3 times to each instance.

3.2 Methodology

To solve the presented problem, the first step was to develop the programming codes of the mentioned algorithms (LS, SA, DBA, Palmer and CDS) in VBA with Microsoft Excel. Then, based on the literature review, a study of the best parameters of each metaheuristic was carried out, to enhance the capabilities of each algorithm. After finding the best parameters of each algorithm, they were applied 5 times to an instance (except LS with Palmer and CDS that would converge to the same solution) to understand which one had a better performance in terms of smaller makespan, faster convergence rate and lower process time. After that, the algorithms under study were implemented to 30 instances of the same size. The processing times were randomly found between the values of 1 and 100.

3.3 Parameterization

Having the algorithms implemented it became important to understand the best parameters for the problem at hand. For this, several combinations were tested, considering the values found in the literature. This range of values, as well as the step, can be seen in Table 2. Note that for each combination 5 iterations were performed, for a total of 1500 iterations.

Table 2. Parameterization values.

Simulated annealing			Bat algorithm	
Parameter	Range	Step	Parameter	Combinations
T_0	[50,450]	100	fmin - fmax	(2 - 4) (4 - 8) (6 - 12) (8 - 16) (10 - 20)
L	[100,200]	50		
Alpha	[0.8,0.99]	0.05	Bat * stop criterion	5 - 1000 10 - 500; 15 - 333 20 - 250;
Stop criterion	[2000,5000]	1000		

Note: 5 more instances for the SA best parameters were performed with stop criterion = 10000

In the results analysis, two essential criteria were used, lowest average of the 5 values for a given parameter combination and iteration with the lowest minimum. Thus, for SA, the parameters that generated best average values were alpha = 0.95, L = 150, T0 = 50 and stopping criterion = 10000 while for DBA, the best parameters were fmin = 4, fmax = 8, bats = 10 and stopping criterion = 500.

3.4 Detailed Instance Analysis

With the results obtained in the parameterization, the algorithms mentioned in Subsect. 3.2 were used, applying each method 5 times to an experimental instance in order to draw conclusions for the rest of the study (except LS with Palmer/CDS that always converges to the same local optimum). The results obtained for the average of the objective functions as for the computation times can be seen in Table 3.

Table 3. Problem results.

	Local search		Simulated annealing		Discrete bat algorithm	
	Makespan	Time	Makespan	Time	Makespan	Time
Standard	9743	19	9396	30	9209	115
With Palmer	9526	9	9290	29	9208	116
With CDS	9283	12	9252	30	9151	117
With palmer + CDS					9151	117

Thus, analysing the results obtained, it can be inferred that, as expected, Local Search is the fastest method, but also the one that generates the worst solutions. Nevertheless, when it starts from a solution obtained through CDS it obtains a good result, even better than those obtained with standard SA, in a shorter time interval. In the case of SA, the values were in fact better than those obtained by LS but, starting from the same initial solution obtained by CDS, superior results were expected, considering the number of iterations to be significant (Stop Criterion = 5000). In the case of DBA, the results were indeed the best obtained but, although the number of solutions generated was equal to that of SA (since the multiplication between the number of bats by the Bat Algorithm's stopping criteria was equal to SA's stopping criteria, the result took 4 times longer). Therefore, to try to understand if the convergence of the bat algorithm would not be superior, and therefore needing less iterations (and consequently less processing time), the evolution of the best solution of both metaheuristics were evaluated. In order to make the comparison fairer, the same initial solution was used, obtained through the CDS method, whose results can be seen in Fig. 1, where the red line represents DBA and the black line SA algorithm.

In fact, with the data presented, it can be seen that the DBA is clearly superior in terms of convergence for this example, reaching in few iterations (60, which is equivalent to a stopping criterion of 6 for 10 bats), a superior solution to the Simulated Annealing for 5000 iterations, and its final makespan in less than 1000 iterations (100 in the stopping criteria for 10 bats). Given these results, it was decided that in the study for the 30 instances the performance of Bat Algorithm with CDS would also be evaluated, with 125 as the stopping criteria and 10 bats.

3.5 30 Instances Results

Before presenting the results obtained for the 30 instances, it is necessary to recall that the central limit theorem states that if you have a population with mean μ and standard deviation σ and take sufficiently large random samples from the population, then the

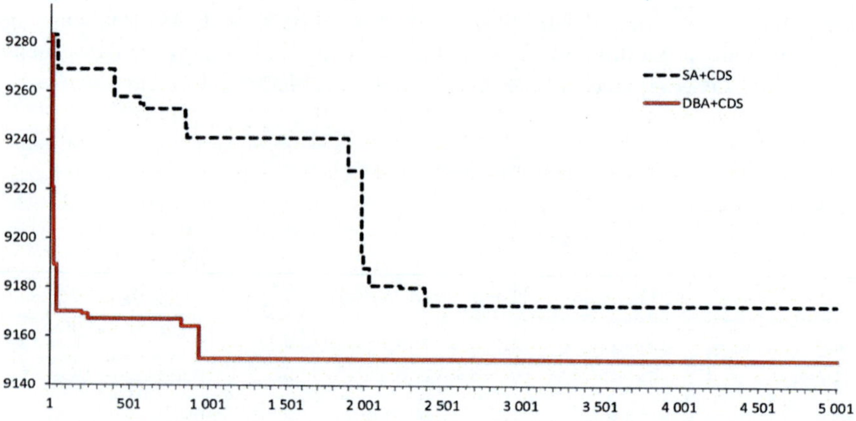

Fig. 1. Evolution of the best makespan for SA + CDS and DBA + CDS.

distribution of the sample means will be approximately normally distributed [30]. An essential component of the Central Limit Theorem is that the average of the sample means will be the population mean. Therefore, instances follow an approximately normal distribution and can therefore be used to assess solution-seeking behaviour, furthermore, allow the use of statistical analyses that require data following a normal distribution. Moving on to the results obtained, 3 iterations were performed for each of 4 methods (LS, SA, DBA and DBA with CDS for the initial solution of one of the bats) for each of the 30 instances. In view of these values, the relative deviations averages of each method in relation to the best value were calculated so that comparisons could be made. These values enabled the values shown in Fig. 2 to be drawn up.

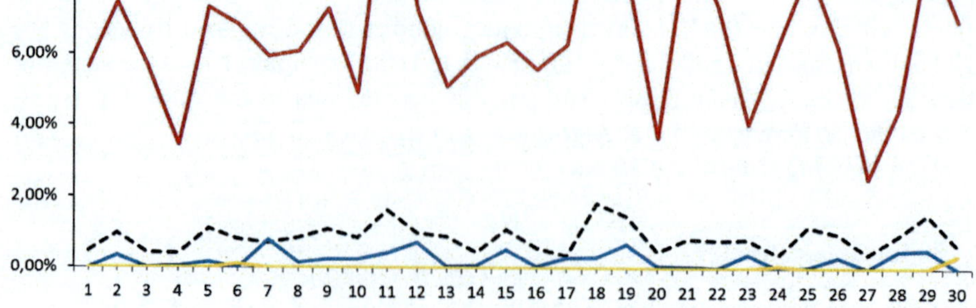

Fig. 2. Relative deviations from each algorithm in relation to the best.

Effectively, as can be inferred, the results obtained by DBA were superior to those obtained by SA (grey and orange lines) for an equivalent number of iterations (500 stopping criteria and 10 bats and 5000 stopping criteria, respectively). Nevertheless, with the use of CDS for an initial solution of one bat the results were even better, reaching the best solution found in 27/30 instances (0.08%, 0.1%, 0.36% in the remaining ones). However, although the visualized results seem sufficient to draw conclusions, a statistical evaluation is necessary. To do so, a one-way ANOVA will be used, which is an extension of Student's T hypothesis test, to analyse the means of multiple populations. Thus, the means will be μLS for Local Search and μSA, μDBA and μDBACDS by the same logic for the others, representing the average relative deviations from the best solution in each of the 30 instances. The hypotheses analysed by ANOVA are:

H0: $\mu LS = \mu SA = \mu DBA = \mu DBACDS$.

H1: There is a difference between the averages of the relative deviation in relation to the optimums obtained.

The result of the ANOVA is presented in the Table 4 and as it is possible to verify, that statistical evidence was found to reject the null hypothesis, with 95% confidence, that is, it is not possible to affirm that the results obtained have similar values (Pr 0.000).

Table 4. ANOVA of the mean of the relative deviation of the algorithms.

	Df	Sum of Squares	Mean square	F Value	Pr(>F)
Between algorithms	3	0.08751	0.029170	255.6	< 2e-16**
Within algorithms	116	0.01324	0.000114		
Total	119	0.10075	0.029284		

As from the analysis performed there is at least one population that performed differently from the others, it is necessary to perform a Post Hoc test, in this case the Scheffe Test to understand which ones led to this result. This test can be seen in Table 5, which will locate pairs that have an average relative deviation from the best solution that can be considered similar.

Thus, analysing the data from the Scheffe Test present in Table 5, it cannot be concluded with 95% confidence that the deviations of the relative solutions are similar, and therefore, there are statistically proven differences between the various methods. Thus, considering the pval values, it can be concluded that the LS has a different average of the others, in this case worse (as expected) but more importantly proves that there is a difference already significant between Bat Algorithm with CDS and SA, also proven by the p value. Finally, besides the differences in obtaining the best values of the objective function, it was also important to understand the differences in terms of computing time. To this end, the average processing times can be seen in Table 6.

As can be seen, not only is the DBA with CDS superior to SA, but it also manages to take almost the same processing time, further highlighting its overall quality.

Table 5. ANOVA of the mean of the relative deviation of the algorithms.

		Mean difference	pval
LS	SA	0.0571	<2e−16***
	DBA	0.0633	<2e−16***
	DBA+CDS	0.0655	<2e−16***
SA	LS	−0.0571	<2e−16***
	DBA	0.0062	0.1708
	DBA+CDS	0.0084	0.0289*
DBA	LS	−0.0633	<2e−16***
	SA	−0.0062	0.1708
	DBA +CDS	0.0022	0.8874
DBA+CDS	LS	−0.0655	<2e−16***
	SA	−0.0084	0.0289*
	DBA	−0.0022	0.8874

Table 6. Average processing times.

	LS	SA	DBA	DBA + CDS
Time (in seconds)	8,7	29,9	116,4	31,4

4 Conclusions

Once the study was concluded, it was possible to compare and draw conclusions between the various algorithms. The objective was to demonstrate the superiority of the Bat Algorithm for discrete optimization problems, in comparison with the Local Search heuristic and Simulated Annealing, both in its normal form and based on constructive methods like Palmer or CDS. For this, parameterization was performed, followed by a detailed application analysed for an instance. After that, and adjusting the study to the results found, some methods were applied to 30 equally distributed instances, in order to perform a statistical analysis, which included an ANOVA analysis and a Post-Hoc test, in this case Scheffe's Test. In fact, this analysis led to the conclusion of the inferiority of Local Search in relation to the others and the superiority of the Bat Algorithm starting from CDS in relation to the normal SA, both in terms of results obtained and also in terms of computation that were similar. Thus, the objectives of the paper were achieved, believing that this algorithm is a viable and efficient option for discrete problems, being one more possibility to be used in this evolutionary process of solving flowshop scheduling problems. This study raises then the possibility of a high panoply of future work that can pass through the comparison of this approach with other population metaheuristics, starting also from constructive heuristics; the use of more methods to obtain the initial solution and the analysis of these hybrid algorithms, such as the addition of the Nawaz heuristic or the application of these methods to other discrete problems.

Acknowledgement. This work was supported by national funds through the FCT—Fundação para a Ciência e Tecnologia through the R&D Units Project Scopes: UIDB/00319/2020, and EXPL/EME-SIS/1224/2021.

References

1. Arrais, A., Varela, M.L.R., Putnik, G.D., Ribeiro, R.A., Machado, J., Ferreira, L.: Collaborative framework for virtual organisation synthesis based on a dynamic multi-criteria decision model. Int. J. Comput. Integr. Manuf. **31**(9), 857–868 (2018). https://doi.org/10.1080/0951192X.2018.1447146
2. Varela, M.L.R., Aparício, J.N., Silva, S.C.: A web-based application for manufacturing scheduling. In: Hamza, M.H. (Eds.) Proceedings of the IASTED International Conference on Intelligent Systems and Control, Salzburg; Austria, 25–27 June 2003, pp. 400–405 (2003). ISBN: 0889863555
3. Sousa, R.A., Varela, M.L.R., Alves, C., Machado, J.: Job shop schedules analysis in the context of industry 4.0. In: 2017 International Conference on Engineering, Technology and Innovation: Engineering, Technology and Innovation Management Beyond 2020: New Challenges, New Approaches, ICE/ITMC 2017 - Proceedings, 2018-January, pp. 711–717 (2018). https://doi.org/10.1109/ICE.2017.8279955
4. Gangala, C., Modi, M., Manupati, V.K., Varela, M.L.R., Machado, J., Trojanowska, J.: Cycle time reduction in deck roller assembly production unit with value stream mapping analysis. In: Rocha, Á., Correia, A.M., Adeli, H., Reis, L.P., Costanzo, S. (eds.) Recent Advances in Information Systems and Technologies. AISC, vol. 571, pp. 509–518. Springer, Cham (2017). https://doi.org/10.1007/978-3-319-56541-5_52
5. Trojanowska, J., Żywicki, K., Varela, M.L.R., Machado, J.M.: Shortening changeover time - an industrial study. In: The Proceedings of the 2015 10th Iberian Conference on Information Systems and Technologies (CISTI 2015), pp. 1–6 (2015). https://doi.org/10.1109/CISTI.2015.7170373
6. Vieira, G.G., Varela, M.L.R., Putnik, G.D., Machado, J.M., Trojanowska, J.: Integrated platform for real-time control and production and productivity monitoring and analysis. Roman. Rev. Precis. Mech. Optics Mechatron. **2016**(50), 119–127 (2016)
7. Reddy, M.S., Ratnam, C., Agrawal, R., Varela, M.L.R., Sharma, I., Manupati, V.K.: Investigation of reconfiguration effect on makespan with social network method for flexible job shop scheduling problem. Comput. Indust. Eng. Elsevier **110**, 231–241 (2017). https://doi.org/10.1016/j.cie.2017.06.014
8. Lambert, A.J.D.: Optimal disassembly of complex products. Int. J. Prod. Res. **35**(9), 2509–2524 (1997). https://doi.org/10.1080/002075497194633
9. Kang, J.G., Xirouchakis, P.: Disassembly sequencing for maintenance: a survey. Proc. Inst. Mech. Eng. Part B: J. Eng. Manuf. **220**, 1–4 (2006). https://doi.org/10.1243/09544054JEM596
10. Slama, I., Ben-Ammar, O., Masmoudi, F., Dolgui, A.: Disassembly scheduling problem: literature review and future research directions. IFAC-PapersOnLine, **52**, 1–7 (2019). https://doi.org/10.1016/j.ifacol.2019.11.225
11. Aydemir-Karadag, A., Turkbey, O.: Multi-objective optimization of stochastic disassembly line balancing with station paralleling. Comput. Indust. Eng. **65**, 1–12 (2013). https://doi.org/10.1016/j.cie.2013.03.014

12. Zhong, L., Youchao, S., Gabriel, O.E., Haiqiao, W.: Disassembly sequence planning for maintenance based on metaheuristic method. Aircraft Eng. Aerosp. Technol. **83**, 1 (2011). https://doi.org/10.1108/00022661111131221
13. Voß, S.: Meta-heuristics: the state of the art. In: Nareyek, A. (ed.) Local Search for Planning and Scheduling. LNCS (LNAI), vol. 2148, pp. 1–23. Springer, Heidelberg (2001). https://doi.org/10.1007/3-540-45612-0_1
14. Osman, I.H., Kelly, J.P.: Meta-heuristics: an overview. In: Osman, I.H., Kelly, J.P. (eds.) Meta-Heuristics, pp. 1–21. Springer, Boston (1996). https://doi.org/10.1007/978-1-4613-1361-8_1
15. Ram, S.K.M.: Steepest Descent Method (2019). https://doi.org/10.1007/978-981-15-0894-3_6
16. Yang, X.-S.: A new metaheuristic bat- inspired algorihm. In: Gonzalez, J.R., Pelta, D.A., Cruz, C., Terrazas, G. (eds.) Nature Inspired Cooperative Strategies for Optimization (NICSO). SCI, vol. 284, pp. 65–74. Springer, Heidelberg (2010). https://doi.org/10.1007/978-3-642-12538-6_6
17. Yang, X.S.: Bat algorithm: Literature review and applications. Int. J. Bio-Inspir. Comput. **5**(3), 141–149 (2013). https://doi.org/10.1504/IJBIC.2013.055093
18. Luo, Q., Zhou, Y., Xie, J., Ma, M., Li, L.: Discrete bat algorithm for optimal problem of permutation flow shop scheduling. Sci. World J. (2014). https://doi.org/10.1155/2014/630280
19. Chen, C. L., Vempati, V. S., Aljaber, N.: An application of genetic algorithms for flow shop problems. Eur. J. Oper. Res. **80**, 389–396 (1995). https://doi.org/10.1016/0377-2217(93)E0228-P
20. Baker, K.R.: Introduction to Sequencing and Scheduling. Wiley, Nova York (1974)
21. Hundal, T.S., Rajgopal, J.: An extension of palmer's heuristic for the flow shop scheduling problem. Int. J. Prod. Res. **26**(6), 1119–1124 (1988). https://doi.org/10.1080/00207548808947922
22. Prado, A.S., Souza, S.R.: Problema de alocação de salas em cursos universitários: um estudo de caso. In: Anais do XLVI Simpósio Brasileiro de Pesquisa Operacional (2014)
23. Rose, J., Klebsch, W., Wolf, J.: Temperature measurement and equilibrium dynamics of simulated annealing placements. IEEE Trans. Comput. Aid. Des. Integr. Circuits Syst. **9**, 253–259 (1990). https://doi.org/10.1109/43.46801
24. Wodecki, M., Bożzejko, W.: Solving the flow shop problem by parallel simulated annealing. In: Wyrzykowski, R., Dongarra, J., Paprzycki, M., Waśniewski, J. (eds.) Parallel Processing and Applied Mathematics. LNCS, vol. 2328, pp. 236–244. Springer, Heidelberg (2002). https://doi.org/10.1007/3-540-48086-2_26
25. Aarts, E., Korst, J.: Simulated annealing 2.1 Introduction of the algorithm. In: Simulated Annealing: Theory and Application (1987). https://link-springer-com.ezproxy2.library.colostate.edu/content/pdf/10.1007%2F978-94-015-7744-1_2.pdf
26. Reis, P. C.: Ferramenta de Apoio ao Escalonamento da Produção, Porto (2020)
27. Ogbu, F.A., Smith, D.K.: The application of the simulated annealing algorithm to the solution of the n/m/Cmax flowshop problem. Comput. Oper. Res. **17**, 243–253 (1990). https://doi.org/10.1016/0305-0548(90)90001-N
28. Park, M.W., Kim, Y.D.: A systematic procedure for setting parameters in simulated annealing algorithms. Comput. Oper. Res. **25**, 207–217 (1998). https://doi.org/10.1016/S0305-0548(97)00054-3
29. Lee, C.-Y., Lee, D.: Determination of initial temperature in fast simulated annealing. Comput. Optim. Appl. **58**(2), 503–522 (2013). https://doi.org/10.1007/s10589-013-9631-y
30. Kwak, S.G., Kim, J.H.: Central limit theorem: the cornerstone of modern statistics. Korean J. Anesthesiol. **70**(2), 144 (2017)

Geometrical Metrology in Food Product Measurements

M. Wieczorowski[1](✉), N. Swojak[1], M. Szelewski[2], A. Pereira[3], T. G. Mathia[4], and B. Gapinski[1]

[1] Poznan University of Technology, 5 M.Sklodowska-Curie Square, 60-965 Poznan, Poland
michal.wieczorowski@put.poznan.pl
[2] ITA Sp. k, 104 Poznanska Street, 60-185 Poznan, Poland
[3] University of Vigo, 36310 Vigo, Spain
[4] École Centrale de Lyon, 36 Av. Guy de Collongue, 69134 Écully, France

Abstract. The paper presents research and scientific considerations to determine the usefulness of modern measurement systems including computer tomography and machine vision systems for mapping the geometry and dimensional measurements of products in the food industry. Due to the variable characteristics of food products, their control during the production process (active metrology) increasingly requires the search for new measurement solutions. These criteria are met by measurement devices combined with classification, machine learning and artificial intelligence systems for automated feature analysis. The measurement methodology was developed, necessary system components were described and IT procedures were applied to enable correct measurements of the assumed product characteristics. Additionally, the factors influencing the errors of measurements obtained with the described systems are indicated, supported by statistical analysis of the obtained results. The study confirmed the suitability of modern measurement systems for product inspection in the food industry.

Keywords: Food products · Coordinate metrology · Machine vision systems

1 Introduction

Measurement technology in the field of length and angle metrology and the devices working in it are mostly intended for applications in mechanical, electrical and civil engineering [1, 2]. They mostly concern objects made of metallic materials [3, 4] and plastics [5, 6], since they are the basic materials for industry [7, 8]. But sometimes situations arise where the items being tested concern completely different areas of life. An example of this is food inspection and measurement. These are mostly related to testing of mechanical properties [9, 10]. Nevertheless, measurements of geometrical features in certain situations are also very important for food products. Among the discussed situations, they are mainly related to batch preparation on the production line or the necessity to match the dimensions to the packaging.

The analysis of products moving on the production line is important both at the start-up stage of the production process and during production. Economic effects are

J. Machado et al. (Eds.): icieng 2022, LNME, pp. 173–186, 2023.
https://doi.org/10.1007/978-3-031-09382-1_15

important here, so that each too large quantity of material for a given product generates an unjustified increase in production costs. On the other hand, too little material is also a reject, i.e. a product that does not meet the expectations of the recipient, also in visual terms (material loss). The increasing level of automation in production, associated with the need to eliminate man and his mistakes [11] and the use of robots [12] and drones [13], forces more and more advanced control methods, implemented without human involvement, providing information about the production on an ongoing basis. This application, in turn, includes the need to adapt the product to the package or window through which it must pass. In this case, any excess in geometric dimensions will cause the automatic transport cycle to block and require the intervention of a supervisor. Here too, of course, the food product cannot be allowed to be too small in terms of dimensions, as this would result in a complaint from the customer. Guided by the above considerations, the paper presents examples of applications of geometric measurements for food products. The measurement methodology was developed, necessary system elements were made and IT procedures were developed.

2 Measurement Technique

Modern coordinate metrology is based on collecting coordinates of points from a surface and processing them to produce results of interest to the user. The combination of metrology with classification systems, machine learning and artificial intelligence for automated feature analysis is also gaining popularity [14]. A characteristic of many modern instruments is the increasing measurement range and the gradual expansion of measurement possibilities to other scales. In the presented study, the macro scale was transitioning to the meso scale, making computed tomography and focal differentiation microscopy among the selected measurement techniques, although the use of a laser profile sensor was also considered [15]. A detailed description of how these techniques work is provided in the literature, both in terms of tomography [16, 17] as well as regarding surface analysis techniques [18, 19].

Computed tomography (CT) is an X-ray imaging method allowing for the acquisition of images of examined objects. Results obtained from the CT enable both the verification of geometry [20, 21] and the examination of structure [22]. The quality of acquired images reach even micro scale [23, 24]. Possibility of their further processing and obtaining particular measurement and accuracy data – to a large degree – depend on a scanned material density, thereby its susceptibility to the X-ray penetration. Problems during the object testing, accounted for the low quality of acquired images that make it impossible to adequately interpret measurement results, arise in the case of X-ray scanning of both low- and high-density materials. Food items definitely belong to the low density ones [25], for which the literature is relatively sparse. Therefore, the authors decided to present their own concept of increasing the ability to reproduce geometric quantities of low-density materials using single-sided coatings of specimen.

Machine vision systems are also increasingly being incorporated into measurement and inspection work [26, 27]. It represents computer image processing systems, supported by the use of additional input/output modules and networks of connections between components [28]. The correct choice of the type of illumination and the appropriate software filters play a crucial role in obtaining the appropriate image quality for analysis [29]. Vision systems are most commonly found in manufacturing areas, but are increasingly boldly entering various areas of life, often not associated with them at all [30]. They are eagerly used in image recognition applications (usually with the use of artificial intelligence) and in measurement systems. Each of the implemented applications of the vision system very often requires an individual approach and designing an application for a specific task [31]. Due to increasing pressure on the need to offer customers the most homogeneous batches of products and increased inspection of equipment and mechanical parts during technological processes, measurements of overall dimensions and geometric conditions based on image analysis methods have found their wide application in the field of mechanical and manufacturing engineering, also as an extension of optical scanning techniques [32]. Measurements of geometric features used in the broad field of mechanical engineering are mainly carried out on objects with uniform and clear edges. The same cannot be said for the inspection of food products. In this sector of economy, it is the shape that is one of the most frequently evaluated aspects during the product classification process.

An outline detection system based on image processing was used in their work by ElMasry et al. [33] using it to separate irregularly shaped potatoes from correct vegetables. In the application, it proved to be very important to select correct, uniform illumination to avoid spots and reflections that can hide or alter important features on the surface of the potatoes. Gongal et al. [34] also wrote about the development of automatic fruit harvesting based on shape and color in their publication. They proved that the accuracy of vision systems in detecting and locating fruits mainly depends on the uncertain and changing lighting conditions in the field environment. A very large number of papers also describe the application of image analysis in the food industry for the detection of distinct enhanced edges. This effect was used in their study by Li et al. [35] who developed a system to detect cracks on the surface of chicken eggs with 100% correctness in sorting products based on the presence or absence of damage on the shell. Here again, machine vision can provide support for scanning systems [36]. In another experiment, Nashat et al. [37] developed a system for crack detection of cakes or baked goods characterized by uneven color distribution and characteristic surface texture. Taking into account the above considerations, it was decided to develop an application for the evaluation of the overall dimensions of a product characterized by uneven edges and a problematic top surface, as well as an application for determining the surface geometry of wafers.

3 Application of Coordinate Measurement Technique

An edible wafer analysis was chosen as the task using the coordinate measurement technique. A Vtomex s computed tomograph was used to obtain images in a relatively fast way. In the course of the test it turned to be material, which CT did not cope with in a satisfactory way. This was predominantly due to the lack of CT ability to find the limit between the air and the measured material. Based on that, we decided to increase contrast between the studied element and its background using unilateral coatings. The effect of the above-mentioned considerations was presented in Fig. 1.

As it can be seen above, the use of a unilateral coating on the measured object enabled separation of individual peaks representing either material or background. According to

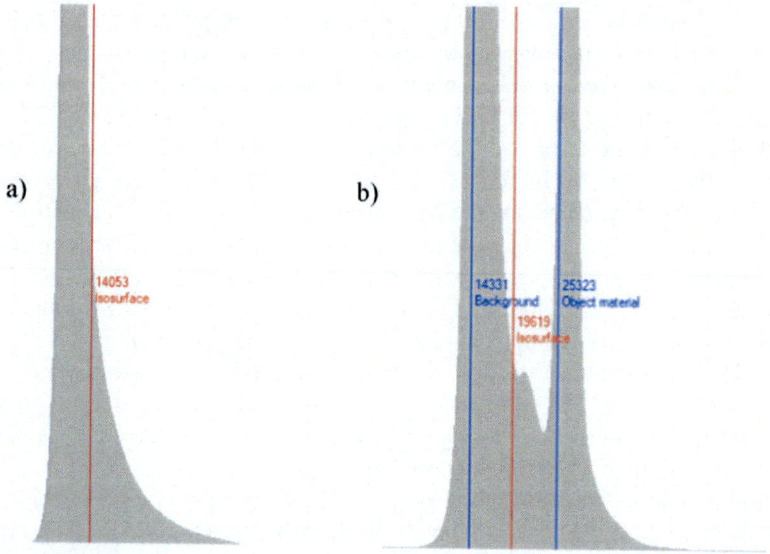

Fig. 1. Contrast between the studied material and background for an object without coating (a) and with coating (b).

that, CT measurements for low-density material were realized, using transparent spray lacquer, black spray paint and paraffin as coatings. A coating-free side of the wafer was examined for shape and geometry point of view. An Infinite Focus G5 microscope was used for that purpose, as a reference device. It gives very good images, but obtaining a large field of view, comparable with a CT is much slower. The whole analysis was performed in a way that is used for surface topography [38]. As an example, a pseudo color view of the coating-free wafer surface, leveled using a least squares method is presented on Fig. 2. For this particular surface maximum height Sz value was equal to 1,35 mm, while ten point height S10z reached 1,04 mm. These values give a good impression about the depth of the structure on the surface of wafer. Valleys occupy a slightly larger part of the surface, which is reflected in a positive skewness value of 0,07, as well as a larger value of maximum peak height Sp of 691 μm than maximum pit height - 662 μm. The nature of the surface strongly deviating from that described by normal distribution is also reflected by the kurtosis value of 1,46. The repeatability of the pattern was made using profile curves converted into a series of profiles. This was depicted on Fig. 3.

From this series a mean profile was evaluated (Fig. 4).

Next, for series of profiles step height calculation was performed. The results of this action was shown in Fig. 5. The same procedure was also applied for a mean profile. The heights calculated from it are shown in Fig. 6. The 3D model of a coating-free wafer was than superimposed onto the surface image originated from focus-variation microscope that was considered a reference data. The results are presented in Fig. 7. Based on the analysis of the images presented above, it can be perceived that geometrical deviations in relation to the reference surface for non-coated wafer were higher at the bottom

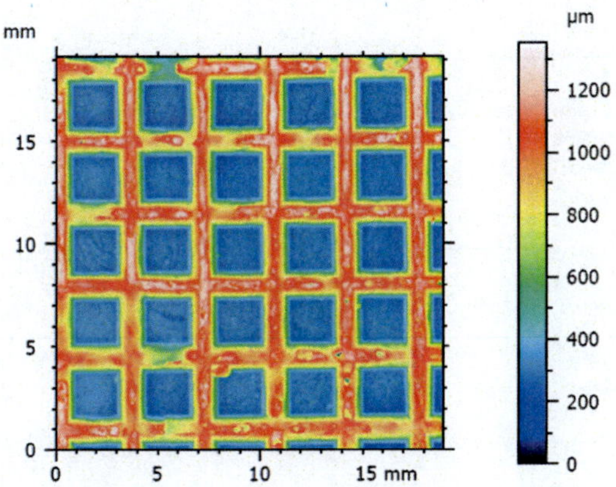

Fig. 2. A pseudo color view of the coating-free wafer obtained with a focus variation microscope.

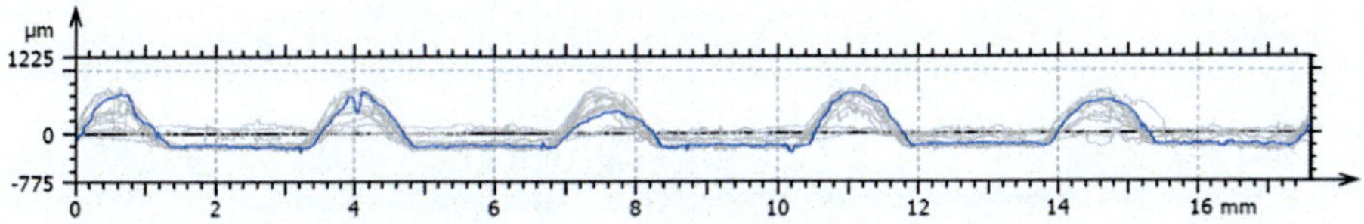

Fig. 3. Profile curves converted into a series of profiles for a coating-free wafer.

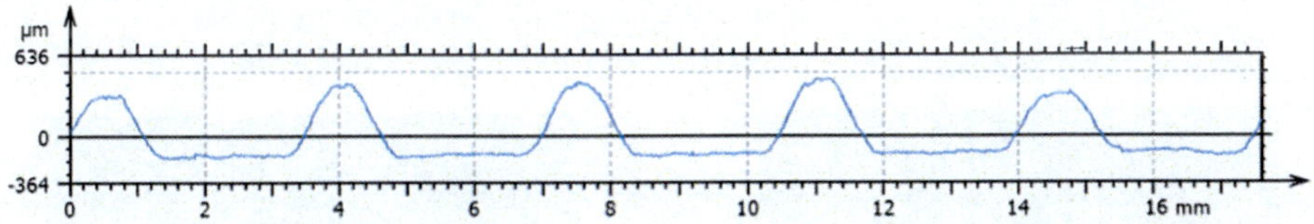

Fig. 4. A mean profile evaluated from a series of profiles for a coating-free wafer.

edges of central rasters, and their values reached approx. 0,3 mm in terms of the order of magnitude. In the case of wafer which bottom side was coated with paraffin, the performance was better. Geometrical deviations at the borders significantly decreased to about 0,05 mm.

At the next stage, it was necessary to check whether similar relationships connected to the improvement of geometry reproduction occurred in the event of lacquer or spray paint coating application onto the bottom side of a low-density element. The results are presented in Fig. 8.

Based on the analysis of the above-mentioned data, we concluded that coating the bottom side of wafer both with lacquer and wax worsens the object geometric properties in relation to the assumed reference surface. This may come from spray-derived moisture effect on the studied element and changes in its dimensions. In order to check if the precision of the shape reproduction would improve, the bottom side of the objects presented above was coated with an additional wax coating. Unfortunately, it did not exert a required effect that would lead to the improvement of wafer geometrical properties.

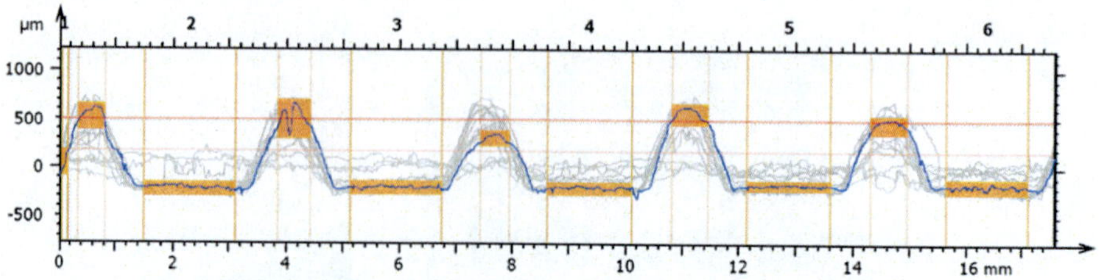

Parameters	Unit	Step 1	Step 2	Step 3	Step 4	Step 5	Step 6
Width	mm	0.116	1.60	1.60	1.50	1.46	1.47
Maximum depth	µm	552	748	734	733	688	715
Mean depth	µm	460	704	695	693	667	670

Fig. 5. Step height calculation for a series of profiles (a coating-free wafer).

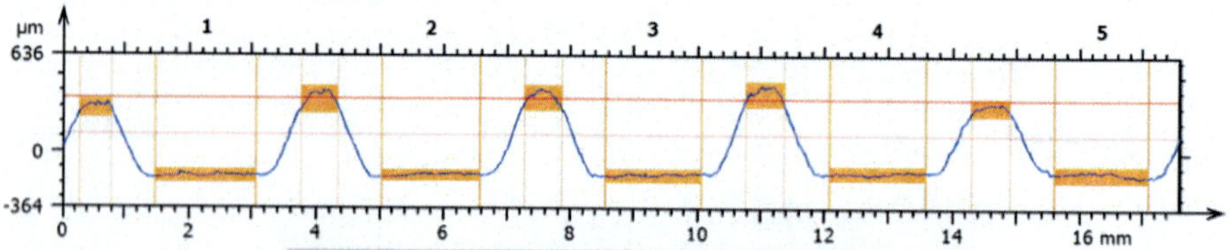

Parameters	Unit	Step 1	Step 2	Step 3	Step 4	Step 5
Width	mm	1.57	1.55	1.52	1.52	1.50
Maximum depth	µm	522	513	515	501	509
Mean depth	µm	506	498	495	483	478

Fig. 6. Step height calculation for a mean profile (a coating-free wafer).

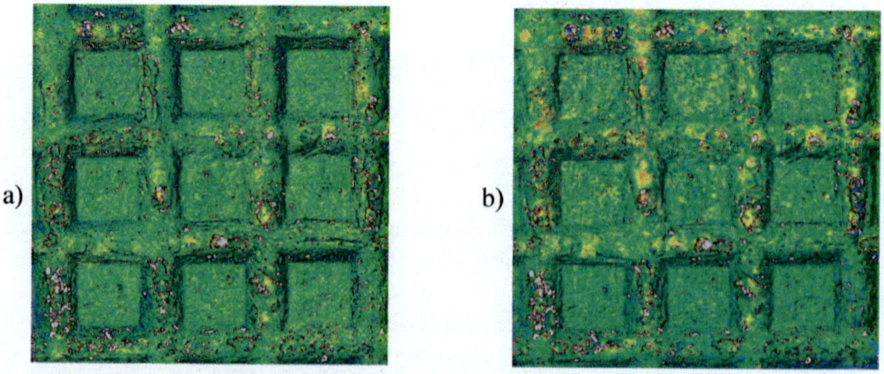

Fig. 7. A geometric deviation image for coating-free wafer (a) and for wax-layer coated wafer (b) in relation to the reference surface.

The findings indicate the existence of a relationship between introducing coatings and the enhancement of accuracy in the reproduction of low-density object shapes. Paraffin turned out to be the best in this respect. In contrast to lacquer and paint, paraffin did not destroy wafer structure, however it generated a considerable reduction in geometrical deviations in relation to the assumed reference surface. In the case of shining surface

a) b)

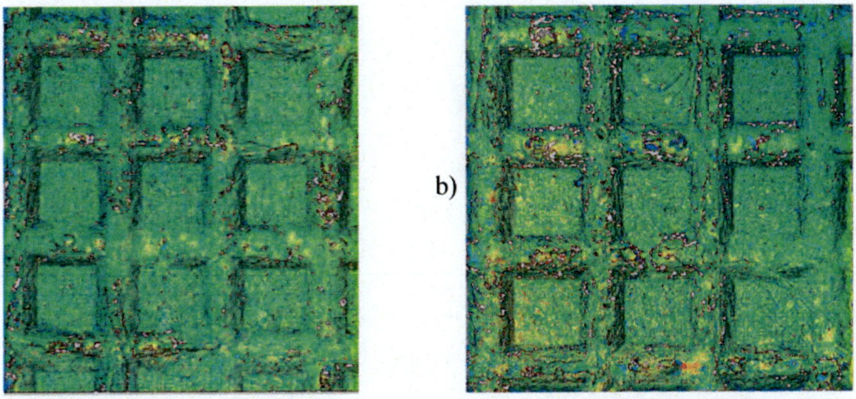

Fig. 8. A geometric deviation image for lacquer (a) and for paint (b) coated wafer.

measurements using optical scanners, it is necessary to tarnish the measured area. Perhaps a similar approach should be made for CT measurements of materials relatively resistant to X-rays, where it would be necessary to use higher-density coatings. In the course of further investigations we are going to test the impact of other, more user-friendly coatings on the accuracy of shape reproduction in the CT. It is also essential to examine whether the presence itself in the CT-affected space of an object that is aimed at enhancing contrast between a low-density material and background is sufficient for an increase in the reproduction ability of geometrical sizes, excluding a necessity to coat studied surfaces with a layer of an additional material.

4 Application of Machine Vision

Due to the fact that the design of a vision system that is reliable in application requires the recognition of many aspects [39], the main focus of the research was on the preparation of as much data as possible about the measurement object and the identification of basic problems occurring during the inspection of a given product using a dedicated image processing program. Incorrect discernment in the aforementioned subsections is the most common problem of malfunctioning of a vision system.

The object of study is bread made of rye. Learning about and analyzing the characteristics of the selected item was aimed at determining the appropriate tools for measurement used by the vision system, as well as the selection of filters and lighting. Using industrial cameras and the proposed control program, the overall dimensions (length L, height H, and width W) of the molded bread have been measured. The measurement is carried out at the widest parts of the object and the range of variation of dimensions can be ± 2 mm. Figure 9 shows an example of the test object along with the determination of the values of the overall dimensions.

Bread is an object with uneven edges. In addition, the upper surface of the bread is covered with white flour and there are cracks formed during the baking process. Therefore, unambiguous measurement of its dimensions can be difficult to achieve

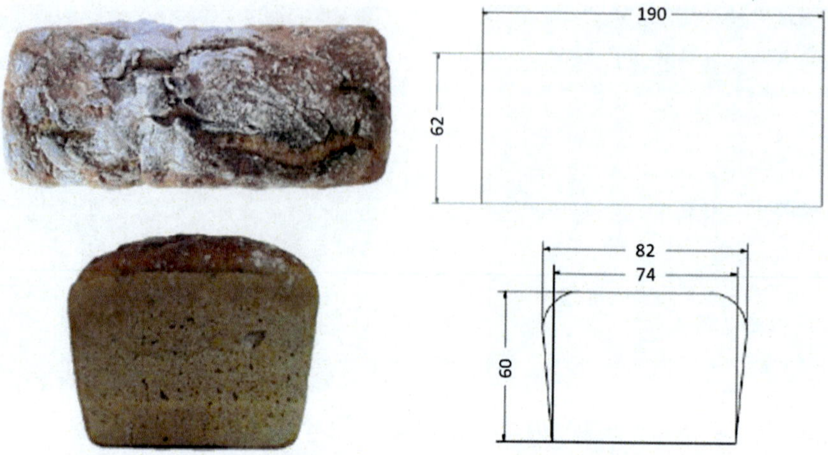

Fig. 9. The overall dimensions of the test object.

both with traditional measuring methods and modern optical systems while maintaining automation of the entire production process.

During the work on the system the method of measurement, types of filters, tools and lighting were chosen to obtain correct and repeatable results. The most problematic issue when analyzing images of the top surface of bread is the presence of white flour spots on the surface. It should be mentioned that the analysis of items using a machine vision system requires that inspections of items be performed against a homogeneous background in order to avoid noise that introduces distortions in the image processing and provides a contrast to the object being examined. The machine vision system does not distinguish between flour and a white background and therefore erroneously detects edges between which distance measurements should take place.

In order to obtain clear edges of the examined object on the analyzed image, it is advisable to apply appropriate filters (Sobel, Prewitt, Forward, Central) enabling sharpening of the object outline. The Sobel and Prewitt filters very accurately represent the course of the edges of the test object and are therefore considered to be the best for use in the described application. As in the case of the previous program as a result of the problematic bread top surface the edges were incorrectly approximated and thus the measurement result turned out to be erroneous.

Taking into account the considerations presented above, two methods were developed to enable the measurement of bread dimensions. This allowed the elimination of errors occurring during image analysis due to the problematic surface of the object. The first one is based on the use of a real filter in the form of milky foil, which, placed at a certain height between the camera and the object under investigation, makes white flour spots and cracks invisible in the image, while the edges of the object are still recognized by the software. The disadvantage of this method is that the outline of the object is blurred, so that the outside dimensions may differ from the actual ones. However, this fact can be taken into account when developing a formula that converts the number of pixels measured by the system into units for measuring the distance between two edges. The results of the applied solution are shown in Fig. 10.

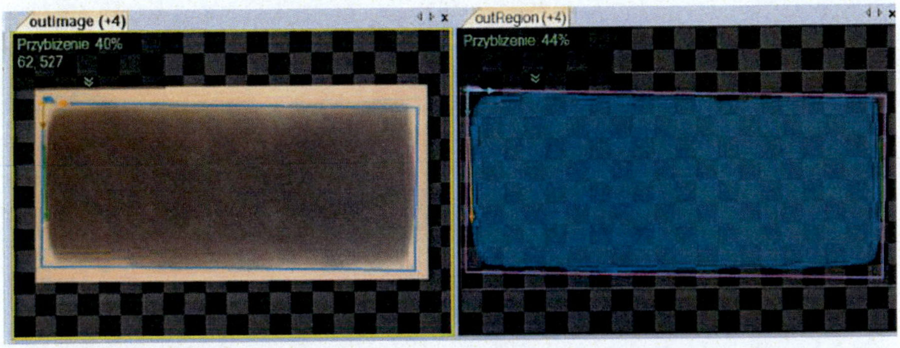

Fig. 10. Eliminating the effect of uneven surface on the image analysis process.

Another method involves the use of backlighting. In this case, the image should be overexposed by means of a very intensive light source placed on the underside of the tested product. The overexposure leads to an image that has a very high optical density and an increased contrast between the product and the background. This method is very suitable for machine vision, where it is necessary to inspect an object with uneven structure and avoid errors occurring during image analysis introduced by surface defects. The backlight is also very beneficial for the detection of clear edges of the outline of the examined object. Figure 11 shows an image of the top surface of the bread obtained by using backlighting and an example of the visualization of the size measurement results displayed on the HMI.

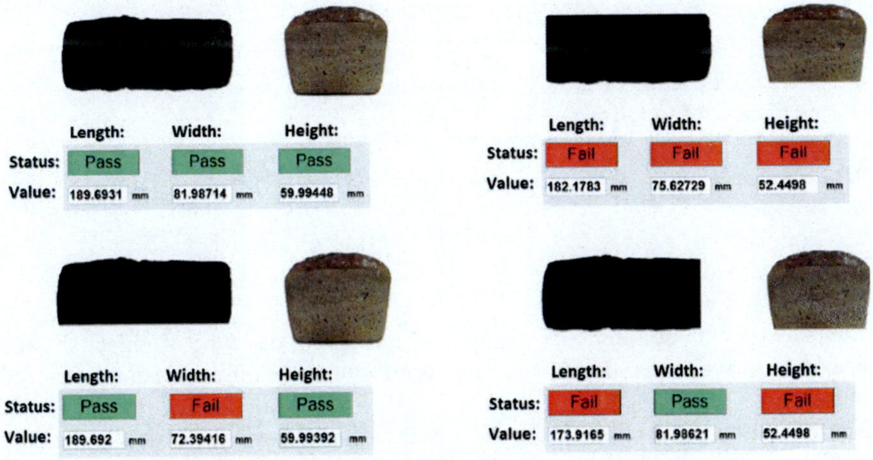

Fig. 11. An example of the user panel of the presented application

The order of measurement uncertainty that can be obtained using machine vision systems is primarily related to the parameters of the selected camera and the accuracy and magnification of the optical system used. The images were taken with a camera with a total resolution of 16 MPx (4608 × 3456 px) and a CCD matrix size of 3½ inch (6,17 × 4,55 mm). The value of the distance between two edges given in pixels has to be converted into the unit of length expressed in millimeters in the next step. The calibration of the system was done on certified glass standards.

To find repeatability of the system, thirty images of bread were taken, each time removed from the measurement setup and placed back in different positions and angles. The obtained images were then processed using a control program. The summary of the performed analysis of the dimensional measurements is shown in Table 1.

Table 1. Summary results of statistical parameter values.

Dimensions	Values of statistical parameters					
	x_m [mm]	σ [mm]	3σ [mm]	V [%]	R [mm]	M_e [mm]
L	190,022	0,046	0,14	0,024	0,25	190,039
W	82,018	0,015	0,018	0,018	0,068	82,015
H	60,034	0,023	0,038	0,046	0,12	60,043
Dimensions	Values of statistical parameters					
	S	K	δ [%]	x_m-$3\sigma < x_{typ} < x_m + 3\sigma$		
L	-0,91	0,49	0,012	$189{,}89 < x_{typ} < 190{,}16$		
W	0,063	-0,61	0,022	$81{,}97 < x_{typ} < 82{,}06$		
H	-0,70	-0,45	0,057	$59{,}95 < x_{typ} < 60{,}12$		

where:
x_m – arithmetical mean
σ – standard deviation of the arithmetic mean
V – coefficient of variation
R – range
M_e – median
S – skewness
K – kurtosis
δ – relative error
x_{typ} – interval with measured value

Comparison with conventional methods of measuring this type of object showed good agreement at a level satisfactory to the manufacturer (the difference did not exceed 0,5 mm).

The largest range of results around the arithmetic mean was recorded in the case of length measurement (the largest range of variation) of the object, while the smallest took place in the case of width measurement (the smallest range of variation). This is confirmed by the values of the range and the coefficient of variation, which complement the knowledge on the variation of the observed values of the variables. The above conclusion can be justified by the fact that the range of measurement results increases as the measured distance increases and in the absence of the use of proper lighting to sharpen the object edges. The median indicates that when measuring the length and height of an object, the middle values of the measurement are greater than the arithmetic mean. In the case of width, these values are most similar to each other. The interpretation of kurtosis and skewness gives knowledge about the shape of the obtained distribution. The value of the skewness coefficient above "0" (width) indicates a right-skewed distribution, while values below "0" (length, height) indicate a left-skewed distribution. The value of kurtosis determined for length measurements informs about obtaining leptokurtic distribution (a lot of results are close to the mean), while in the case of width and height about platykurtic distribution (more extreme values). The greatest number of outliers is found for the width measurement.

5 Conclusions

The examples of food product measurements presented in this paper show that this issue is not easy and requires a specific approach. It is very often different depending on the analyzed product. Nevertheless, in the era of widespread automation and replacement of human work by automatic systems, such activities become a necessity.

The use of the coordinate technique has demonstrated first of all the very good suitability of the focus variation microscope for the characterization of the surface of food products. This measurement technique is very well suited as a reference. For faster measurements with lower accuracy, a CT scanner can be used, but the specifics of its operation may require additional steps (surface coating). On an example of a wafer which was about 1 mm thick, the differences between these two device do not exceed 10–15% in height.

It would be of course interesting to investigate how X-rays affect the properties of measured foods and whether there is any effect on the person eating these foods. There are X-ray based devices working in line e.g. to inspect whether there are any unwanted parts in plastic bottles with ketchup or mayonnaise. These products are later bought by consumers. On the airports there are devices verifying bags, including grocery products, that afterwards are eaten by passengers. A lot depends on power of the lamp and time of exposure. For our research the test was more a first part inspection, so the method was not used in 100% control.

When designing measurement applications based in its structure on machine vision systems, it is necessary to carry out a very precise and reliable recognition of a potential measured object and find out what problems may occur during the process of image analysis. On this basis, in the next stage, the executive and control elements should be selected with appropriate parameters, meeting the requirements of a given system. Performing measurement applications where the measurement is to be performed with a specified uncertainty, it is necessary to select a matrix with appropriate accuracy and resolution and an optical system with a specified magnification. In the presented measurement application, the selection of the correct illumination guaranteeing elimination of processing errors resulting from flour stains and cracks on the top surface of the bread proved to be the most important. The system proved to be able to eliminate a loaf of bread with wrong dimensions, with repeatability range not exceeding 0,25 mm in length, 0,07 in width and 0,12 in height.

References

1. Isa, M.A., Lazoglu, I.: Design and analysis of a 3D laser scanner. Measurement **111**, 122–133 (2017). https://doi.org/10.1016/j.measurement.2017.07.028
2. Majchrowski, R., Grzelka, M., Wieczorowski, M., Sadowski, Ł, Gapiński, B.: Large area concrete surface topography measurements using optical 3D scanner. Metrol. Measur. Syst. **22**(4), 565–576 (2016). https://doi.org/10.1515/mms-2015-0046
3. Gapiński, B., et al.: Use of white light and laser 3d scanners for measurement of mesoscale surface asperities. In: Diering, M., Wieczorowski, M., Brown, C.A. (eds.) Advances in Manufacturing II. LNME, pp. 239–256. Springer, Cham (2019). https://doi.org/10.1007/978-3-030-18682-1_19
4. González, J., Rodríguez, I., Prado-Cerqueira, J-L., Diéguez, J.L., Pereira, A.: Additive manufacturing with GMAW welding and CMT technology. Proc. Manuf. **13**, 840–847 (2017). https://doi.org/10.1016/j.promfg.2017.09.189
5. Gapinski, B., Wieczorowski, M., Grzelka, M., Alonso, P.A., Tomé, A.B.: The application of micro computed tomography to assess quality of parts manufactured by means of rapid prototyping. Polimery/Polymers **62**(1), 53–59 (2017). https://doi.org/10.14314/polimery.2017.053
6. Guerra, M.G., Gregersen, S.S., Frisvad, J.R., De Chiffre, L., Lavecchia, F., Galantucci, L.M.: Measurement of polymers with 3D optical scanners: evaluation of the subsurface scattering effect through five miniature step gauges. Measur. Sci. Technol. **31**(1), 015010 (2019). https://doi.org/10.1088/1361-6501/ab3edb
7. Rokicki, P., et al.: Rapid prototyping in manufacturing of core models of aircraft engine blades. Aircr. Eng. Aerosp. Technol. **86**(4), 323–327 (2014). https://doi.org/10.1108/AEAT-10-2012-0192
8. Ribeiro, F.: 3D printing with metals. Comput. Control Eng. J. **9**(1), 31–38 (1998). https://doi.org/10.1049/cce:19980108
9. Sorensen, J.N., Edelenbos, M., Wienberg, L.: Drought effects on green pea texture and related physical-chemical properties at comparable maturity. J. Am. Soc. Horticult. Science **128**(1), 128–135 (2003). https://doi.org/10.21273/JASHS.128.1.0128
10. Hopkins, D.L., Toohey, E.S., Kerr, M.J., van de Ven, R.: Comparison of two instruments (G2 Tenderometer and a Lloyd Texture analyser) for measuring the shear force of cooked meat. Animal Prod. Sci **51**, 71–76 (2010). https://doi.org/10.1071/AN10136
11. Butlewski, M., et al.: Safety and Reliability: Methodology and Applications - Proceedings of the European Safety and Reliability Conference ESREL 2014, pp. 1101–1106. CRC Press, London (2015)
12. Wan, Z., Lai, L., Yin, X., Mao, J., Zhu, L.: Robot line structured light vision measurement system: light strip center extraction and system calibration. Optical Eng. **60**(11), 114102. https://doi.org/10.1117/1.OE.60.11.114102
13. Wieczorowski, M., Swojak, N., Pawlus, P., Pereira, A.: The use of drones in modern length and angle metrology. In: Śniatała, P., Iyengar, S.S., Bendarma, A., Klósak, M. (eds.) Modern Technologies Enabling Safe and Secure UAV Operation in Urban Air-space, pp. 125–140. IOS Press (2021). DOI: https://doi.org/10.3233/NICSP210013
14. Eastwood, J., Newton, L., Leach, R., Piano, S.: Generation and categorisation of surface texture data using a modified progressively growing adversarial network. Precis. Eng. **74**, 1–11 (2022). https://doi.org/10.1016/j.precisioneng.2021.10.020

15. Swojak, N., Wieczorowski, M., Jakubowicz, M.: Assessment of selected metrological properties of laser triangulation sensors. Measur. J. Int. Measur. Confeder. **176**, 109190 (2021). https://doi.org/10.1016/j.measurement.2021.109190
16. Carmignato, S., Dewulf, W., Leach, R. (eds.): Industrial X-Ray Computed Tomography. Springer, Cham (2018). https://doi.org/10.1007/978-3-319-59573-3
17. Kruth, J.P., Bartscher, M., Carmignato, S., Schmitt, R., De Chiffre, L., Weckenmann, A.: Computed tomography for dimensional metrology. CIRP Ann. Manuf. Technol. **60**, 821–842 (2011). https://doi.org/10.1016/j.cirp.2011.05.006
18. Mathia, T.G., Pawlus, P., Wieczorowski, M.: Recent trends in surface metrology. Wear **271**(3–4), 494–508 (2011). https://doi.org/10.1016/j.wear.2010.06.001
19. Subbarao, M., Choi, T.: Accurate recovery of three-dimensional shape from image focus. IEEE Trans. Pattern Anal. Mach. Intell. **17**, 266–274 (1995)
20. Gapinski, B., Wieczorowski, M., Marciniak-Podsadna, L., Dybala, B., Ziolkowski, G.: Comparison of different method of measurement geometry using CMM, optical scanner and computed tomography 3D. Proc. Eng. **69**, 255–262 (2014). https://doi.org/10.1016/j.proeng.2014.02.230
21. Hiller, J., Maisl, M., Reindl, L.M.: Physical characterization and performance evaluation of an x-ray micro-computed tomography system for dimensional metrology applications. Measur. Sci. Technol. **23**(8), 085404 (2012). https://doi.org/10.1088/0957-0233/23/8/085404
22. Maszybrocka, J., Gapiński, B., Dworak, M., Skrabalak, G., Stwora, A.: Modelling, manufacturability and compression properties of the CpTi grade 2 cellular lattice with radial gradient TPMS architecture. Bull. Polish Acad. Sci. Techn. Sci. **61**(4), 719–727 (2019). https://doi.org/10.24425/bpasts.2019.130181
23. Hamrol, A., Ciszak, O., Legutko, S., Jurczyk, M. (eds.): Advances in Manufacturing. LNME, Springer, Cham (2018). https://doi.org/10.1007/978-3-319-68619-6
24. Townsend, A., Pagani, L., Scott, P., Blunt, L.: Areal surface texture data extraction from X-ray computed tomography reconstructions of metal additively manufactured parts. Precis. Eng. **48**, 254–264 (2017). https://doi.org/10.1016/j.precisioneng.2016.12.008
25. Vainberg, É., Kalinchev, V., Sviridova, Z.: Comparing the effectiveness of radiography and x-ray computer tomography for controlling low-density ceramics. Refract. Ind. Ceram **26**, 199–202 (1985)
26. Sergiyenko, O.Y., Tyrsa, V.V.: 3D optical machine vision sensors with intelligent data management for robotic swarm navigation improvement. IEEE Sens. J. **21**(10), 11262–11274 (2021). https://doi.org/10.1109/JSEN.2020.3007856
27. Banda, T., Jie, B.Y.W., Farid, A.A., Lim, C.S.: Machine vision and convolutional neural networks for tool wear identification and classification. In: Ab. Nasir, A.F., Ibrahim, A.N., Ishak, I., Mat Yahya, N., Zakaria, M.A., P. P. Abdul Majeed, A. (eds.) Recent Trends in Mechatronics Towards Industry 4.0. LNEE, vol. 730, pp. 737–747. Springer, Singapore (2022). https://doi.org/10.1007/978-981-33-4597-3_66
28. Hornberg, A.: Handbook of Machine and Computer Vision. University of Applied Sciences Of Esslingen, Germany (2017)
29. Vernon, D.: Machine Vision: Automated Visual Inspection and Robot Vision. Prentice Hall, Cambridge (1991)
30. Holton, C.W.: Automated Inspection of microparts requires vision. Vision Syst. Des. **8**, 1–4 (2007)
31. Kopparapu, S.K.: Lighting design for machine vision application. Image Vis. Comput. **7**, 720–726 (2006)
32. Rekas, A., Kaczmarek, T., Wieczorowski, M., Kucharski, D., Marciniak-Podsadna, L.: Analysis of tool geometry for the stamping process of large-size car body components using a 3D optical measurement system. Materials **14**(24), 7608 (2021). https://doi.org/10.3390/ma14247608

33. Elmasry, G., Cubero, S., Molto, E., Blasco, J.: In-line sorting of irregular potatoes by using automated computer based machine vision system. J. Food Eng. **1–2**, 60–68 (2012)
34. Gongal, A., Amatya, S., Karkee, M., Zhang, Q., Lewis, K.: Sensors and systems for fruit detection and localization. Comput. Electron. Agric. **116**, 8–19 (2015)
35. Li, Y., Dhakal, S., Peng, Y.: A machine vision system for identification of micro- crack in egg shell. J. Food Eng. **1**, 127–134 (2012)
36. Wieczorowski, M., Gapiński, B., Swojak, N.: The use of optical scanner for analysis of surface defects. Ann. DAAAM Proc. Int. DAAAM Symp. **30**(1), 76–85 (2019). https://doi.org/10.2507/30th.daaam.proceedings.010
37. Nashat, S., Abdullah, A., Abdullah, M.Z.: Machine vision for crack inspection of biscuits featuring pyramid detection scheme. J. Food Eng. **120**, 233–247 (2014)
38. Pawlus, P., Reizer, R., Wieczorowski, M.: Functional importance of surface texture parameters. Materials **14**(18), 5326 (2021). https://doi.org/10.3390/ma14185326
39. Hyoung- Keun, P.: Development of machine vision monitoring system for semiconductor package sorter. Int. J. Control Autom. **4**, 63–72 (2006)

Analysis and Validation of a CFD Simulation of the Wind Through a Horizontal Axis Wind Turbine as an Actuator Disc with a Porous Jump Condition

Diogo Silva[1], João Silva[2](✉), Paulo Pinto[3], Senhorinha Teixeira[4], and José Teixeira[2]

[1] Department of Mechanical Engineering, University of Minho, 4800-058 Guimarães, Portugal
[2] MEtRICs Research Centre, University of Minho, 4800-058 Guimarães, Portugal
js@dem.uminho.pt
[3] MEGAJOULE, 4465-171 São Mamede de Infesta, Portugal
[4] ALGORITMI Research Centre, University of Minho, 4800-058 Guimarães, Portugal

Abstract. Wind energy plays a crucial role in reducing climate change and it is expected to grow in the coming decades. Due to the increase in energy demands, wind turbines have increased to greater sizes. As such, evaluating the wake produced by them has become a way to maximize their respective power output when studying different wind turbine arrangements in wind parks. Thenceforth, numerical simulations have become valuable assets in this industry. The objective of this work is to analyse the wind flow on a flat terrain through a horizontal axis wind turbine with the use of a CFD procedure developed with ANSYS Fluent. This technique implements the k-ε turbulence model and custom inlet equations to describe a neutral atmospheric boundary layer. The rotor was considered as an actuator disc, modelled with the use of a porous jump boundary condition. The solution presented comforting results in terms of decrease in wind velocity and pressure drop at the location of the rotor, as well as an increase of turbulence due to the obstacle in the flow. The numerical model was validated with the use of mathematical wake models, and it presents reasonable results in the far wake of the stream, with a ≈5% accuracy.

Keywords: ANSYS fluent · Actuator disc · CFD · Horizontal axis wind turbine · Porous jump · RANS · Wake · Wind energy · Wind flow

1 Introduction

The global energy consumption is growing every year and, with climate changes becoming a problem of prime concern, wind energy represents a vital role in mitigating both the energy crisis and the environmental problems. Wind power has received substantial attention for the past two decades and it is now one of the most significant renewable energy sources. As wind energy plays a key role in mitigating climate changes, it is expected to grow substantiality in the upcoming decades. The increase in wind energy

J. Machado et al. (Eds.): icieng 2022, LNME, pp. 187–199, 2023.
https://doi.org/10.1007/978-3-031-09382-1_16

demand has led to an increase of the wind rotors to drastic sizes since, the bigger the turbine, the larger will be the energy output because the energy harvested is proportional to the swept area of the rotor. Wind turbine sizes have increased in last decades and they are expected to grow even more, up to diameters three times the length of a Boeing 747–8 airplane.

The wakes produced by wind turbines are a very important topic to study when trying to maximize the production of a wind farm. As such, understanding the behavior and development of turbine wakes is a crucial aspect when trying to maximize power production. To perform this kind of research, many depend on Computational Fluid Dynamics (CFD) analysis and commercial software that allows to study these types of atmospheric flows and evaluate different wind turbine arrangements and study their wake impacts.

Some general and comprehensive reviews on wind turbine and wind energy topics were performed by Sawant et al. [1] and by O'Brien et al. [2], considering many key aspects such as wind turbine aerodynamics, CFD simulation techniques and wake modelling. Parente et al. [3] were able to develop an improved turbulence model for the numerical simulation of neutral Atmospheric Boundary Layer (ABL) flows, with arbitrary sets of fully developed inlet conditions for a flat terrain. Their results indicate the capability of the approach on ensuring a homogeneity of velocity and turbulent quantities throughout an empty computational domain. Chang et al. [4] presented a CFD methodology to simulate ABL flows by implementing a RANS turbulence model over a flat terrain. This is a common approach in the wind industry when trying to model the ABL and produces consistent steady-state results. Hargreaves and Wright [5] were capable of implementing a neutral ABL over a flat terrain using commercially available CFD software by applying modelling suggestions given by Richards and Hoxey [6] which were later revisited by Richards and Norris [7]. Nedjari et al. [8] studied the wind turbine's wake evolution in wind farms over both complex and flat terrains, modelling the turbine with an actuator disc model, yielding satisfactory wake results. Tang et al. [9] carried out wind tunnel tests to determine the wake characteristics of a three-blade horizontal axis wind turbine, such as velocities as turbulence. With this procedure, it was possible to perform a verification on the decrease of the oncoming flow velocity upon contact with the wind turbine as well as a significant increase in turbulence intensity. Uchida et al. [10] employed a new Large Eddy Simulation (LES) CFD porous disc approach to predict the changes of the wind velocity downstream the wind turbine. This model is validated with the use of a Reynolds-averaged Navier–Stokes (RANS) turbulence model and proved appropriate to reproduce the flow phenomenon. Neunaber et al. [11] performed a small-scale experimental study of a wind flow through a model wind turbine in which a detailed turbulence analysis was performed, giving important results for turbulence modelling in wakes. Regarding analytical wake models, Brusca et al. [12] and Shakoor et al. [13] detail both the Jensen and the Larsen models and review them in terms of accuracy and simplicity. They conclude that the Larsen model is in good agreement with experimental values while the Jensen model is only able to identify the mean velocity deficit of the wake.

This study focuses on creating a CFD procedure capable of analyzing the atmospheric wind flow on a flat terrain and through a horizontal axis wind turbine. For this, two main modelling practices are deployed: a method to model a neutral ABL and a technique to implement a simplified wind turbine rotor in the domain. In Sect. 2, the numerical model is presented, in which it is shown the turbulence model that was used as well as how the modelling of the neutral ABL and the rotor were achieved. Section 3 contains all the set up data and needed values for the simulation, such as computational domain, boundary conditions, solver configurations and discretization method. In Sect. 4 the results of the CFD study are discussed and a validation study is performed. Section 5 comprises all the conclusions taken from this work as well as some possible future work.

2 Numerical Model

In order to develop a CFD model capable of simulating a neutral atmospheric wind flow through a wind rotor, some parameters and procedures were considered. First, when studying the aerodynamics of a wind turbine, the energy extracting phenomenon can be described with the use of a simple approach called the actuator disc theory [14]. Second, some assumptions used in the development of the research need to be addressed, such as neutral atmospheric stability, incompressible and isothermal flow, one wind direction and intensity and absence of buoyancy, canopy model [15] and Coriolis force [16].

2.1 Turbulence Modelling

The most common turbulence model in CFD practices involving wind flows is through the use of the Reynolds Averaged Navier-Stokes equations, or RANS model, as it is considered a good tradeoff between accuracy and computational cost [16].

Since this analysis revolves around modelling an external atmospheric flow, the Standard k-ε with some modifications was utilized to account for a neutral atmospheric flow. As such, in Table 1, the constants used by the Standard model as well as those used in the modified atmospheric model are displayed.

Table 1. Turbulence model Constants [16].

	C_μ	$C_{\varepsilon 1}$	$C_{\varepsilon 2}$	σ_k	σ_ε
Standard	0.090	1.44	1.92	1.00	1.300
Modified	0.033	1.44	1.92	1.00	1.835

For the case in hand, the value of σ_ε was calculated using Eq. (1) and assuming a value of k= 0.4 (von Kármán constant) [5].

$$\sigma_\varepsilon = k^2/(C_{\varepsilon 2} - C_{\varepsilon 1})\sqrt{C_\mu} \tag{1}$$

2.2 Neutral Atmospheric Boundary Layer Modelling

Richards and Hoxey [6] concluded that "in order to adequately model the atmospheric surface layer the boundary conditions, turbulence model and associated constants must be consistent". These authors also made the following assumptions when attempting to model a homogeneous ABL [5]:

- The velocity profile is of the log law type;
- The vertical velocity is zero;
- The pressure is constant in the vertical and in the streamwise direction;
- The turbulent kinetic energy (k) and the dissipation rate (ε) satisfy their conservative equations.

With these considerations, Richards and Hoxey [6] recommend that the inlet of the domain requires an assignment of velocity and turbulence, given as profiles of mean velocity (u), turbulent kinetic energy (k) and turbulent dissipation rate (ε), represented with the Eqs. (2) to (4) [5, 6]:

-

$$\text{Logarithmic velocity profile}: \quad u_x = \frac{u_*}{\kappa} ln\left(\frac{y + y_0}{y_0}\right) \tag{2}$$

-

$$\text{Turbulent kinetic energy}: \quad k = \frac{u_*^2}{\sqrt{C_\mu}} \tag{3}$$

-

$$\text{Turbulent dissipation rate}: \quad \varepsilon = \frac{u_*^3}{\kappa(y + y_0)} \tag{4}$$

where u_* is the friction velocity, k is the von Kármán constant, y is the height and y_0 is the roughness length.

2.3 Rotor Modelling

The rotor of the horizontal axis wind turbine is modelled using the Actuator Disk Model (ADM), which is a model based on the Linear Momentum Theory and it has become one of the most widely used methods for simplified studies of the flow field in horizontal axis wind turbines. With this model, the aerodynamic effects of the turbine rotor are represented by a pressure drop over an infinitely thin disc with the same swept area as the real rotor. As such, there is no need to fully resolve the blades of the turbine, being that the turbine disc is modelled as a porous material that supports pressure but not velocity discontinuities. The ADM allows for a simple but robust representation of the wind turbine that allows for an accurate calculation of the flow in the far wake with low computational costs [17].

To represent the discontinuity caused by the wind turbine, this disc is defined as a porous jump boundary condition. The porous jump is a simplified 1D momentum sink based on velocity (pressure-drop) characteristics. The pressure change is shown in Eq. (5).

$$\Delta p = -\left(\frac{\mu}{\alpha}u_1 + C_2\frac{1}{2}\rho u_1^2\right)\Delta m \tag{5}$$

where ρ is the air density, α is the permeability, C_2 is the pressure jump coefficient, Δm is the depth of the porous media and u_1 is the velocity normal to the porous face (velocity at the actuator disc).

3 Set Up Data

In order to implement and perform the analyses previously explained, one must first understand the inputs of the problem as well as some set-up considerations. As such, the following parameters are considered:

- Air density: $\rho = 1.225\ \text{kg/m}^3$;
- Gravity turned on in the y-direction as $-9.81\ m/s^2$;
- Wind velocity: $y_{ref} = 55\ \text{m} = \rightarrow u_{ref}\ 8.5\ \text{m/s}$;
- The friction velocity is calculated with the use of the Eq. (2): $= u_*\ 0.39\ \text{m/s}$;
- The terrain of the domain is assumed to be a flat grassy plain: $y_0 = 0.01\ \text{m}$ [18];
- Dynamic viscosity: $\mu = 1.7894 \times\ 10^{-5}\text{kg/m} \cdot \text{s}$.

3.1 Wind Turbine V42–600 kW

The wind turbine that was selected to serve as model for this study was a Vestas V42–600 kW. This wind turbine's rotor has a hub height of 55 m and a diameter of 42 m, with a swept area of 1,357 m^2. More specifications about this wind turbine, such as the power curve and the power coefficient plot, can be found at [19] and [20].

Since it was assumed a wind velocity of 8.5 m/s at hub height, the chosen wind turbine displays a power coefficient (C_p) of 0.432. As such, it is possible to calculate the axial induction factor (a) with the use of Eq. (2), resulting in $a = 0.1492$.

$$C_P = 4a(1-a)^2 \tag{6}$$

With the value of the axial induction factor, it is also possible to predict the wind velocity at the wind turbine's rotor (u_1), with the use of Eq. (7). translating into a wind speed at the actuator disc of 7.2 m/s.

$$u_1 = u_0(1-a) \tag{7}$$

3.2 Computational Domain and Boundary Conditions

For this specific case, in which the ABL flow is being modelled in a flat terrain, the computational domain consists in a rectangular cartesian coordinate system, as shown in Fig. 1, and has the following dimensions: x-direction = 2,000 m; y-direction = 300 m; z-direction = 500 m; rotor located at x = 800 m. Special attention was given to the height of the domain in order of guaranteeing it to be high enough so that the logarithmic wind profile is capable of fully developing itself. For the inlet of the domain, a "velocity-inlet" condition was applied with Eq. (2) characterizing the velocity and Eqs. (3) and (4) describing the turbulence. Both sides as well as the top part of the domain were selected as "symmetry". At the downstream end of the fetch, the outlet face was created as a pressure-outlet boundary. At the bottom of the domain, a "wall" condition was used with a no slip shear condition and the standard roughness model. Many considerations were taken into modelling an accurate ground representation. Blocken explains some of these ground concerns in [21] and [22] as well as Hargreaves in [5]. The rotor of the wind turbine is represented as an actuator disc, and it is solved with the use of a porous jump boundary condition.

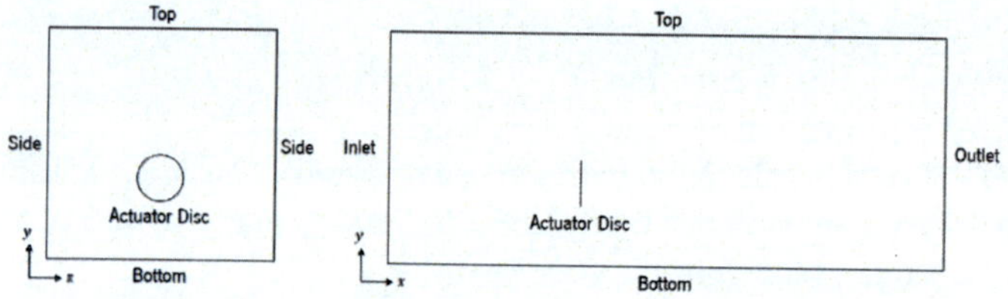

Fig. 1. Computational domain of the CFD simulation.

3.3 Solver and Discretization

The unsteady RANS governing equations, integrating the turbulence and ADM model, are applied in the numerical calculation performed by the CFD software ANSYS Fluent 2020 R2.

The coupled method is employed for the pressure-velocity coupling because of its implicit nature, being that it is more stable than the segregated solver and it will produce results, insuring both accuracy and convergence of the solution.

In terms of discretization, the Green-Gauss Node-Based approach was chosen to compute the gradients as it provides a better accuracy than the default cell-based scheme [23]. For the pressure, the PRESTO! was the scheme deployed since it is the most adequate for cases using porous media [23]. In terms of momentum, turbulent kinetic energy and turbulent dissipation rate, the Second Order Upwind scheme was applied in order to achieve a high-order accuracy. The monitoring is set at 1 x 10^{-5} in all residuals (continuity, x-velocity, y-velocity, z-velocity, k and ε).

4 Results and Discussion

4.1 Mesh

The mesh, containing approximately 4 million elements, is shown in Fig. 2 in which is seen the refinements near the actuator disc as well as near the ground of the domain. The whole domain was divided in various sections to produce a structured mesh with hexahedral elements.

A grid dependency test was performed, in which a coarser mesh with a 15% decrease in the number of divisions and a finer mesh with a 15% increase in the number of divisions were tested. This resulted in a variation of the results of the pressure field and, due to this, the normal mesh was utilized for the simulation.

For the simulation of the wind flow through an actuator disc, an analysis regarding the velocity profiles, the pressure drop and the turbulent kinetic energy was performed.

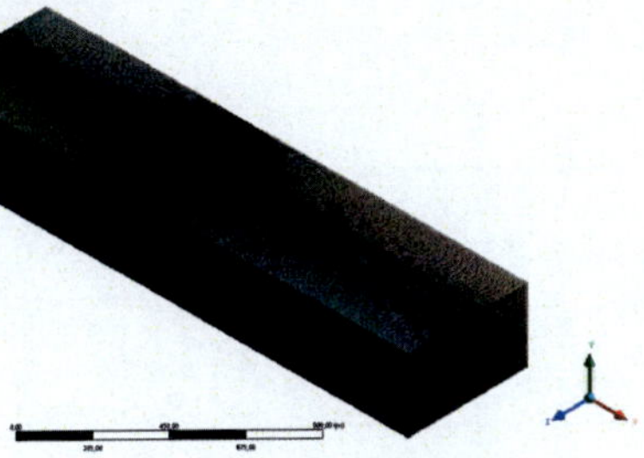

Fig. 2. Mesh of the simulation case

4.2 Velocity Variation

In terms of velocity, the expecting trend is to observe a drop in velocity as the wind approaches the rotor, followed by a downstream recovery of the flow to its free-stream values. Figure 4 shows that the simulation agrees with the expected trend. The velocity at the hub height takes the value of 8.5 m/s as modelled and, as the flow approaches the actuator disc, the velocity of the wind starts to decrease. Right downstream of the wind turbine, the flow continues to slow down until it reaches the minimum velocity value until it starts to recover to its original free-stream velocity. This recovery is more accentuated right after the position of the rotor and the recovery rate gradually reduces as the flow moves further away from the disc.

The CFD model was capable of maintaining the free-stream velocity as well as to establish a velocity at the actuator disc that follows what was expected, within a very small deviation.

Figure 3-(a) shows the velocity contours seen from the side view of the domain and Fig. 3-(b) from the top view of the domain. The contours show the decrease in velocity upon the actuator disc as well as the recovery of the wake right after the rotor. This helps to understand the impact and changes of the wind flow caused by placing a wind turbine in its region.

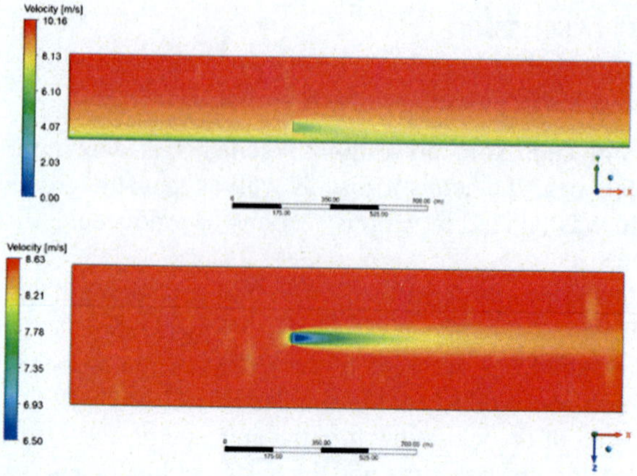

Fig. 3. Velocity contours of the simulation at: (**a**) plane *xy*, showing the view from the side of the domain; (**b**) plane *xz*, showing the view from the top of the domain

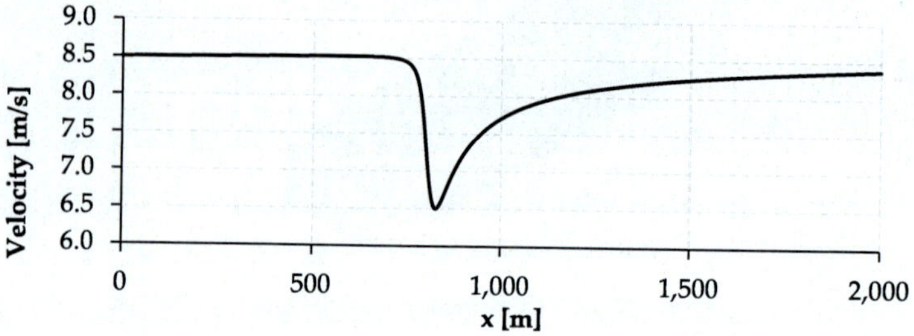

Fig. 4. Velocity along the center line of the actuator disc

4.3 Pressure Field

The rotor of a wind turbine acts as an obstruction to the flow, causing an abrupt pressure drop. Considering the 1D linear momentum theory, it is possible to calculate the theoretical pressure drop using Eq. (5) and compare it to the one resulted from the CFD simulation. The theoretical pressure drop is 22.5 Pa and the pressure drop resulted from the simulations is 22.8 Pa, corresponding to an error of 1.3%. Thus, the pressure variation was well interpreted, and the simulation can accurately represent the drop in pressure of the flow on the impact with the actuator disc.

Figure 5 displays the pressure distribution along the centre line of the rotor. As expected, the pressure increases as the flow moves closer to the turbine since it slows down the incoming wind and more particles gather in front of the blades. Upon the impact with the rotor, the pressure suddenly drops as energy is extracted from the wind. As the wind moves further away from the actuator disc, the pressure gradually recovers to its original atmospheric value, as observed.

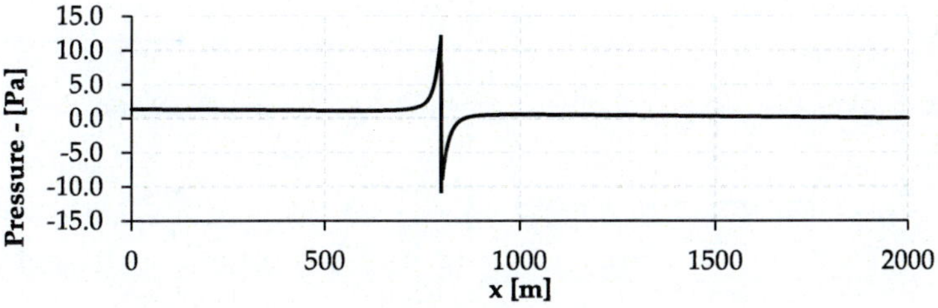

Fig. 5. Pressure drop along the center line of the actuator disc.

4.4 Turbulent Kinetic Energy

The turbulent kinetic energy is the mean kinetic energy per unit mass related to eddies in a turbulent flow. As such, it is an important variable to analyse since it allows to measure the intensity of turbulence.

Figure 7 depicts the evolution of the turbulent kinetic energy along the domain of the flow around one actuator disc. After passing through this obstruction to the flow, the velocity of the oncoming wind decreases while the turbulence intensity of the flow increases significantly. This phenomenon happens since the actuator disc modelled as a porous media acts as a momentum sink, causing an unbalance in the flow and increasing the turbulence.

The turbulent kinetic energy allows to understand the evolution of turbulence, permitting to understand the development of instability due to the flow hitting the actuator disc. As such, in Fig. 6-(a) and 6-(b) the turbulent kinetic energy contours for the side view and for the top view of the domain, respectively, are shown. It is important to note that this is a simplified case with various assumptions, meaning that in a real case, the turbulence would vary in a more chaotic way because of a rotating wake and other considerations. It can also be mentioned that there is a higher variation in the higher part of the wake in Fig. 6-(a) since the velocity of the wind at this point is slightly higher than at the bottom of the disc because of the wind velocity logarithmic profile.

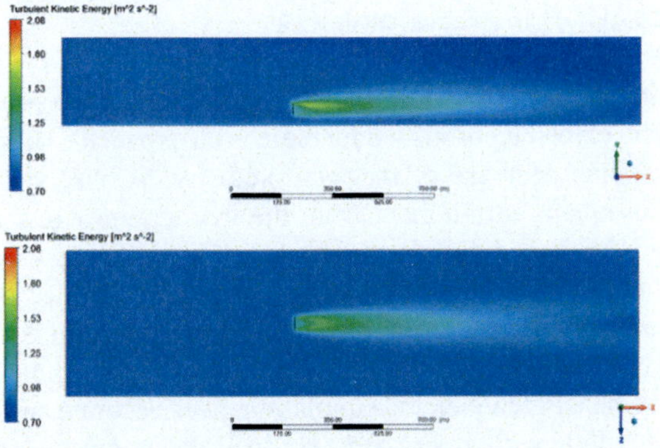

Fig. 6. Turbulent kinetic energy contours of the simulation at: **(a)** plane *xy*, showing the view from the side of the domain; **(b)** plane *xz*, showing the view from the top of the domain

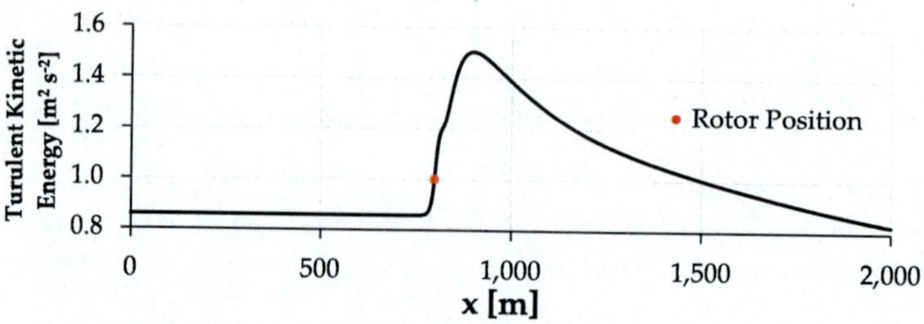

Fig. 7. Turbulent kinetic energy along the center line of the actuator disc.

4.5 Validation Procedure

The validation of the results is a very important step in order to check if these are accurate and realistic. Usually, this step is performed using experimental data, such as data from a wind tunnel or from meteorological masts. Because there was no availability for this type of data for this research, an initial validation process is performed using mathematical and theoretical models. These allow to verify if the basic equations are implemented correctly and thus if the code works in more simple cases. With this in mind, in order to validate the simulations, the CFD model is compared with the Jensen and the Larsen models. More about these models is provided in [12, 24] and [25].

These models are only applicable for the far wake of the wind flow, and they are not capable of accurately representing the near wake. Nevertheless, this is not problematic since the goal of this study is to analyse the far wake of wind turbines and their overall impact on the wind flow. Figure 8 shows the plot of the evolution of the wind velocity performed by the CFD simulation as well as predicted by the Jensen model and the Larsen model.

In terms of the Jensen model, it shows that the near wake region presents different results, with an associated velocity deficit of around 20%. When approaching the far wake, it is shown that the error decreases to a value below 1%, which is very reassuring.

Regarding the Larsen Model, the velocity deficit starts by being very large in the near wake and, for the downstream region of the flow, the deficit is less than 5% and tending to around 3%, which is an acceptable value.

Just downstream (near wake) of the porous jump media, both the Jensen and Larsen models show a significant deviation from the CFD results. However, such behaviour is unrealistic. As the flow exits the porous media there is an expansion of the cross-section area resulting in a sudden decrease of the wind velocity, as observed from the CFD results. However, subsequent entrainment from the free stream wind, a recovery process occurs resulting in an increase of the flow velocity approaching its upstream value. Such feature is captured by the CFD model as opposed to the simpler models proposed by Jensen and Larsen, as observed in the plot in the Fig. 8, resulting in a discrepancy in the near wake in both models. With these two different wake model analyses, it is possible to affirm that the CFD model and simulation gives accurate and realistic results for the far wake of the flow, with a very low error and deviation, making it capable of correctly modelling the wake effects provoked by the wind flow upon the contact with an actuator disc.

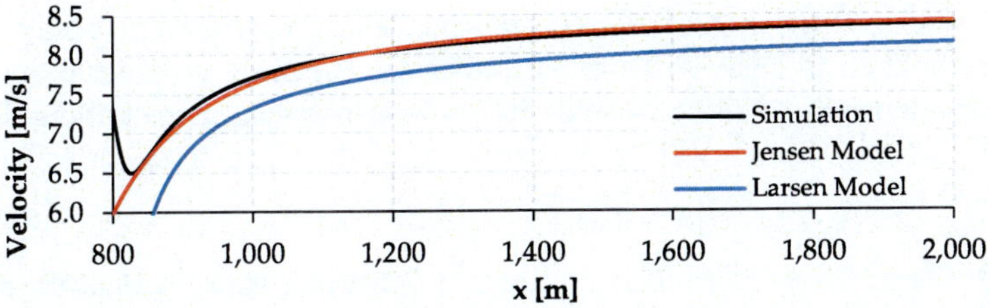

Fig. 8. Comparison between the velocity results from the Jensen model, Larsen model and from the CFD simulation.

5 Conclusions

In this research work it was investigated the effects that a wind turbine would cause in a wind flow. This paper presented a CFD procedure which allowed for a simple and direct analysis of the wake produced by a wind turbine. The rotor of the wind turbine was implemented with the use of a porous jump boundary condition acting as an actuator disc and the flow was assumed as a neutral atmospheric wind flow, with the wind assuming a logarithmic profile. A 3D steady state flow was assumed and the RANS equations along with the Custom Atmospheric k-ε turbulence model were solved.

The results and key findings led to the following conclusions:

- The ABL was well maintained throughout all the domain;
- This CFD procedure proved to be capable of predicting both the near and far wakes of a wind turbine;
- The RANS turbulence model turned out to be a good and versatile approach when modeling the atmospheric boundary layer;
- The porous jump boundary condition was a reliable modelling technique which gave reasonable results in terms of the far wake, allowing to perform a quick approximation of the velocity and turbulence in this zone of the wake. This can be a valuable simulation tool with a good balance between accuracy and computational cost;
- The expected decrease in the wind speed and the pressure drop were detected at the location of the rotor and the values were very comforting;
- The used approach was compared to LES models and rotating frame modelled rotors. This analysis showed similar behaviors and tendencies in terms of wake and velocity variation but also proved the simplicity of the used procedure since it couldn't replicate the rotational aspect of the flow as well as eddies created due to the motion of the blades;
- This CFD simulation was compared with results of the Jensen and Larsen wake models, which shows an agreement within 5% in the far wake. However, the differences are substantial in the near wake region where the referred models are expected to perform poorly. On the other hand, the CFD simulation can capture with accuracy the behavior of the flow field. A more comprehensive evaluation would require experimental data.

In this entire work, a flat terrain was assumed for the bottom of the domain. One possible future development could be the implementation of a complex terrain and consequently analyse the best locations on which to build a wind turbine. This procedure can also be utilized to analyse the impacts of an upstream turbine wake on another downstream wind turbine and thus study its performance due to positioning. It would also be interest to perform a validation process using data from a real or experimental case.

Acknowledgments. This work has been supported by FCT within the R&D Units Project Scope UIDB/00319/2020 (ALGORITMI) and R&D Units Project Scope UIDP/04077/2020 (MEtRICs).

References

1. Sawant, M., Thakare, S., Rao, A.P., Feijóo-Lorenzo, A.E., Bokde, N.D.: A review on state-of-the-art reviews in wind-turbine- and wind-farm-related Topics. Energies **14**, 2041 (2021). https://doi.org/10.3390/en14082041
2. O'Brien, J.M., Young, T.M., O'Mahoney, D.C., Griffin, P.C.: Horizontal axis wind turbine research: a review of commercial CFD, FE codes and experimental practices. Prog. Aerosp. Sci. **92**, 1–24 (2017). https://doi.org/10.1016/j.paerosci.2017.05.001
3. Parente, A., Gorlé, C., van Beeck, J., Benocci, C.: Improved k–ε model and wall function formulation for the RANS simulation of ABL flows. J. Wind Eng. Ind. Aerodyn. **99**, 267–278 (2011). https://doi.org/10.1016/j.jweia.2010.12.017
4. Chang, C.-Y., Schmidt, J., Dörenkämper, M., Stoevesandt, B.: A consistent steady state CFD simulation method for stratified atmospheric boundary layer flows. J. Wind Eng. Ind. Aerodyn. **172**, 55–67 (2018). https://doi.org/10.1016/j.jweia.2017.10.003
5. Hargreaves, D.M., Wright, N.G.: On the use of the k–ε model in commercial CFD software to model the neutral atmospheric boundary layer. J. Wind Eng. Ind. Aerodyn. **95**, 355–369 (2007). https://doi.org/10.1016/j.jweia.2006.08.002
6. Richards, P., Hoxey, R.: Appropriate boundary conditions for computational wind engineering models using the k-ε turbulence model. J. Wind Eng. Ind. Aerodyn. **46–47**, 145–153 (1993). https://doi.org/10.1016/0167-6105(93)90124-7
7. Richards, P.J., Norris, S.E.: Appropriate boundary conditions for computational wind engineering models revisited. J. Wind Eng. Ind. Aerodyn. **99**, 257–266 (2011). https://doi.org/10.1016/j.jweia.2010.12.008
8. Daaou Nedjari, H., Guerri, O., Saighi, M.: CFD wind turbines wake assessment in complex topography. Energy Convers. Manag. **138**, 224–236 (2017). https://doi.org/10.1016/j.enconman.2017.01.070
9. Tang, H., Lam, K.-M., Shum, K.-M., Li, Y.: Wake effect of a horizontal axis wind turbine on the performance of a downstream turbine. Energies **12**, 2395 (2019). https://doi.org/10.3390/en12122395
10. Uchida, T., et al.: A new wind turbine CFD modeling method based on a porous disk approach for practical wind farm design. Energies **13**, 3197 (2020). https://doi.org/10.3390/en13123197
11. Neunaber, I., Hölling, M., Stevens, R.J.A.M., Schepers, G., Peinke, J.: Distinct turbulent regions in the wake of a wind turbine and their inflow-dependent locations: the creation of a wake map. Energies **13**, 5392 (2020). https://doi.org/10.3390/en13205392
12. Brusca, S., et al.: On the wind turbine wake mathematical modelling. Energy Proc. **148**, 202–209 (2018). https://doi.org/10.1016/j.egypro.2018.08.069
13. Shakoor, R., Hassan, M.Y., Raheem, A., Wu, Y.-K.: Wake effect modeling: a review of wind farm layout optimization using Jensen's model. Renew. Sustain. Energy Rev. **58**, 1048–1059 (2016). https://doi.org/10.1016/j.rser.2015.12.229

14. Manwell, J.F., McGowan, J.G., Rogers, A.L.: Wind Energy Explained: Theory, Design, and Application. 2nd ed. Wiley, Hoboken (2009). ISBN: 9780470015001
15. Blocken, B., van der Hout, A., Dekker, J., Weiler, O.: CFD simulation of wind flow over natural complex terrain: case study with validation by field measurements for Ria de Ferrol, Galicia. Spain. J. Wind Eng. Ind. Aerodyn. **147**, 43–57 (2015). https://doi.org/10.1016/j.jweia.2015.09.007
16. Radünz, W.C., Mattuella, J.M.L., Petry, A.P.: Wind resource mapping and energy estimation in complex terrain: a framework based on field observations and computational fluid dynamics. Renew. Energy **152**, 494–515 (2020). https://doi.org/10.1016/j.renene.2020.01.014
17. Avramenko, A., Agafonova, O., Sorvari, J., Haario, H., Avramenko, Y.: Fast Numerical Modelling Method for Wind Flow Investigation Based on Depth-Averaged Equations, pp. 21–29 (2017)
18. Burton, T., Jenkins, N., Bossanyi, E., Sharpe, D., Graham, M.: Wind Energy Handbook, 3rd edn. Wiley, Hoboken (2021)
19. A/S, V.W.S. Vestas V42 - 600,00 kW. https://en.wind-turbine-models.com/turbines/109-vestas-v42#marketplace. Accessed 24 Sep 2021
20. Mollestad, M.K.: Butinge Wind Power Plant - Pre-engineering. Steinkjer (1997)
21. Blocken, B., Stathopoulos, T., Carmeliet, J.: CFD simulation of the atmospheric boundary layer: wall function problems. Atmos. Environ. **41**, 238–252 (2007). https://doi.org/10.1016/j.atmosenv.2006.08.019
22. Blocken, B., Carmeliet, J., Stathopoulos, T.: CFD evaluation of wind speed conditions in passages between parallel buildings—effect of wall-function roughness modifications for the atmospheric boundary layer flow. J. Wind Eng. Ind. Aerodyn. **95**, 941–962 (2007). https://doi.org/10.1016/j.jweia.2007.01.013
23. Ansys Inc. ANSYS FLUENT Theory Guide. http://www.afs.enea.it/project/neptunius/docs/fluent/html/th/main_pre.htm
24. Katic, I., Højstrup, J., Jensen, N.O.: A simple model for cluster efficiency. In: Proceedings of the EWEC1986, Roskilde, vol. 1, pp. 407–410 (1987)
25. Larsen, G.C.: A simple stationary semi-analytical wake model, Risø-R-1713(EN), Risø National Laboratory, Roskilde, Denmark (2009)

Eutectic Modification of Al-Si Alloys by Using $AlSr_{10}$

Iryna Hren(✉), Sylvia Kusmierczak, Nataša Naprstková, and Roman Horky

Faculty of Mechanical Engineering, Jan Evangelista Purkyně University in Ústí nad Labem, Pasteurova 7, 400 01 Ústí nad Labem, Czechia
iryna.hren@ujep.cz

Abstract. The eutectic modification of the AlSi7Mg0.3 alloy with Sr allows for an effective refinement of the eutectic in aluminium alloys. The additive introduced as a modifier must have the following requirements: forming sufficiently strong adsorption bonds with melt modifying atoms; sufficient stability in the aluminium melt at high temperatures without changing the chemical composition. The aim of the realized experiment is first of all to investigate the distribution and influence of strontium in the structure of Al-Si alloys. Subsequently, an experiment was carried out on the quality evaluation of effect of the modifier on the structural integrity in AlSi7Mg0,3 alloy. Elemental distribution of Sr in the microstructure of AlSi7Mg0,3 alloy was determined by scanning electron microscope with EDS analyser.

Keywords: Aluminium alloy · Modification · Electron microscopy

1 Introduction

The modification of hypoeutectic alloy is used as one of the methods of refining the crystals of eutectic silicon. The addition of a modifier in an alloy, such as strontium, leads to a change in eutectic silicon up to filamentous morphology. This technology is used for casting light metal alloys in many industries such as aerospace, automotive and military. The AlSi7Mg0.3 hypoeutectic alloy was prepared for the automotive industry with the aim of improving and maintaining the mechanical properties for the resulting castings [1].

One of the methods of determining the mechanical properties of materials is the modification process. During the modification process, elements such as sodium (Na), antimone (Sb) and strontium (Sr) are used. Antimony (Sb) and sodium (Na) can cause occupational health and safety problems, as well as problems with exposure time, respectively. On the another hand strontium (Sr) modification is the most popular method of converting coarse and elongated grains into a finer morphology, which leads to improved mechanical properties. In addition, with the help of Sr modification, it is possible to achieve a better material flow. [2].

Anson [3] and Stucky [4] described, in their works, the dependence of the distance of the secondary dendrite arm on the mechanical properties of hypoeutectic alloy. The

J. Machado et al. (Eds.): icieng 2022, LNME, pp. 200–208, 2023.
https://doi.org/10.1007/978-3-031-09382-1_17

addition of Sr changes the morphology of Si from a coarse plate form to a thin or fibrous form, resulting in an increase in the mechanical properties of the material.

Smith [5] in his article has examined the effect of the morphological changes (transformation) of the eutectic silicon phase. The effect of strontium distribution at the atomic level was proved by experiment. It was found that Sr-Al-Si segregation inhibits and limits the growth of eutectic Si phase.

An article of Crosley and Mondolfo [6] focused on the development of microstructure of modified Al-Si alloys. The fact is that the modification process is accompanied by changes in the alloy structure, such as phase redistribution. Changes in nucleation and eutectic growth dynamics depending on modifier type were presented in this paper.

Backerud [7] considered the effect of overheating of modified alloys in his article. The low superheat temperature leads to the breakdown of iron-rich β-needles. In this article, it was proposed to increase the superheat temperature to an optimum temperature t = 750 °C, which resulted in a uniform phase distribution and thus an increase in the strength of the material.

Tavitas-Medrano in his study [8] considered the effect of the addition of strontium and pressure increase on the microstructure and defects of castings made of AlSi7Mg0.3 alloy during low-pressure casting. As a result of the study, he found that the Sr content in the alloy and the casting at a higher pressure leads to a decrease in the porosity of the castings.

Birol's [9] work dealt with the effect of adding strontium to the AlSi7Mg0.3 alloy to investigate the microstructure and mechanical properties of it. It was the use of Sr in the amount of 0.01 to 0.02 wt. % and its effect on the resulting alloy structure, to provide a very fine-grained matrix and fine fibrous eutectics. The results of the experiment showed that strontium modified not only eutectic Si particles, but also β-Al_5FeSi needles, which exhibit a morphology very similar to the eutectic Si phase morphology and which are very detrimental to the mechanical properties of the casting. On the other hand, with increasing Sr additives (0.02 to 0.04 wt%), the content of these particles gradually disappears.

The modification is a stochastic phenomenon, the purpose of which is to achieve the longest possible duration of action of the modifier on the alloy while maintaining the high mechanical properties of the material [10–12]. Therefore, extensive research and analysis of the modified AlSi7Mg0.3 alloy was performed. The head point of this work was to investigate and analyse the influence of the modifier on the mechanical properties of the material. Prolonging the modification time and maintaining the mechanical properties of the material can be done by examining all factors affecting the process.

2 Experiment and Results

The samples were made from AlS7Mg0.3 hypoeutectic alloy which subsequent chemical composition in weight percent is 7.1% Si, 0.17% Fe, 0.05% Cu, 0.1% Mn, 0.32% Mg, 0.08% Ti, Al-balance. This hypoeutectic alloy has very good machinability, electrical and thermal conductivity, good casting properties [13–18]. The disadvantage is low mechanical properties. Analysis of literary data [14–19] and production experience have shown that in order to achieve the required level of technological, mechanical and applied properties of secondary aluminum alloys several technological operations have to be applied at all stages of the metallurgical process from waste and scrap sorting, refining and modification to heat treatment of the alloys. It is most commonly used in the automotive and aerospace industries [19, 20].

2.1 Experimental Material

The microstructure of the sample without modifier is shown in Fig. 1, which is typically hypoeutectic alloy [21–25]. The structure consists of α-solid solution and silicon in the form of coarse hexagonal plates, which appear, in the metallographic ground plane, as sharp-edged or partially rounded irregular needles of varying size in the range of 10–40 μm.

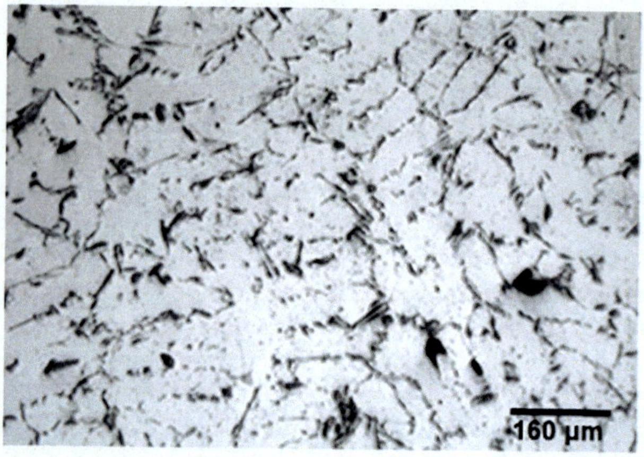

Fig. 1. Microstructure of AlSi7Mg0, 3 alloy, mag. 100x

2.2 Alloy Modification

The most widespread process of changing the morphology of hypoeutectic alloy is modification. This process consists in influencing the crystallization process with added substances in order to change the shape and type of the excluded eutectic. The casting parameters according to a standard ČSN EN 1371–1, i.e. melting point 750^{0}C, were used for the starting alloy. Thereafter, strontium modifier AlSr10 was added to the melt and casting was performed over a time interval from 1 to 2 h of modification. A total of 12 castings were produced for the experiment. Four of them were unmodified and the rest were modified with strontium. The remaining 8 castings was prepared as follows: four of them were cast after 1 h of modification and another 4 - after 2 h of modification.

The microstructure of the sample after the addition of the modifier is shown in Fig. 2, where eutectic silicon is eliminated in the form of rounder and finer grains than unaltered in the alloy.

The samples were subjected to microscopic and static analysis to check the casting process and assess the expected effect of the modification on the alloy structure. EDS analysis of modified samples was also performed to verify modification in the alloy.

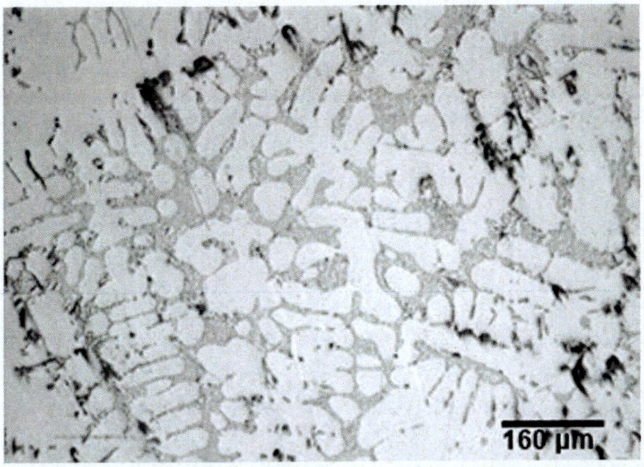

Fig. 2. Microstructure of AlSi7Mg0,3 + AlSr10 alloy, mag. 100x

2.3 Microscopic Analysis

To assess the microstructure all experimental samples were prepared manually. For the final mechanical-chemical polishing was used an OPS suspension from Struers and then etched with 10% phosphoric acid. The structure of the resulting castings was evaluated using the LEXT OLS 3100 confocal laser microscope. Figure 3 shows the structure of a modified alloy casting where eutectic silicon is eliminated in the form of rounded needles.

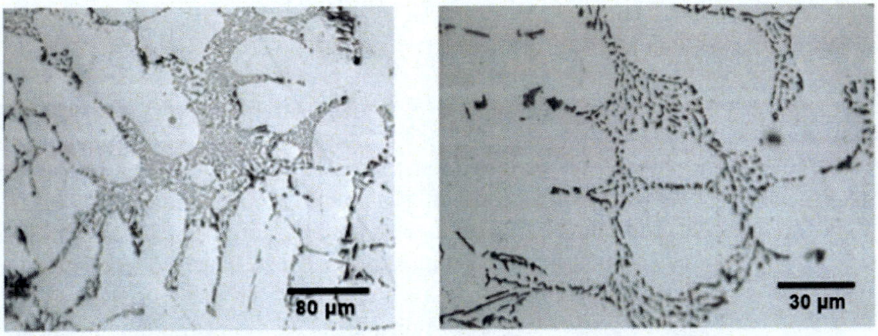

a) mag. 200x b) mag. 500x

Fig. 3. Structure of the modified sample after 1 h of modification

Figure 4 shows the modified alloy structure after 2 h of modification. It can be seen here that, after the addition of strontium to the AlSi7Mg0.3 alloy, the size of the eutectic silicon needles increases, resulting in the expansion of the plates of the eutectic phases. During this process, several thin plates are also combined to form long needle-like Si phases.

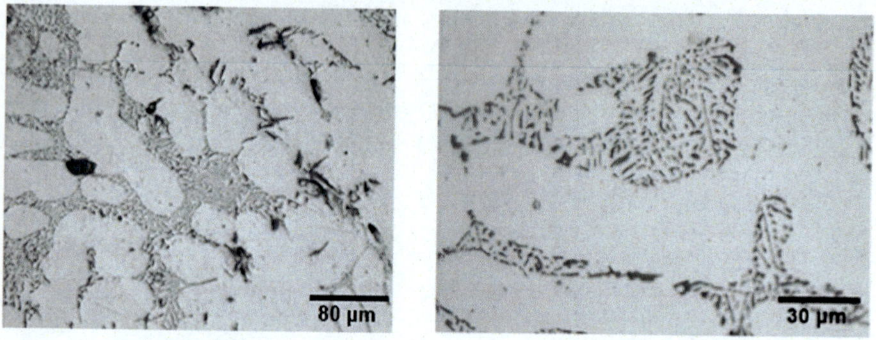

a) mag. 200x b) mag. 500x

Fig. 4. Structure of the modified sample after 2 h of modification

It can also be seen that some phase is slightly stretched. That is, the strontium modifying effect disappears over time. Otherwise, microscopic analysis confirms compliance with theoretical assumptions where modification with a certain amount of Sr results in a change in eutectic silicon morphology.

2.4 Scanning Electron Microscopy and EDS Analysis

To investigate the microstructure of an alloy sample was used an electron microscope type Tescan VEGA3. The chemical composition of the microstructure structural components was determined by Energy Dispersive Spectrometry (EDS) via an EDS analyser Bruker.

Figure 5a shows the area where strontium was found. Elemental maps of Si and Sr distribution show that strontium particles are present on the surface of silicon particles (Fig. 5b). Analysis of the chemical composition in these places, which covers more than 95% of the area of all samples showed that this area consists of 35.5 wt. % of silicon, 63.2 wt. % of aluminium, 0.9 wt. % of magnesium and 0.4 wt. % of strontium. Elemental Si and Sr distribution maps show that strontium particles are present on the surface of silicon particles, and this finding is confirmed by areal EDS analysis in the area.

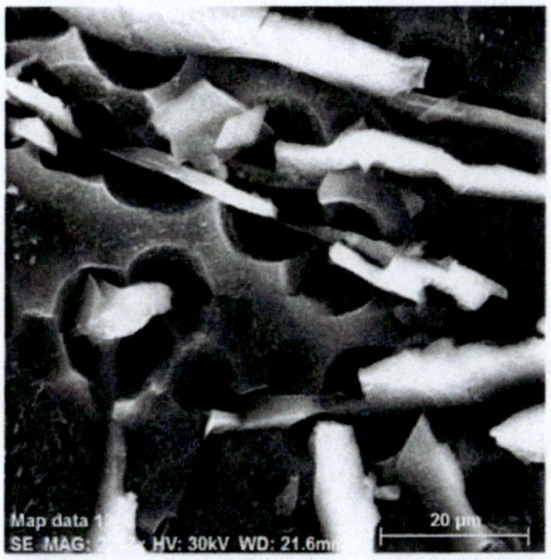

a) microstructure of AlSi7Mg0,3 + AlSr10

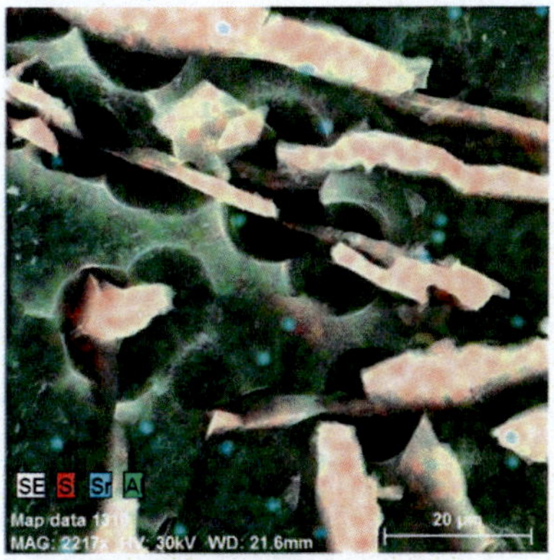

b) EDS map of Sr atoms in AlSi7Mg0,3 alloy surface

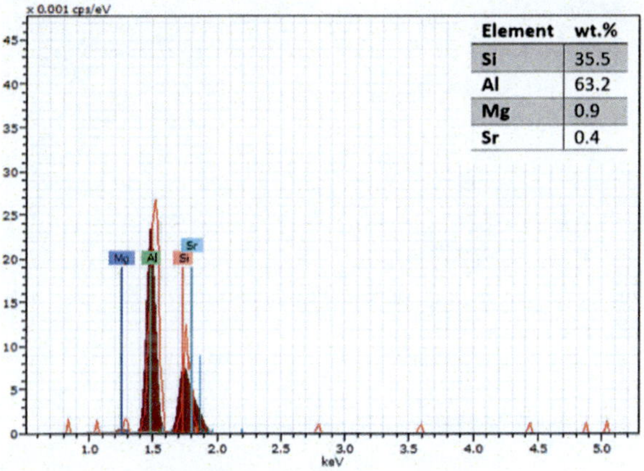

Element	wt.%
Si	35.5
Al	63.2
Mg	0.9
Sr	0.4

c) Recording of EDS area analysis

Fig. 5. Microstructure of sample

The microstructure of the sample after 2 h of modification are shown on Fig. 6a. EDS analysis of the Sr atom map on the AlSi7Mg0.3 alloy surface is shown in Fig. 6b. The presence of strontium on the surface of eutectic silicon particles has been demonstrated by microstructure examination by scanning electron microscope. Quantification of results from the indicated area is shown in Fig. 6c. Analysis of the chemical composition in these places, showed that this area consists of 36.5 wt. % of silicon, 61.5 wt. % of aluminium, 0.8 wt. % of magnesium and 1.2 wt. % of strontium.

This observation is confirmed by the literature sources that describe the modification of Al-Si alloys by strontium, which during the crystallization process surface-bound to the crystallizing nuclei of eutectic silicon. This fact consequently significantly influences the resulting morphology of the silicon particles.

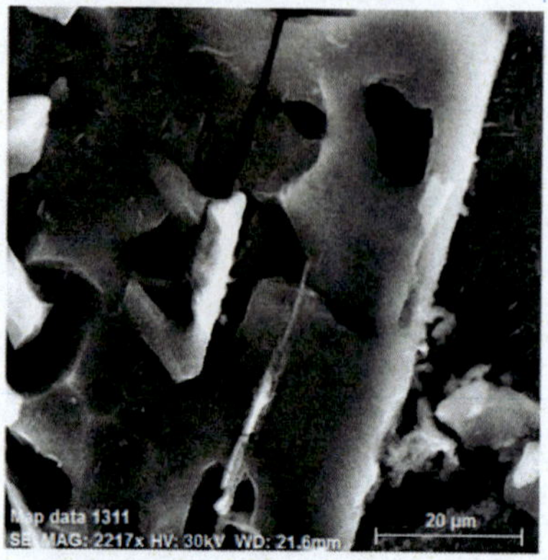

a) microstructure of AlSi7Mg0,3 + AlSr10

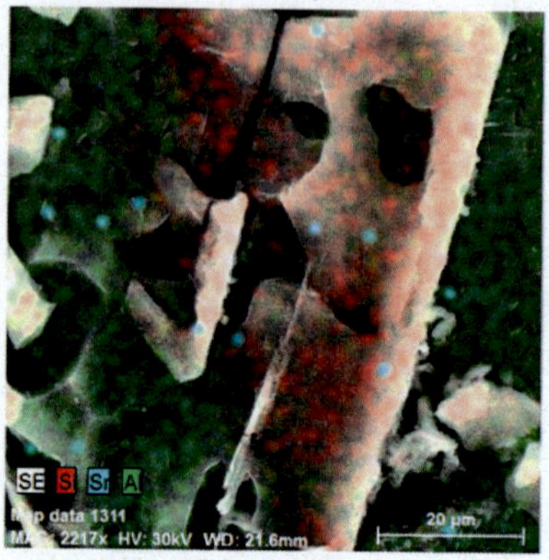

b) EDS map of Sr atoms in AlSi7Mg0,3 alloy surface

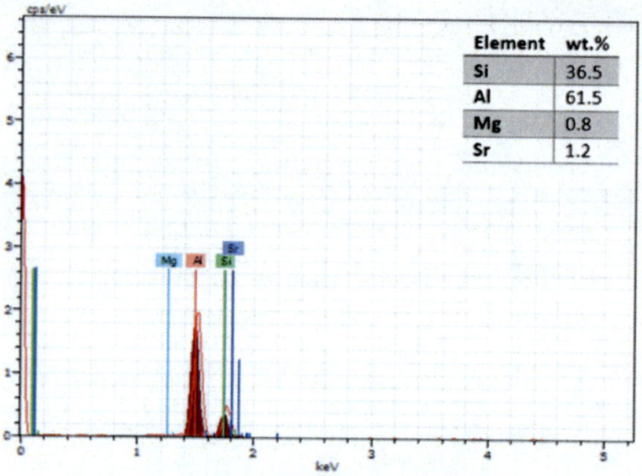

c) Recording of EDS area analysis

Fig. 6. Microstructure of sample

3 Conclusion

The purpose of the experiment was to investigate the distribution and influence of strontium in the structure of Al - Si alloys. During the experiment, 12 samples were produced from the AlSi7Mg0.3Sr alloy, which were subsequently subjected to a detailed investigation, in which the effect of strontium on the structure of the material was studied. As a result of the conducted analyses, the following conclusions were made:

- the addition of strontium causes a significant change in the morphology of eutectic silicon already at a concentration of 0,04 wt% Sr,
- the presence of strontium on the surface of eutectic silicon particles has been demonstrated when examining the microstructure of the samples,
- EDS analysis confirmed that during the modification process strontium binds to silicon thus affecting its morphology,

- the addition of Sr to the AlSi7Mg0.3 alloy, the size of the eutectic silicon needles increases, resulting in the expansion of the plates of the eutectic phases.

On the basis of the above conclusions, it can be stated that the strontium modification has no negative influence on the morphology of the hypoeutectic AlSi7Mg0.3 alloy. The analyses carried out in this article are part of a large experiment where additional analyses will be carried out in the future for a more detailed analysis of strontium compounds with silicon by using surface analysis like XRD.

Acknowledgement. Supported by the project SG 48207 15 2014–48 01 at UJEP.

References

1. Glazoff, M., Khvan, A., Zolotorevsky, V., Belov, N., Dinsdale, A.: Casting Aluminum Alloys: Their Physical and Mechanical Metallurgy, 2nd edn. Butterworth-Heinemann, Cambridge (2018)
2. Derin, S., Aybarç, U., Birol, Y.: Optimization of Strontium addition for the modification of AlSi7Mg0.3 Alloy. In: 18th International Metallurgy & Materials Congress UCTEA Chamber of Metallurgical & Materials Engineers, pp. 626 –629 (2016)
3. Anson, J.P., Stucky, M., Gruzleski, J.E.: Effect of Sr concentration on microporosity formation in A356 aluminium alloy. AFS Trans. **109**, 1–16 (2001)
4. Anson, J.P., Stucky, M., Gruzleski, J.E.: Effect of Sr modification on nucleation and growth of microporosity during solidification of Al-7% Si foundry alloy. AFS Trans. **108**, 419–426 (2000)
5. Smith, R.W.: Modification of aluminum–silicon alloys. In: Proceedings of the Conference on The Solidification of Metals, pp. 224–236. London, Brighton (1967)
6. Crosley, P.B., Mondolfo, L.F: The modification of aluminum-silicon alloys. Modern Castings **46**, 89–100 (1966)
7. Backerud, L., Chai, G., Tamminen, J.: Solidification characteristics of aluminum alloys. Foundry Alloys **2**, 25–38 (1990)
8. Tavitas-Medrano, F.J., Gruzleski, J.E., Samuel, F.H., Valtierra, S., Doty, H.W.: Effect of Mg and Sr-modification on the mechanical properties of 319-type aluminum cast alloys subjected to artificial aging. Mater. Sci. Eng., A **480**(1–2), 356–364 (2008)
9. Ibrahim, M.F., Abdelaziz, M.H., Samuel, A.M., Doty, H.W., Samuel, F.H.: Effect of rare earth metals on the mechanical properties and fractography of Al–Si-based alloys. Int. J. Metalcasting **14**, 1–17 (2019)
10. Derin, S., Aybarc, U., Birol, Y.: Effect of strontium addition on microstructure and mechanical properties of AlSi7Mg0.3 alloys. Inter Metalcast **11**, 688–695 (2017)
11. Timelli, G., Caliari, D., Rakhmonov, J.: Influence of process parameters and Sr addition on the microstructure and casting defects of LPDC A356 alloy for engine blocks. J. Mater. Sci. Technol. **32**(6), 515–523 (2016)
12. Shivkumar, S., Ricci, S., Jr., Apelian, D.: Influence of solution parameters and simplified supersaturation treatments on tensile properties of A356 alloy. AFS Trans. **98**, 913–922 (1990)
13. Hren, I., Svobodova, J.: Fractographic analysis of strontium-modified Al-Si alloys. Manuf. Technol. **18**(6), 900–905 (2018)
14. Farahany, S., Ourdjini, A., Idrsi, M., Shabestari, S.G.: Evaluation of the effect of Bi, Sb, Sr and cooling condition on eutectic phases in an Al-Si-Cu alloy (ADC 12) by in situ thermal analysis. Thermochim. Acta **559**, 59–68 (2013)

15. Hren, I., et al.: Research of the influence on the modification of beryllium in Al–Si alloy. Metallofiz. Noveishie Tekhnol **40**(12), 1637—1647 (2018)
16. Haghdadi, N., Zarei-Hanzaki, A., Abedi, H.R., Sabokpa, O.: The effect of thermomechanical parameters on the eutectic silicon characteristics in a non-modified cast A356 aluminium alloy. Mater. Sci. Eng. A **549**, 93–99 (2012)
17. Adachi, M.: Modification of hypo-eutectic and eutectic Al-Si system casting alloys. J. Japan Inst. Light Metals **34**(6), 361–373 (1984)
18. Telang, Y.P.: Process variables in Al-21 Si alloys refinement. Mod. Cast. **43**(5), 232–240 (1963)
19. Helmanova, M.: The influence of solidification rate on the spatial distribution of eutectic silicon in technical silumins. Prakt. Metallogr. **12**, 59–68 (1975)
20. Kissing, R.J., Wallace, J.F.: Refinement of aluminum-silicon alloys. Foundry **91**(4), 74–79 (1963)
21. Grazieski, J.E.: The art and science of modification: 25 years of progress. AFS Trans. **100**, 673–683 (1992)
22. Bates, A.P.: Hypereutectic aluminum-silicon alloys: a review of published information. Metallurgia **61**(364), 70–79 (1960)
23. Gobrecht, J.: The influence of alloying elements on the duration of modification of Na and Sr in Al-Si cast alloys. Giesserei **65**, 158–164 (1978)
24. Emadi, D., Gruzleski, J.E.: Effects of casting and melt variables on porosity in directionally solidified Al-Si alloys. AFS Trans. **102**, 307–312 (1994)
25. Nogita, K., Dahle, A.K.: Eutectic growth mode in Sr, Sb and P modified hypoeutectic Al-Si foundry alloys. Mater. Trans. **42**(3), 393–396 (2001)

Flow Visualizations in a PDMS Cerebral Aneurysm Biomodel

Andrews Souza[1], Maria Sabrina Souza[2], Conrado Ferrera[3], Hélder Puga[4], Rui Lima[1,5](✉), and João Ribeiro[2,6]

[1] METRICS, University of Minho, Guimarães, Portugal
rl@dem.uminho.pt
[2] Instituto Politécnico de Bragança, Bragança, Portugal
[3] Depto. de Ingeniería Mecánica, Energética y de los Materiales and Instituto de Computación Científica Avanzada, Universidad de Extremadura, Badajoz, Spain
[4] CMEMS, University of Minho, Guimarães, Portugal
[5] CEFT, Faculty of Engineering of the University of Porto, Porto, Portugal
[6] CIMO, Instituto Politécnico de Bragança, Bragança, Portugal

Abstract. Cerebral aneurysm is an abnormal dilatation of the blood vessel which affects a high percentage of the worldwide population. One way to investigate this pathology is using *in vivo* techniques, but these types of experiments have a high cost and low reproducibility. Thus, to understand the local hemodynamics of brain aneurysms, it is imperative to manufacture *in vitro* models that simulate real brain aneurysms. These biomodels are suitable for experimental testing, as well as for evaluating and validating computational models. In this work, was manufactured a biomodel of a cerebral aneurysm made by polydimethylsiloxane (PDMS), combining rapid prototyping technology with a PDMS gravity casting process. Experimental flow visualizations were performed at different flow rates. The flow visualizations results have shown that there is a transition from laminar to turbulent flow for a flow rate near 6 ml/min. The proposed PDMS biomodels have shown the ability to perform flow visualizations and have the potential to help the development and validation of computational models.

Keywords: Cerebral aneurysm · In vitro tests · PDMS flow biomodels

1 Introduction

A cerebral aneurysm is an abnormal dilatation of a blood vessel [1]. Its appearance is caused by a combination of physical and biological factors such as the weakening of the arterial wall with increasing age [2], atherosclerotic changes, hypertension and smoking. This set of factors alters blood flow and, consequently, the vessel wall. Therefore, it is extremely important to obtain more information on how the different flow regimes affect blood vessels and brain aneurysms.

In vivo experiments are an effective way of investigating this pathology, but besides the ethical issues, these types of experiments are expensive and have low reproducibility

J. Machado et al. (Eds.): icieng 2022, LNME, pp. 209–215, 2023.
https://doi.org/10.1007/978-3-031-09382-1_18

[3]. *In silico* studies have been a widely used tool in recent years and have achieved substantial progress in the study of aneurysms. However, the limitation of numerical studies is the need for experimental validation, according to recommended international guidelines [4]. Thus, for a better understanding of the pathology and to study the blood flow inside aneurysms, it is necessary to develop *in vitro* models. These biomodels are suitable for refining endovascular treatments and validating theoretical and computational models [5].

In vitro models can be idealized or patient-specific [6]. Currently, the 3D printing technique has been used for the construction of the initial mold of the vessel lumen and polydimethylsiloxane (PDMS) is the material commonly used for the manufacture of biomodels. This material is transparent, biocompatible and can replicate the geometry of veins [7, 8]. Due to its advantages, PDMS has been used in a wide variety of applications in the biomedical field [9–17]. Pinho et al. [6] developed a biomodel of aneurysms in PDMS, based on clinical data, and this model was used for experimental studies using the digital image correlation technique and *in silico* tests [2, 18]. However, this model did not allow visualization of the flow within the cerebral aneurysm.

In order to improve the existing model, in this work, we propose a simple process to manufacture an *in vitro* PDMS cerebral aneurysm biomodel by combining rapid prototyping technology with a PDMS gravity casting process. Additionally, the model allows the visualization of the different behaviors of blood flow within the aneurysm. Thus, this work reports the acquisition of the real geometry of the cerebral aneurysm by using a prototyping technique to manufacture the mold and the biomodel in PDMS. It was also performed an experimental work of flow visualizations at different flow rates.

2 Materials and Methods

2.1 Geometry and Manufacturing Process

The geometry of the *in vitro* biomodel was obtained by performing image treatment from angiography images from a real cerebral aneurysm. The use of this type of geometry is essential to study the flow phenomena that occur within an aneurysm more realistically. Figure 1 illustrates the general measurements of the cerebral aneurysm model obtained from angiography images.

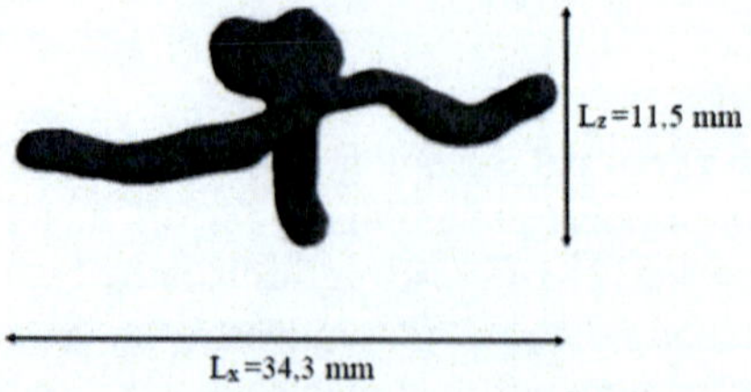

Fig. 1. General measurements of the cerebral aneurysm model.

The treated angiography images were converted into STL format and manufactured by means using a 3D printer model ProJet 1200 (3D Systems, USA). The geometry was printed with the Visijet FTX Green resin as shown in Fig. 2. This process was chosen because the SLA (stereolithography) technique has good accuracy and high print resolution.

Fig. 2. Cerebral aneurysm model produced by the 3D Systems ProJet 1200 printer.

2.2 *In Vitro* PDMS Aneurysm Biomodel

To obtain the biomodel, the PDMS was prepared with a ratio of 10:1, i.e., 10 g of the PDMS base polymer correspond to 1 g of the curing agent. To perform the mixture, it was used a vacuum pump to remove the bubbles and after that, the physical model was placed in an acrylic box and the PDMS was poured into the mold by means of a gravity casting process. The curing period of PDMS was 48 h. In order to obtain the lumen of the blood vessel, the mold was placed in a glass beaker with acetone to remove the printed material. The described process is shown in Fig. 3.

Fig. 3. Process of obtaining the transparent biomodel. (a) curing the PDMS in the acrylic box; (b) biomodel in PDMS with the printed model; (c) transparent PDMS biomodel with a cerebral aneurysm to perform the fluid experiments.

2.3 Flow Visualizations Setup

In the flow visualization tests, it was used a high-speed video microscopy system consisting of an inverted microscope (IX71, Olympus, Japan) combined with a high-speed camera (Photron FASTCAM SA3) and an objective lens (ZEISS, 2.5 ×). To visualize the movements of the particles inside the cerebral aneurysm, it is necessary that the working fluid has the same value as the PDMS refractive index, which is 1.412. A particulate blood analog fluid composed of 61% glycerol and 39% water (w/w) was developed [19, 20] and this fluid has the same refractive index as PDMS, which allows the visualization of the particles. 0.1% of suspended Polymethylmethacrylate (PMMA) particles of 60 μm (diameter) were added to the fluid. The experiments were performed at room temperature (T = 20 ± 2 °C). Figure 4 illustrates the experimental setup used to perform the flow visualizations. To process the images previously recorded was used the ImageJ software being the flow behavior automatically analyzed by the Z Project plugin [21].

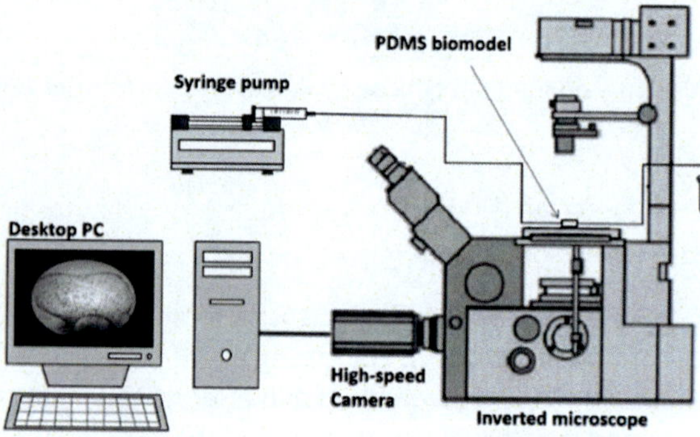

Fig. 4. Setup for *in vitro* experimental tests.

3 Results and Discussion

The developed cerebral aneurysm biomodels, in PDMS, proved to be suitable for experimental flow assays, managing to reproduce the geometry of the blood vessel and with good transparency.

To perform the flow visualizations, the PDMS biomodel was placed under the microscope and the analog fluid was injected into the biomodel through a syringe pump with a constant flow rate. In this study, it was tested different flow rates, i. e., 3 mL/min and 6 mL/min, in order to visualize the moment of transition of the flow behavior and the beginning of recirculation in the aneurysmal sac. The recorded images, shown in Figs. 5 and 6, were processed in the ImageJ software by using the Z Project plugin. With this plugin, the PMMA trajectories captured by a video record were converted automatically into one single image as is shown in Figs. 5 and 6.

From these flow experiments, it was shown that the manufactured PDMS biomodels were suitable to perform hemodynamic studies. Qualitatively, it is possible to observe (Figs. 5 and 6) that at the regions close to the channel inlet and outlet, typical laminar flows were observed for all the tested flow rates. However, when the trace particles get closer to the neck of the aneurysm, the flow tends to suffer a deviation into the direction of the aneurysm and as a result, it was observed the formation of vortices at a flow rate of 6 mL/min. In contrast, at a lower flow rate of 3 mL/min, no recirculation was observed.

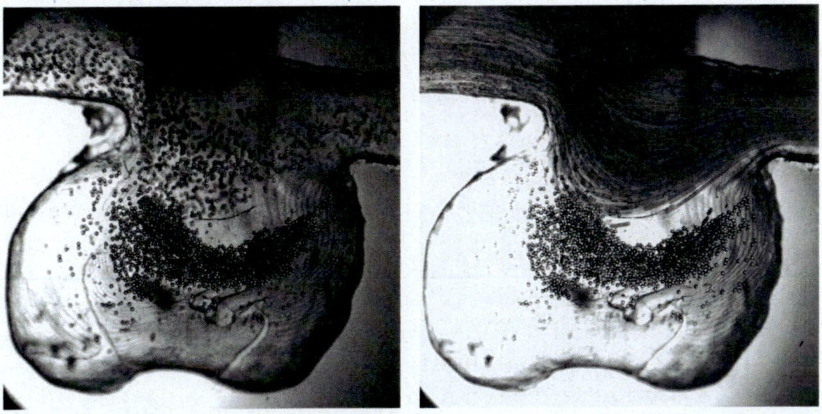

Fig. 5. Flow visualizations of the blood analogue fluid recorded from a high-speed camera at a flow rate of 3 mL/min of an individual recorded image (right side); treated by the Z Project plugin (left side).

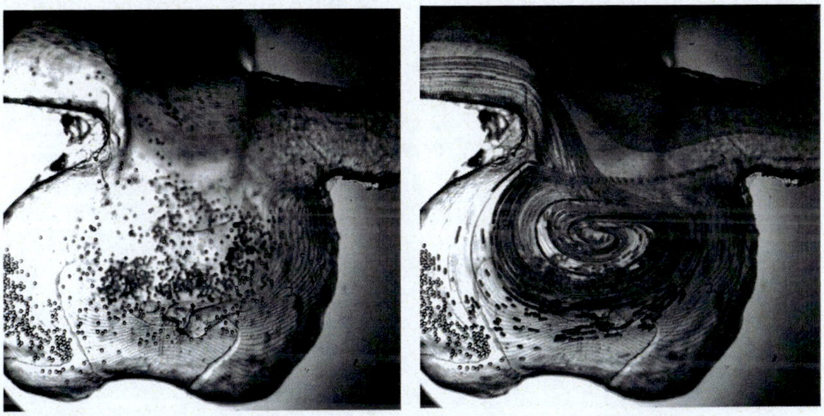

Fig. 6. Flow visualizations of the blood analog fluid recorded from a high-speed camera at a flow rate of 6 mL/min of an individual recorded image (right side); treated by the Z Project plugin (left side).

Hence, these qualitative results indicate that there is a transition from laminar to turbulent flow at a flow rate of about 6 ml/min. The experiments also demonstrate the potential for classifying blood flow patterns within the brain aneurysm. According to the study developed by Cebral et al. [22, 23] who showed that ruptured brain aneurysms are more likely to have complex flow patterns, unstable flow, concentrated inlet jets and small impact regions. Through the final images obtained, we can visualize a single recirculation region indicating a simple flow pattern. The change in flow behavior indicates flow instability. It is also possible to observe that only a part of the flow enters the aneurysmal sac, characterizing it as concentrated inlet jets. Although the flow rate used in the tests was lower than the actual flow rates of a cardiac cycle, the technique used in this study indicates great potential and additional tests are in progress and will be published in due course.

4 Conclusion

It has been reported that the brain aneurysm is caused by a set of factors that are still not well understood. Although many numerical flow studies have been carried out still the majority of these models need to be validated. With the difficulty of performing *in vivo* assays, the *in vitro* study of aneurysms has proved to be an efficient alternative with great potential to validate numerical studies, test new medical devices and improve surgical techniques. However, in order to carry out an *in vitro* test, it is necessary to build biomodels that allow the visualization of the particulate flow. The proposed PDMS biomodels have the potential to be used as *in vitro* models and to study hemodynamics within a brain aneurysm, through the visualization of the flow. By using this flow visualization technique, it will be possible to identify the kinds of flow and to relate them to the flow patterns that the literature indicates as possible causes of aneurysm rupture. The used technique also shows great potential to assist in the development and validation of computational models.

Acknowledgements. The authors acknowledge the financial support provided by Fundação para a Ciência e a Tecnologia (FCT), through the projects EXPL/EME-EME/0732/2021, PTDC/EEI-EEE/2846/2021, funded by NORTE 2020, PORTUGAL2020, and FEDER. This work was also supported by Fundação para a Ciência e a Tecnologia (FCT) under the strategic grants UIDB/04077/2020, UIDB/00690/2020, UIDB/04436/2020 and UIDB/00532/2020. Andrews Souza acknowledges the PhD scholarship 2021.07961.BD attributed by FCT. Partial support from the Junta de Extremadura through Grants No. GR21091 and IB20105 (partially financed by FEDER funds) is gratefully acknowledged.

References

1. Rodriguez-Régent, C., et al.: Non-invasive diagnosis of intracranial aneurysms. Diagn. Interv. Imaging. **95**, 1163–1174 (2014)
2. Cardoso, C., Fernandes, C.S., Lima, R., Ribeiro, J.: Biomechanical analysis of PDMS channels using different hyperelastic numerical constitutive models. Mech. Res. Commun. **90**, 26–33 (2018)
3. Bosman, W.M.P.F., et al.: Aortic customize: an in vivo feasibility study of a percutaneous technique for the repair of aortic aneurysms using injectable elastomer. Eur. J. Vasc. Endovasc. Surg. **40**, 65–70 (2010)
4. Berg, P., Saalfeld, S., Voß, S., Beuing, O., Janiga, G.: A review on the reliability of hemodynamic modeling in intracranial aneurysms: why computational fluid dynamics alone cannot solve the equation. Neurosurg. Focus. **47**, 1–9 (2019)
5. Bosman, W.M.P.F., et al.: The effect of injectable biocompatible elastomer (PDMS) on the strength of the proximal fixation of endovascular aneurysm repair grafts: an in vitro study. J. Vasc. Surg. **52**, 152–158 (2010)

6. Pinho, D., Bento, D., Ribeiro, J., Lima, R., Vaz, M.: An in vitro experimental evaluation of the displacement field in an intracranial aneurysm model. In: Flores, P., Viadero, F. (eds.) New Trends in Mechanism and Machine Science. MMS, vol. 24, pp. 261–268. Springer, Cham (2015). https://doi.org/10.1007/978-3-319-09411-3_28
7. Hassler, C., Boretius, T., Stieglitz, T.: Polymers for neural implants. J. Polym. Sci. Part B Polym. Phys. **49**, 18–33 (2011)
8. Lima, R., et al.: In vitro blood flow in a rectangular PDMS microchannel: experimental observations using a confocal micro-PIV system. Biomed. Microdev. **10**, 153–167 (2008)
9. Maitz, M.F.: Applications of synthetic polymers in clinical medicine. Biosurf. Biotribol. **1**, 161–176 (2015)
10. Miranda, I., et al.: Properties and applications of PDMS for biomedical engineering: a review. J. Funct. Biomater. **13**, 2 (2022)
11. Ariati, R., Sales, F., Souza, A., Lima, R.A., Ribeiro, J.: Polydimethylsiloxane composites characterization and its applications: a review. Polymers (Basel). **13**, 4258 (2021)
12. Carvalho, V., et al.: 3d printing techniques and their applications to organ-on-a-chip platforms: a systematic review. Sensors **21**, 7 (2021)
13. Souza, A., et al.: 3D manufacturing of intracranial aneurysm biomodels for flow visualizations: low cost fabrication processes. Mech. Res. Commun. **107**, 103535 (2020)
14. Rodrigues, R.O., Lima, R., Gomes, H.T., Silva, A.M.T.: Polymer microfluidic devices: an overview of fabrication methods. U.Porto J. Eng. **1**, 67–79 (2015)
15. Souza, R.R., et al.: Recent advances on the thermal properties and applications of nanofluids: From nanomedicine to renewable energies. Appl. Therm. Eng. 201 (2022)
16. Sadek, S.H., Rubio, M., Lima, R., Vega, E.J.: Blood particulate analogue fluids: a review. Materials (Basel). **14**, 16–21 (2021)
17. Carvalho, V., et al.: In vitro stenotic arteries to perform blood analogues flow visualizations and measurements: a review. Open Biomed. Eng. J. **14**, 87–102 (2020)
18. Rodrigues, R.O., Pinho, D., Bento, D., Lima, R., Ribeiro, J.: Wall expansion assessment of an intracranial aneurysm model by a 3D Digital image correlation system. Measurement **88**, 262–270 (2016)
19. Pinho, D., Campo-Deaño, L., Lima, R., Pinho, F.T.: In vitro particulate analogue fluids for experimental studies of rheological and hemorheological behavior of glucose-rich RBC suspensions. Biomicrofluidics **11**, 054105 (2017)
20. Campo-Deaño, L., Dullens, R.P.A., Aarts, D.G.A.L., Pinho, F.T., Oliveira, M.S.N.: Viscoelasticity of blood and viscoelastic blood analogues for use in polydymethylsiloxane in vitro models of the circulatory system. Biomicrofluidics **7**, 034102 (2013)
21. Carvalho, V., et al.: Manual and automatic image analysis segmentation methods for blood flow studies in microchannels. Micromachines **12**, 1–20 (2021)
22. Cebral, J.R., Mut, F., Weir, J., Putman, C.M.: Association of hemodynamic characteristics and cerebral aneurysm rupture. Am. J. Neuroradiol. **32**, 264–270 (2011)
23. Cebral, J.R., Castro, M.A., Burgess, J.E., Pergolizzi, R.S., Sheridan, M.J., Putman, C.M.: Characterization of cerebral aneurysms for assessing risk of rupture by using patient-specific computational hemodynamics models. Am. J. Neuroradiol. **26**, 2550–2559 (2005)

What Times Should Be Used for fNIRS and EEG Protocol?

Iara Margolis[1,2](✉), Rute Silva[1,3], Diego Pinal[4], Nelson Costa[1], and Bernardo Providência[2]

[1] University of Minho, Guimarães/Braga, Portugal
iaramr@gmail.com, ncosta@dps.uminho.pt
[2] Lab2PT and ALGORITMI Center, University of Minho, Guimarães/Braga, Portugal
providencia@arquitetura.uminho.pt
[3] Fashion Design anda ALGORITMI Center, University of Minho, Guimarães/Braga, Portugal
[4] CIPsi, University of Minho, Braga/Braga, Portugal
diego.pinal.fdez@psi.uminho.pt

Abstract. fNIRS (Functional near-infrared spectroscopy) and EEG (Electroencephalogram) are two technologies that measure brain activity. Recently expanded to new fields, such as Design. These technologies are beginning to be applied simultaneously in the laboratory or in real life, as they allow researchers to collect distinct data - electrical activity in the form of brain waves and the concentration of oxygenated and deoxygenated hemoglobin (HbO and HbR), respectively. The purpose of this study is to present a review on the event timing applied individually to optimize the usage of both tools, and a reflection on how a study protocol should be designed for the integration of both technologies into a hybrid fNRIS-EEG recording. The results show that the purpose of the study and the type of stimulus or task to be done by the participants, must be considered since they determinate the events timing in each step of the protocol. Event timing includes longer durations when there is interaction between the participant and the stimulus, or for completion of a task. Additionally, the resting period for fNIRS signals baseline correction must be carefully considered for the convergence on the use of both technologies in the same protocol.

Keywords: Neurocognition · fNIRS · EEG · Event timing protocol · Design · Affective · Emotion

1 Introduction

The area of human brain activities is still a field to be explored. The real understanding of how the human being captures and processes information can be aided with some technologies such as fNIRS (Functional near-infrared spectroscopy) and EEG (Electroencephalogram).

The brain is composed of two hemispheres (right and left), each divided into four lobes - frontal, parietal, temporal and occipital. When electrical activity is synchronized in large populations of neurons, the electrical potentials are summed, which over time

J. Machado et al. (Eds.): icieng 2022, LNME, pp. 216–227, 2023.
https://doi.org/10.1007/978-3-031-09382-1_19

produces a wave-like signal when recorded at the scalp, that is called brain waves [1]. Electroencephalography (EEG) is a non-invasive tool that allows the detection of electrical activity through electrodes placed on the surface of the scalp. It is a tool of high temporal resolution and is exceptional for the study of cognitive processes, since it allows the capture of neural activity in the period when cognition occurs, that is, in milliseconds. With advances in neuroscience, EEG makes it possible to record the brain waves of individuals when they are exposed to a given stimulus. Thus, the EEG technique has been applied beyond the clinical forum, in different research areas, such as Design. Here, for instance, it provides evaluation's possibility of the emotional reaction of individuals to a given stimulus. Damasio [2] reinforces the importance of studying the brain with respect to the influence of emotions on the decision-making process. In the evaluation of human emotions, the brain's response is usually analyzed through frequency bands. Frequently, refers to the speed of electrical oscillations and is measured in cycles per second - Hertz (Hz). Frequencies are usually categorized into five types - from highest to lowest: Gamma (30–80 Hz), Beta (13–30 Hz), Alpha (8–13 Hz), Theta (4–7 or 8 Hz) and Delta (0–4 Hz) [3]. EEG has a low spatial resolution. However, it does not allow to precisely locate the associated cortical sources [4]. Another major limitation is the sensitivity to electrical noise, which requires users to employ methods to decrease interference with the recording system.

Whereas fNIRS is a non-invasive, safe, easily applicable, noise resilient, and motion tolerant optical imaging method that measures brain activity through hemoglobin oxygenation and deoxygenation [5–10]. In fact, the technology measures the change in oxygenation against a baseline [11]. However, one of its limitations is the spatial resolution, being able to measure only the hemodynamic activities coming from the brain surface [6, 7]. Balconi and Venutelli [12] argue that it has a higher spatial resolution than EEG, despite its temporal resolution being far more limited. This technology has been used to analyze emotional responses, especially in the prefrontal cortex [6]. In its application, lighting and some blockages (such as hair or skin oiliness), may cause interference with the optical capture.

Both technologies are safe and non-invasive and can be used in a complementary and simultaneous way. And their results can help in the human understanding of the physiological perspective when facing several problems and challenges, including emotional issues. Providing their different time resolution, this paper aimed to provide event timing parameters to the combined use of these technologies, to assist in assembling protocols for the simultaneous recording using the two technologies.

2 Methodology

The research was made from two perspectives: fNIRS and EEG. Initially the goal was to analyze the protocols used in previous studies to analyze design and emotion, with emphasis on products. However, a search in 13 journals focused on design and in CAPES platforms, yielded no results within the delimitations drawn from product design. Therefore, the analysis was made on 9 articles with EGG applications of the use of product's design and, regarding the fNIRS, 33 articles were selected that used this technology to access affective responses in healthy adults, and that referred to event timing with numerical values. For EEG, the search consisted of three keywords - "product design", "EEG" and "emotion". As for the fNIRS the words searched were "fNIRS" and "affective".

All articles were read in entire, event timing, stimulus, and protocol information tabulated in Excel, and handled manually. Worth mentioning that a distinction was made between pauses and rest periods. A pause was understood here as the time between blocks, and a rest the inter-stimulus interval. It should be noted that this study considered the pre-baseline term for actions taken before the baseline period. To compare the studies, it was necessary to parameterize them, which includes categorizing what was the baseline period, the stimulus, and the rest period. However, not all papers made this information clear. Sometimes it was understood that the baseline period would be a moment for resting (even when it was not explicitly said, as in the case of the study [13, 14], or for the studies [15, 16] that started the study with a resting period. Besides, it is important to emphasize that the studies were subdivided into 4 groups, namely:

- Digital: those articles that had as stimuli videos, digital images or digital images with sound. This was the largest group, containing 25 studies [8, 9, 12, 13, 15–35];
- Task: the experiments that include some kind of activities to be developed by the participants, such as playing Sudoku [36], communicating with drums [37], solving arithmetic activities [38, 39] or playing instruments [40];
- Interaction: including the interaction between a user and a software [14], virtual reality [41], a car simulator [42] or a virtual game [43];
- Others: studies that did not fit into the previous categories, such as reflective actions [7, 10], muscular thermal change [44], audio analysis [45], hand clench movement [11], classes [6], physical product analysis [13, 46] or product development [47].

There were 9 studies [13, 14, 30–35, 46] using only EEG and 4 studies using both technologies together [8, 12, 17, 39]. It is noteworthy that the study by Stylios and Chen [13] had two moments, one with a physical and another with a digital product, and is therefore in two categories. As well as the study by Soltanlou et al. [39] that offers participants two arithmetic problems: first with one digit and a stimulus presentation time of 10 s, and the other with two-digit problems and a stimulus presentation time of 30 s, both being in the task category.

In the studies that had a varying event timing, we used either the average (when properly referred to in the text, as in the case of [47]) or the longest time used (as in the case where the stimulus timing varied from 16 to 20 s [26]).

Finally, all the event timings were transformed to seconds, which are the units presented in the graphs. Not all studies presented the event timing numerically, despite addressing the condition. Only studies that provided the timing of that analysis were considered. The studies were also separated into those that used EEG, fNIRS, and both technologies together.

3 Results

In the event analysis, timing was used in studies using digital stimuli (Fig. 1), 17 works presented baseline periods and it was possible to notice that a set of 5 studies from the same group of authors [8, 12, 16, 18, 19] used a baseline period of 120 s, both using fNIRS and the combination of fNIRS and EEG equipments. The second longest baseline period was of 60 s [28], also from a study of the previous group of authors. The remaining studies used a baseline period of 15 or less seconds, with 4 studies [15, 23, 24, 33] using between 6.5 and 15 s and 7 studies [16, 20, 21, 30–32, 35] using a baseline period between 0.5 and 2 s. For the rest period, on the other hand, when the stimuli were presented rapidly, the rest tended towards miliseconds, but overall, the rest periods ranged from 0.3 to 2 s [17–19, 22, 23, 30, 34], 3 to 6 s [13, 20, 21, 24] and 8 to 10 s [8, 12, 15, 16, 28]. The longest rest period found was that of the study by Hu et al. [29] who presented videos for a virtual assistant emotional detection and had a rest period duration of 45 s.

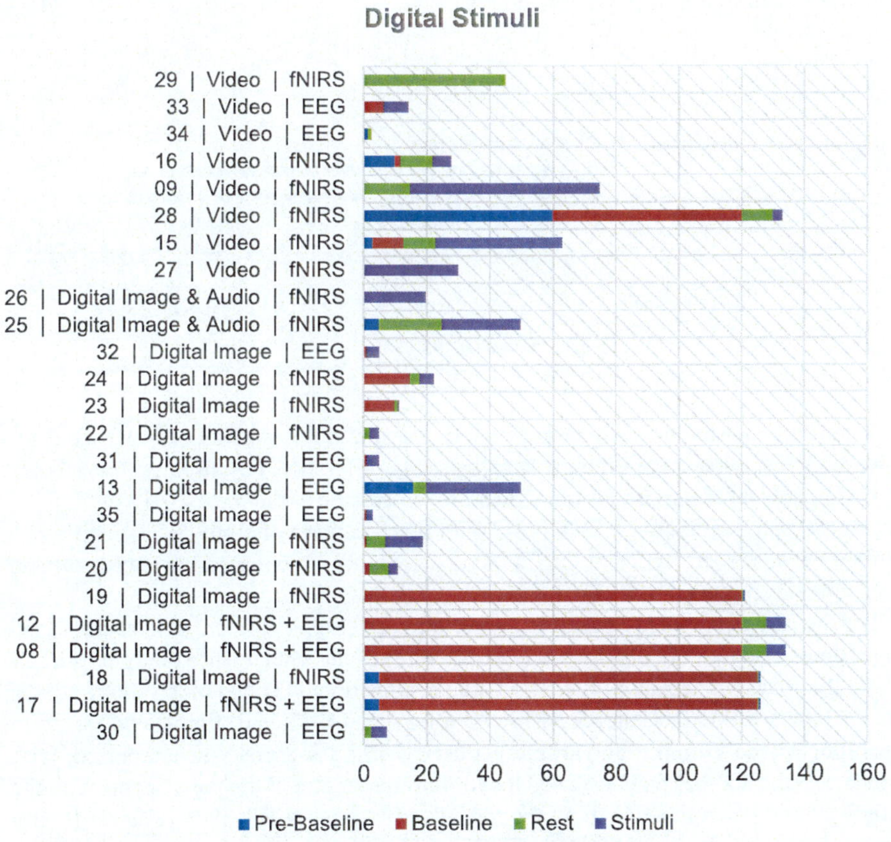

Fig. 1. Digital stimuli timing

Regarding stimuli, 4 studies [17–19, 23] used 0.5 s of stimulus presentation, 11 studies [8, 12, 16, 20, 22, 24, 28, 30–32, 35] had the stimulus presentation time ranging between 2 and 6 s, with two of these studies using videos [16, 28]. And 5 studies using fNIRS [9, 15, 25–27] and 1 using EEG [13] had the stimulus presented for longer than

20 s, with the longest times being 40 and 60 s for videos [9, 15], followed by 30 s of stimulus presentation for videos using fNIRS [27] and for digital images using EEG [13]. It is not shown, but 4 studies had post- stimulus intervals, 3 of them lasting 4–5 s [17, 19, 21] and another one [34] lasting 40 min. There were still 8 studies [8, 17–19, 23, 24, 28, 31] that had pauses between blocks ranging from 5 s to 45 min.

When analyzing interaction studies (Fig. 2), only a work that used EEG presented a baseline period, with 60 s [14], all other articles did not refer to baseline periods. The stimulus presentation timing was quite varied: 3 s [43], 5 s [41], 3 min [14] and 15 min [43]. And 2 studies reported a resting interval, which lasted 3 [43] and 15 s [41], as presented in Fig. 2. It is noteworthy that the 2 car studies used a pre-baseline period, the virtual reality one [41] used 30 min of training, and the simulator study [42] 5 min of training in a different scenario and a certain group with 12 min writing a letter to evoke an emotion before performing the stimulus.

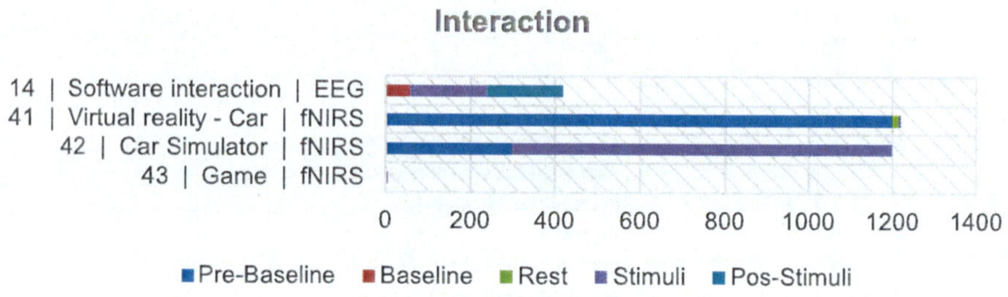

Fig. 2. Interaction timing

Regarding the studies using tasks (Fig. 3), the stimulus lasted between 10 and 50 s. However, in the other event timings there was a big difference. For instance, two studies referred to the performance of arithmetic tasks, where one of them used 0.5 s of baseline period [39] and the other one 20 s [38]. With the exception of study [39] - which was the only one that used fNIRS with EEG in this category - the rest of the studies tended to include a rest interval of 20–40 s.

For the last category (others) it is possible to see (Fig. 4) that in general they used longer times than those in previous categories. The articles with the use of digital stimuli to help the reflexive decision [7, 10] or with pseudowords [45] had much shorter times than the others within this category. Three studies [06, 44, 47] had stimulus presentation times longer than 20 min. One article [7] used 5 min for a pre-baseline period, with simulation; and another one [13] used 8 s for familiarization of the participants with the equipment and environment. About the rest intervals, one fNIRS study [6] used 10 min of rest interval, another one [46] using EEG included 1 min of rest interval and the others [7, 10, 11, 13, 44] between 5 and 20 s. It is to be noted that the muscular thermal change study [44] has a larger stimulus presentation time (20 min) but a short rest interval (10 s) that converges to be used as the baseline period for baseline correction.

Finally, it was also analyzed what was presented in the baseline, pause and rest periods (Table 1). It was possible to notice the use of a fixation cross (a cross in the center of the screen), an empty screen (with some kind of background, such as black,

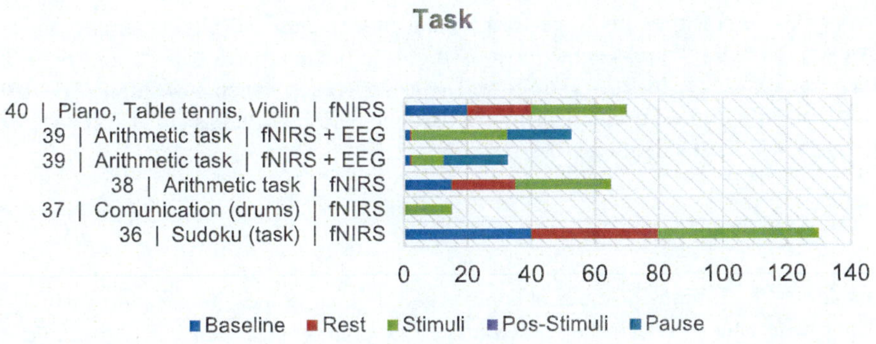

Fig. 3. Tasks timing

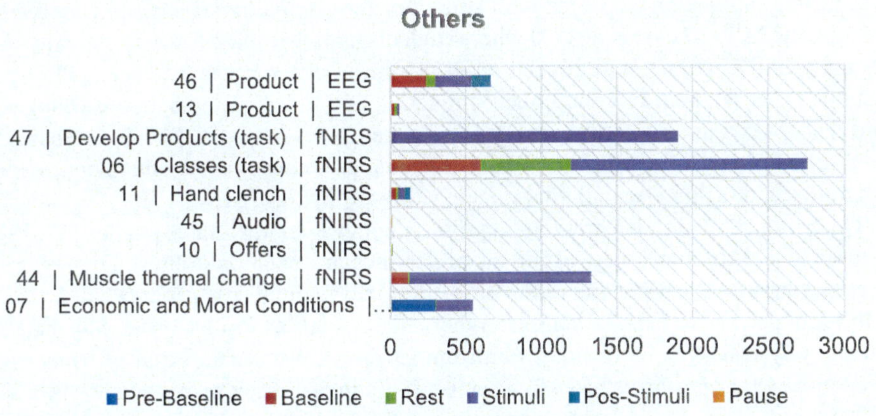

Fig. 4. Others timing

Table 1. Presentation screen in baseline, rest and pause.

Baseline		Rest		Pause	
(1) Cross Mark	9	(1) Cross Mark	15	(1) Blank	3
(2) Eye instructions; (3) Relax/Rest	3	(2) Blank screen	5		
(4) Asterisk; (5) Base product; (6) Calm the mind; (7) Focus on the screen; (8) Neutral condition	1	(3) Black screen with REST word; (4) Blank interval; (5) Closed eyes; (6) Dot; (7) No tasks performance	1		

white, gray), or Instructions to perform ocular movements and blinks, among others. However, the fixation cross is the most used during the rest period, in 11 studies [7, 8, 10, 12, 15, 20, 22, 25, 26, 28, 43], followed by using a black screen, in 4 studies [16–19], all of them with digital stimuli. During the baseline period, 9 studies [11, 15, 16, 21, 22, 30, 32, 35, 45] used a fixation cross.

4 Results Analysis

Considering the timings of the events presented, it was not possible to notice any significant difference between the fNIRS and EEG technologies. But it was possible to notice that the event timings varied considerably in view of the objective of the study. Regarding the type of stimuli - digital and non-digital, the results allow us to observe significant differences. Digital stimuli can be very fast and it is also easier to use screens for baseline, pause and rest periods. The longest event timings were found for studies requiring interaction of people with stimuli or the completion of tasks, being them respectively: Milovanovic et al. [47] where students had to develop 3 accessibility products using TRIZ, Zheng et al. [6] that used as stimuli different class formats, Phillips et al. [44] that worked with muscular contraction and assessed temperature changes, Unni et al. [41] that used virtual reality environment with cars and FakhrHosseini and Jeon [42] who used a car simulator. The shortest event timings, on the other hand, were with millisecond stimuli with digital imaging [17–19, 23].

The contrast of stimulus presentation time is noticeable, in all three studies [17–19] that used visual stimulus relay on 500 ms of stimulus presentation, followed by 300 ms of rest interval for each image presentation, after these stimuli participants had 5 s for performing certain actions. In contrast, other studies such as Phillips et al. [44] had 2 different physical stimuli: 15 min of arm immersion in water and 5 min of isometric contraction. Other protocols had variation in stimuli presentation time, such as the case of Zheng et al. [6] whose classes ranged from 13 to 26 min, or the Burns, et al. study [26] who had videos ranging from 47 to 180 s. In the Milovanovic et al. study [47], students had to develop 3 accessibility products, averaging 7.5 min in the brainstorming part, 11 min in the morphological stage, and 13.3 min in the TRIZ.

When analyzing the EEG use as compared to fNRIS usage, it was possible to see that the fNRIS studies used a longer period of time for the baseline. It can be seen that all the practical cases that apply EEG technology, except the study [14] that considers 60 s, use a baseline period much shorter than 30 s, in contrast to the longer periods in studies that used fNRIS. This issue may be related to the limitation mentioned above, the delay in the blood response. However, the time of the baseline, pause and rest periods did not follow a logic. For example, three studies - from the same groups of authors - [17–19] used a stimulus presentation time of 0.5 s, 0.3 of rest interval and 5 s of pause between blocks and 120 s of baseline period, while the study [23] used 0.5 s of stimulus presentation time, 1 s of rest interval, 30 s in between blocks pause and 10 s of baseline period - all four studies with digital images. Still on digital images, two studies of the same group [08, 12] used 120 s of baseline period, 8 s of rest interval, 6 s of stimulus presentation time, and one of them [8] used 60 s in between blocks pause. Whereas the study [16] which used videos as stimuli, used 2 s as baseline period, 10 s as rest interval, and 6 s of stimulus presentation time. In the arithmetic tasks study [39], the stimulus presentation had two different length of 10 and 30 s, but the rest interval was exactly the same (0.5 s) as was the baseline period (2 s) and the between blocks pause (20 s). The study [38], which also used arithmetic tasks, the stimulus presentation time was 30 s, with 20 s as rest interval and 10 s as baseline period. There is a study with muscular contractions [44] that has its stimulus 120 times longer than the rest, and another study [13] including the manipulation of physical products, which presents a 0.55-fold difference between the stimulus and the rest interval. The analyzed studies that apply the EEG, consider the step – pause in the protocol, but the period of time is self-determined by the participants. The participants were allowed to pause as much as needed [35] and, in the study [31] the pause was performed every 120 stimuli, in a total of 360 stimuli.

EEG allows the brain's electrical activity to be recorded with milliseconds precision. In other words, it enables to identify the changes that occur in the brain when the individual is exposed to a certain stimulus, allowing the process of differences between stimuli to be observed. The stimulus presented can be physical or digital, as reviewed in the present analysis. On the other hand, fNIRS needs a longer time, as found in 8 studies with an explicit time correction, where recording starts a few seconds after stimulus presentation and ends a few seconds after. One study [22] mentioned that it made a correction, but did not say the exact amount of time considered in that correction. 3 studies [29, 36, 44] used a "baseline correction" of 10 s. There were also 7 [15], 5 [23, 39] and 2 s [24] periods used for the baseline correction.

The total time of the experiment is an important factor to consider since the preparation of these tools is usually longer than the time required to complete the experiment and over time individuals begin to feel fatigue and lose concentration on the experimental tasks. Of the nine EEG studies, three refer to the total time of the experiment, being 35 [14], 40 [34] and 45 min [31]. For the fNIRS studies, 11 [7, 10, 23, 26, 28, 29, 37, 39, 41–43] referred to the total time, ranging from 70 s [29] to 50 min [41].

Despite having analyzed the timing variation, this work leaves as a contribution a summary of the two neuroimaging techniques (Fig. 5). Worth mentioning that, except for the stimuli, all the others had at least one article that did not perform the operation. Then was considered for the result flow the shortest time and the longest time reported.

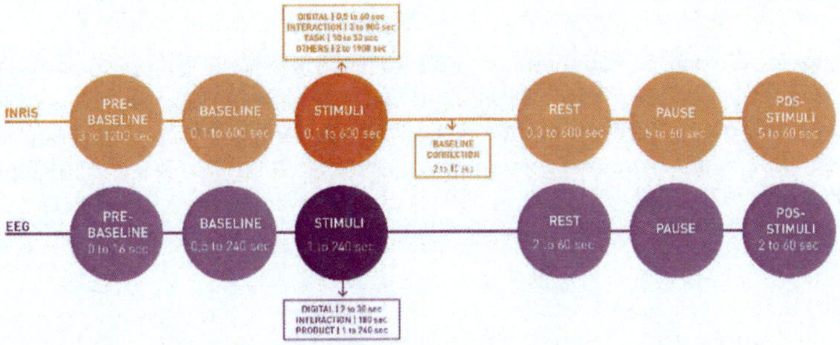

Fig. 5. Result flow – presenting the stages and times of the papers analyzed

5 Conclusion

This article aimed to provide some parameters for choosing the event timing for the different steps of the protocol applied to practical cases with fNIRS and EEG tools. For this, the steps - baseline period, stimulus presentation, rest interval and between blocks pause were considered, as they were the main and most important steps in defining the events timing. A synthesis of the results was presented in Fig. 5.

The design of the protocol requires care and attention. The first point that must be analyzed is the type of stimulus and the participants' tasks, along with the research objective, in order to weigh the event timing for the protocol design. The analysis allowed us to identify that stimulus presentation time varies according to the type of stimulus and the needs of the study. In turn, the baseline period and rest interval converge to a less structured trend in the comparison between the studies analyzed, and may serve as a more consistent reference for future protocols. In the analysis of the total time of the experiment, it is possible to identify that, for the use of both tools - EEG and fNRIS, the time period does not exceed 60 min, being considered a maximum of 50 min [41].

Regarding the event timing defined in the protocol steps, it was not possible to identify significant differences between the use of EEG, fNIRS, or both tools together, except for the baseline period duration when using fNIRS or both tools together. This reinforces the harmonic integration of the two technologies and their respective protocols.

It is important to highlight that the difference in the characteristics of each of these technologies must be taken into consideration. In this sense, understanding the temporal resolution of the two tools is of prime importance, especially in the treatment of the data. In fNIRS it was possible to observe the use of a baseline correction to overcome this limitation of the technology, while the same is not necessary in EEG studies.

This article has the limitation of analyzing the event timing of different studies without accounting for their objectives. However, we sought to look for standards when using different types of stimuli and tasks as a guide for future research. Also, it remains as a suggestion for future work of analysis in a systematic way the use of each of the technologies individually and their unique contributions to the area. It can be concluded that the aforementioned topics were considered in order to provide more security to researchers who wish to use EEG and/or fNIRS, with an emphasis on the event timing in experimental protocols.

Acknowledgment. This work is supported by FCT – Fundação para a Ciência e Tecnologia within the R&D Units Project Scope: UIDB/00319/2020. The authors would also like to thank the European Structural and Investment Funds in the FEDER component, through the Operational Competitiveness and Internationalization Programme (COMPETE 2020) [Project n° 039334; Funding Reference: POCI-01-0247-FEDER-039334] for the Grants of Iara Margolis and Rute Silva.

References

1. Cohen, M.X.: Analyzing Neutal Time Series Data: Theory and Practice. MIT Press, Amsterdam (2014)
2. Damasio, A.: O erro de Descartes: Emoção, razão e o cérebro humano. Temas e Debates, Lisboa (2011)

3. Sauseng, P., Wolfgang, K.: What does phase information of oscillatory brain activity tell us about cognitive processes? Neurosci. Biobehav. Rev. **32**, 1001–1013 (2008)
4. Abiri, R., Borhani, S., Sellers, W., Jiang, Y., Zhao, X.: A comprehensive review of EEG-based brain–computer interface paradigms. J. Neural Eng. **16**(1) (2019)
5. Val-Laillet, D., et al.: Neuroimaging and neuromodulation approaches to study eating behavior and prevent and treat eating disorders and obesity. Neuroimage Clin. **8**, 1–31 (2015)
6. Zheng, L., et al.: Affiliative bonding between teachers and students through interpersonal synchronisation in brain activity. Soc. Cogn. Affect. Neurosci. **15**(1), 97–109 (2020)
7. Vanutelli, M., Meroni, F., Fronda, G., Balconi, M., Lucchiari, C.: Gender differences and unfairness processing during economic and moral decision-making: a fNIRS study. Brain Sci. **10**, 1–15 (2020)
8. Balconi, M., Vanutelli, M.E.: Emotions and BIS/BAS components affect brain activity (ERPs and fNIRS) in observing intra-species and inter-species interactions. Brain Imaging Behav. **10**(3), 750–760 (2015). https://doi.org/10.1007/s11682-015-9443-z
9. Bandaraa, D., Velipasalara, S., Brattb, S., Hirshfield, L.: Building predictive models of emotion with functional near-infrared spectroscopy. Int. J. Hum Comput Stud. **110**, 75–85 (2018)
10. Balconi, M., Fronda, G.: Morality and management: an oxymoron? fNIRS and neuromanagement perspective explain us why things are not like this. Cogn. Affect. Behav. Neurosci. **20**(6), 1336–1348 (2020). https://doi.org/10.3758/s13415-020-00841-1
11. Propper, R., Dodd, K., Christman, S., Brunyé, T.: Relationship between sustained unilateral hand clench, emotional state, line bisection performance, and prefrontal cortical activity: a functional near-infrared spectroscopy study. Brain Cogn. **22**(6), 671–689 (2016)
12. Balconi, M., Vanutelli, M.: Empathy in negative and positive interpersonal interactions. what is the relationship between central (EEG, fNIRS) and peripheral (autonomic) neurophysiological responses? Adv. Cogn. Psychol. **13**(1), 105–120 (2017)
13. Stylios, G., Chen, M.: Psychotextiles and their interaction with the human brain. In: Smart Textiles and Their Applications, pp. 197–239. Woodhead Publishing. Duxford (2016)
14. Chen, Z., Lin, L.: Emotional experience evaluation method of interaction task based on EEG technology. IOP Conf. Ser. Mater. Sci. Eng. **573**(1) (2019)
15. Aranyi, G., Pecune, F., Charles, F., Pelachaud, C., Cavazza, M.: Affective interaction with a virtual character through an fNIRS brain-computer interface. Front. Comput. Neurosci. **10**(70), 1–14 (2016)
16. Crivelli, D., Sabogal Rueda, M.D., Balconi, M.: Linguistic and motor representations of everyday complex actions: an fNIRS investigation. Brain Struct. Funct. **223**(6), 2989–2997 (2018). https://doi.org/10.1007/s00429-018-1646-9
17. Balconi, M., Crivelli, D., Vanutelli, M.: Why to cooperate is better than to compete: brain and personality components. BMC Neurosci **18**(68), 1–15 (2017)
18. Balconi, M., Gatti, L., Vanutelli, M.: When cooperation goes wrong: brain and behavioral correlates of ineffective joint strategies in dyads. Int. J. Neurosci. **128**(2), 155–166 (2017)
19. Balconi, M., Vanutelli, M.: When cooperation was efficient or inefficient. Functional Near-Infrared Spectroscopy evidence. Front. Syst. Neurosci. **11**(26), 1–10 (2017)
20. Gier, N., Strelow, E., Krampe, C.: Measuring dlPFC signals to predict the success of merchandising elements at the point-of-sale: A fNIRS approach. Front. Neurosci. **14**, 1–13 (2020)
21. Giles, G.E., et al.: Habitual exercise is associated with cognitive control and cognitive reappraisal success. Exp. Brain Res. **235**(12), 3785–3797 (2017). https://doi.org/10.1007/s00221-017-5098-x
22. Xie, J., Yang, H., Xia, X., Yu, S.: The influence of medical professional knowledge on empathy for pain: evidence from fNIRS. Front. Psychol. **9**, 1–8 (2018)

23. Yeung, M., Lee, T., Chan, A.: Negative mood is associated with decreased prefrontal cortex functioning during working memory in young adults. Soc. Psychophysiol. Res. **58**(6), e13802 (2021)
24. Yu, W., et al.: Episodic memory encoding and retrieval in face-name paired paradigm: an fNIRS study. Brain Sci. **11**(7), 13–16 (2021)
25. Azhari, A., Rigo, P., Bornstein, M., Esposito, G.: Where sounds occur matters: context effects influence processing of salient vocalisations. Brain Sci. **10**, 1–12 (2020)
26. Burns, S., Barnes, L., Katzman, P., Ames, D., Falk, E., Lieberman, M.: A functional near infrared spectroscopy (fNIRS) replication of the sunscreen persuasion paradigm. Soc. Cogn. Affect. Neurosci. **13**(6), 628–636 (2018)
27. Andreu-Perez, A., et al.: Single-trial recognition of video gamer's expertise from brain hemodynamic and facial emotion responses. Brain Sci. **11**(1), 1–20 (2021)
28. Balconi, M., Crivelli, D., Cortesi, L.: Transitive versus intransitive complex gesture representation: a comparison between execution, observation and imagination by fNIRS. Appl. Psychophysiol. Biofeedback **42**(3), 179–191 (2017)
29. Hu, X., Zhuang, C., Wang, F., Liu, Y.-J., Im, C.-H., Zhang, D.: fNIRS Evidence for recognizably different positive emotions. Front. Hum. Neurosci. **13**, 1–11 (2019)
30. Deng, L., Wang, G.: Application of EEG and interactive evolutionary design method in cultural and creative product design. Comput. Intell. Neurosci. **2019** (2019)
31. Wang, R., Ke, T., Chuang, S., Liu, I.: Sex differences in high-level appreciation of automobile design-evoked gamma broadband synchronization. Sci. Rep. **10**(1) (2020)
32. Zhang, W., Jin, J., Wang, A., Ma, Q., Yu, H.: Consumers' implicit motivation of purchasing luxury brands: an EEG study. Psychol. Res. Behav. Manag. **12**, 913–929 (2021)
33. Guo, F., Li, M., Hu, M., Li, F., Lin, B.: Distinguishing and quantifying the visual aesthetics of a product: an integrated approach of eye-tracking and EEG. Int. J. Ind. Ergon. **71**, 47–56 (2019)
34. Guo, F., Ding, Y., Wang, T., Liu, W., Jin, H.: Applying event related potentials to evaluate user preferences toward smartphone form design. Int. J. Ind. Ergon. **54**, 57–64 (2016)
35. Schoen, F., Lochmann, M., Prell, J., Herfurth, K., Rampp, S.: Neuronal correlates of product feature attractiveness. Front. Behav. Neurosci. **12** (2018)
36. Ashlesh, P., Deepak, K., Preet, K.: Role of prefrontal cortex during Sudoku task: fNIRS study. Transl. Neurosci. **11**(1), 419–427 (2020)
37. Rojiani, R., Zhang, X., Noah, A., Hirsch, J.: Communication of emotion via drumming: dual-brain imaging with functional near-infrared spectroscopy. Neuroscience **3**(10), 1047–1057 (2018)
38. Kakehashi, E., Choi, J., Takekawa, S., Suzuki, A., Kimura, M., Hibino, H.: Physiological and psychological effects of OLED lighting location on office work efficiency. Int. J. Affect. Eng. **17**(2), 99–107 (2018)
39. Soltanlou, M., et al.: Increased arithmetic complexity is associated with domain-general but not domain-specific magnitude processing in children: a simultaneous fNIRS-EEG study. Cogn. Affect. Behav. Neurosci. **17**(4), 724–736 (2017). https://doi.org/10.3758/s13415-017-0508-x
40. Balardin, J., et al.: Imaging brain function with functional near-infrared spectroscopy in unconstrained environments. Front. Hum. Neurosci. **11**, 1–7 (2017)
41. Unni, A., Ihme, K., Jipp, M., Rieger, J.: Assessing the driver's current level of working memory load with high density functional near-infrared spectroscopy: a realistic driving simulator study. Front. Hum. Neurosci. **11**, 1–14 (2017)
42. FakhrHosseini, S.M., Jeon, M.: How do angry drivers respond to emotional music? A comprehensive perspective on assessing emotion. J. Multimodal User Interfaces **13**(2), 137–150 (2019). https://doi.org/10.1007/s12193-019-00300-3

43. Maier, M., Rosenbaum, D., Haeussinger, F., Brüne, M., Fallgatter, A., Ehlis, A.-C.: Disinhibited revenge – an fNIRS study on forgiveness and cognitive control. Front. Behav. Neurosci. **13**, 1–10 (2019)
44. Phillips, K., Verbrigghe, D., Gabe, A., Jauquet, B., Eischer, C., Yoon, T.: The influence of thermal alterations on prefrontal cortex activation and neuromuscular function during a fatiguing task. Environ. Res. Public Health **17**, 1–17 (2020)
45. Gruber, T., et al.: Human discrimination and categorization of emotions in voices: a functional near-infrared spectroscopy (fNIRS) study. Front. Neurosci. **14**, 1–14 (2020)
46. Lee, O., Lee, Y.: Electrophysiological assessment of cushions as attachment objects. Int. J. Cloth. Sci. Technol. **33**(6), 953–965 (2021)
47. Milovanovic, J., Hu, M., Shealy, T., Gero, J.: Characterization of concept generation for engineering design through temporal brain network analysis. Des. Stud. **76**, 1–33 (2021)

Check for updates

Inclusive Platform: GUS for Social Inclusion and Competencies Development in Cerebral Palsy

Gil Dias[1], Filomena Soares[2], Vítor Carvalho[1,3], Armanda Pereira[4], and Tiago Martins[1](✉)

[1] Metrics Research Center, School of Engineering, University of Minho, Braga, Portugal
tiagorsmartins@hotmail.com
[2] Algoritmi Research Center, School of Engineering, University of Minho, Braga, Portugal
[3] 2Ai, School of Technology, IPCA, Barcelos, Portugal
[4] School of Psychology, University of Minho, Braga, Portugal

Abstract. Cerebral Palsy (CP) refers to a set of neurodevelopment disorders associated with motor and functional limitations in individuals, and often accompanied by other impairments. The lack of adapted resources to CP hinders a normal development and social inclusion. Create accessibilities and adapted solutions for this population which promote social inclusion and skills development is necessary. In this paper we propose a conceptual development of an inclusive platform for children with CP and children with normal development. Through Games Used Seriously (GUS), our platform aims to develop skills (motor, cognitive and social) and social interaction, with individual and multiplayer games (e.g. cooperative or competitive games) and communication strategies. Components for increasing of the motivation and engagement with platform is included (e.g. social media information or gamification). Include hardware equipment with biometric sensors and tangible interaction allows more resources to explore motor skills and clinical control. The ability to obtain quantitative indicators in this tool allows to make constant evaluation and to be a source of information for secondary users (parents and therapists). This solution may be an important complementary resource to conventional therapy and can be applied in other environments.

Keywords: Cerebral Palsy · Games Used Seriously · Inclusion platform · Skills development · Social inclusion

1 Introduction

1.1 Cerebral Palsy

Cerebral palsy (CP) is a pathology which is a group of permanent, non-progressive, developmental disorders that mainly affects movement and posture where secondary impairments are associated such a cognitive, language, and visual impairments, and with repercussions in daily activity [1, 2]. The CP affected 17 million of people in the

J. Machado et al. (Eds.): icieng 2022, LNME, pp. 228–239, 2023.
https://doi.org/10.1007/978-3-031-09382-1_20

world, with an estimated prevalence in Europe of 1.5–2.5 children per 1000 live births, being considered the most common cause of childhood physical disability [3–5].

The clinical picture in CP reveals a very wide clinical range and varying degrees of activity limitations on the daily life of youth with CP (e.g., difficulties in normal learning; struggle in their social relationships). Therefore, intervention in CP should promote the autonomy and well-being of youth with CP through the appropriate combination of interventions tailored to the specificity of this conditions [6]. Hence, neurodevelopmental approach should be considered in CP intervention. The goal of this approach is to enhance the children control over the following somatic and mental functions: sensorimotor components, postural control, sensation, perception, memory, attention and executive function [7].

1.2 Rehabilitation

Studies with interventions in CP concluded that effective interventions with children with CP focused on physical interventions (e.g., improve of motor function of hands and feet; overall fitness training), and management of specific health conditions (e.g., spasticity, hip status, and convulsions. Thus, these authors report the importance of context focused therapy and goal-directed training, home-based therapies, and occupational therapies to improve the different capacities of the patients [5].

For supporting this interventions, new approaches have been applied recently to improve the care and rehabilitation of patients with CP. One of this approach is the use of games within the intervention schemes [8–11]. Because their design elements, games can enable constructive, situational, and experiential learning opportunities, enhanced by active game experimentation [12, 13]. For these reasons, games have been applied as a way of promoting the motivations and engagement in participation of patients with CP in their rehabilitation programs [8–11].

1.3 Games

For centuries, games have been used essentially to entertain both children and adults. Altogether, the growth of digital games has led to an increase in interest in their potential to promote some transversal skills [14]. Different studies report which the games can improve different skills, in simultaneous or individual, such as attention, memory, and overall performance in the executive functions [15, 16]. The results of this studies, bring some enthusiasm about the potential of entertainment games as an attractive and fun way to train and improve helpful skills. For this reason, a new category of games emerges, serious games (SG), that can become tools to improve specific skills [17, 18].

The growing interest in SG has created several studies in different contexts, meta-analyses, and systematic reviews. This fact promoted an inconsistency in studies, in application of the different types of the games in interventions. So, some studies took advantage of SG, games designed to particular goal and to promote specific skills, and other based on of commercial-off-the-shelf (COTS) games to improve skills or for rehabilitation purposes (e.g., PlayStation, Wii or Xbox) [19–21]. For this reason, the concept Games utilized seriously (GUS) was proposed for defining "games utilized with serious purposes, regardless of their nature (e.g., entertain vs. serious)" [22, p. 2]. The

COTS/entertainment games are not designed with the purpose serious, for support the acquisition and learning of specific competencies, but when played with serious purpose can fulfil that role. Therefore, the GUS concept stresses the need by focusing on the purpose of the games, and not in their nature (non-entertainment nature vs. entertainment nature) [22].

The features of GUS and its use with a specific application have played an important role in the intervention and in the lives of people with CP. The social participation is important for people with CP because reduces their marginalization and bullying, increases friendships, and allows them to show their peers and skills [23]. Still, social participation in people with CP is reduced, fact directly associated with the CP limitations [24]. One of the promoters of social exclusion is the creation of different environments between children with and without CP [23]. This exclusion causes peers to identify children with CP as fragile and unable to participate in normal activities, thus not having knowledge of the limitations and abilities of their peers, to adapt the rules of the activities [23]. Importantly, the authors revelated in this study which the knowledge of the features allowed obtaining better opportunities for development and a better inclusive environment and developing a network of social contacts.

Thus, the present work aims to present a solution to the problem of social exclusion and development of skills in children with CP. Based on a digital platform and a virtual world, children with or without PC interact with each other through GUS and communication systems. Other features complete the platform with acquisition and data analyses, interaction with secondary users for parental and healthcare control, and engagement strategies.

2 GUS Platform Architecture

In this section we discuss the design of the platform architecture, its different sections, and overview of the current functionality.

2.1 Design Objectives

The service platform design follows several basic principles. First, the platform and associated services should allow access to all the children with CP, having alternatives to the different limitations, and at the same time, still provide interaction between players, with or without CP. Second, this platform with GUS must be enough to support many types of games, with easy access, and every game should be promoting skills development (cognitive, social, and motor). Third, this tool should be a complement and support for therapeutic rehabilitation, giving mobility to some rehabilitation strategies through their implementation in different means (e.g., therapies, home, school). It also should save data for therapeutic analysis, and create other alternative communication between clinical professional, children, and parents. Lastly, the platform should be promoting engagement for rehabilitation and be a good form to give different types of information for users.

The objectives just described are somewhat idealized, and our current prototype does not meet all of them. Nonetheless, our implementation represents a first step toward realizing an GUS platform for social inclusion.

2.2 Logical Platform Architecture

Figure 1 shows the logical relationships of each one of the architecture components.

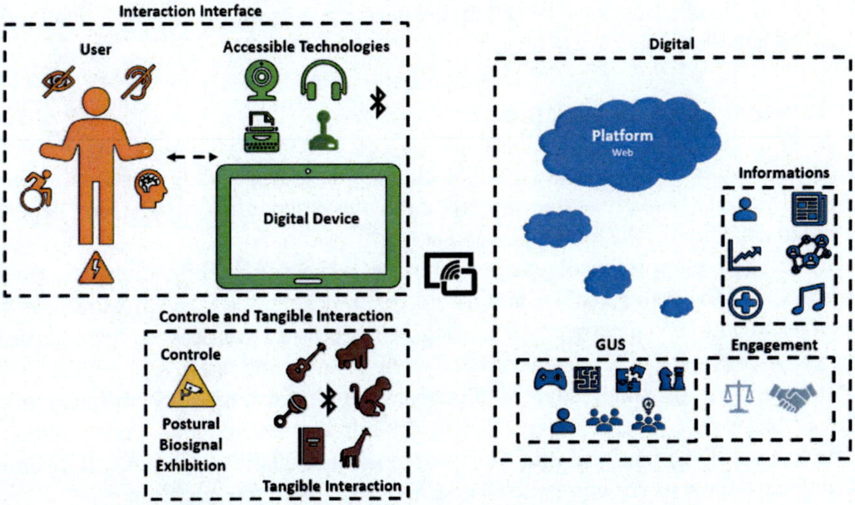

Fig. 1. Logical relationships of each one of the architecture components

Conceptually, the platform may be through a three layers architecture, with: i) user layer, ii) online layer, and iii) control and tangibles elements layers. The first layer represents the interface interaction when player works in a computer/tablet and assistive technologies. In this layer, is important ensure the best form the patient with CP work in the computer/tablet through his compatible assistive technologies. With the contribute of the specialized support (e.g., therapists, rehabilitation engineers) in the implementation, it is intended to guarantee the correct assessment of the limitations and skills of the sick person, to maximize the software potentialities. The online layer encompasses the whole part of the system which represents the digital components as a virtual world, GUS, engagements strategies, and information systems. Through internet, this layer interacts with the other two layers and with the user, and all its software should be considered accessibility guidelines (e.g., WCAG [25]). So, the online platform has a block space for relation between players through different forms, where they can interact using GUS, virtual world in individual or group chat, and internal social media platform. In general, it's intended that the platform complements accessibility, provides GUS and multiple forms of interaction, and a source of information interpretation.

In the end, the use of hardware elements and control strategies, with tangible interaction and sensors, complement the platform allowing to be incorporated component for acquisition information about the player and his interaction with the game, and extend the motor, cognitive and social approach, and increase motivation. With the help of specialized support implementation and taking advantage of internal and external control elements (e.g., timer or biometric sensors), it is intended have clinical information, for analyses or create control of the player integrated in the games. So, tangible interaction with its physical elements intended work motor and sensorial components, important for extend the motor intervention, and work fine and thick motor skills.

In sum, the user works in a computer/tablet and its assistive technologies, if necessary, to access the digital platform integrated in a server. In this, he/she could play the GUS and resort to the other options. Additionally, he/she can profit the tangible elements to extend the intervention.

3 Platform Implementation

Being the CP a neuromotor disorder with a clinical picture very wide range and varying degrees of activity limitations, the success of the implementation this platform depends on a broad capacity for customization and access.

The access to platform involves solid parameter of the compatibilization between the access tool and the web software, and for this, is important the web software consider various personalization parameters. Therefore, it's necessary to assess the characteristics of the possible clients, access solutions and other features and usage scenarios.

Additional, using other resources to increase the platform capacity for intervention makes it possible to extend intervention to the different components and to obtain a better analysis of the platform use. Thus, to implement tangible solutions, acquisition tools and data analysis structure allows new applications for the platform.

3.1 Interface Unit

The interaction interface is the phase of the project that relates the user with the application. This requires he/she has a good knowledge, his/her limitations and ability, forms of accessibility and the environments of use.

Users. As this platform is intended to have several functionalities (e.g., play GUS, clinical considerations, and social interaction), some tips of users are considered. The principal ones are the children with or without CP, and the secondary are parents/caregivers and the healthcare professionals (e.g., therapists, psychologists, others). The children profit the platform for playing GUS, creating social interaction, and accessing to different information. The parents/caregivers have role of controlling and observing the activities on the platform and consulting clinical assessments. In turn, the healthcare professionals regulating access and ensuring the safe use of the platform, the controlling and observation the activities, and clinical appraisal.

In this way, to achieve a more secure utilization of the platform, each child user account must be associated with a caregiver account. In the case of children with CP, is necessary to associate the healthcare professional account, to regulate the parameters of use, to have access to consulting clinical data and extended therapeutic intervention.

Accessibility. The children with CP can have different limitations and capacities that need to be considered. It is essential to orient the design (assistive technologies and accessible software) according to the motor (topographic, motor type, functional classification) and secondary impairments (cognitive, visual, auditive and communicative considerations, and epilepsy). Therefore, is important planning the implementation of assistive technology based on three general considerations: i) individual considerations (e.g., motor or cognitive abilities); ii) environments considerations (e.g., personal and infrastructure involvement); and iii) integrator considerations (e.g., failure mode). For design of the accessible software is important to create alternatives for different limitations, thus following the guidelines of WCAG [25].

Usage Scenarios. The use of the platform out of the GUS is important because components with social and cognitive skills, and motivational factor can be explored. The avatar and virtual world for interaction are present in every platform [25]. This strategy is implemented to promote interaction between clients, to reduce the opportunities to specific groups generation and promote the interaction in a random way. Moreover, the circulation of an avatar in the virtual world allows a longer time for the client, which helps to develop skills in using assistive technology and cognitive skill.

In normal case, the movement in virtual world allows to go to specific points in the virtual world for selecting the options (e.g., GUS spot or settings spot). If for some reason (e.g., preference) the user does not want to go through the virtual world to select the desired options, the platform also allows access to the options by direct access, with specific buttons or commands.

Personalization and Security Parameters. The personalization and security parameters play a fundamental role in the successful implementation of an inclusive solution for CP. To this end, healthcare professionals (e.g., therapist, rehabilitation engineer) have an important responsibility, in definition of these parameters. Thus, for each patient with CP account is also necessary to associate the healthcare professional account, to adapt the parameters, to promote changes of use when pertinent and to have access to consulting and editing clinical data.

The aim of this strategy is to ensure that the platform is as safe and efficient as possible. The regular monitoring of the child by the health professional ensures that the use of the platform does not harm the children health and optimizes their resources and of the platform. Beyond that, the clinical data analyses allow to create a continuous assessment, to inform parents and caregivers regularly, and to adjust the platform to the needs over time.

3.2 Cloud Unit

The cloud unit, where affiliation strategies are promoted, integrates the base elements for users playing GUS, accessing to information and social interaction. It has a wide range of options interesting the clients with different ages, tastes, objectives and needs.

Games Unit. The games in this platform have role to create opportunities for children to enjoy and at the same time work on serious components. To do this, each game to be incorporated must meet some criteria, be accessible to all players so that no one is pre-vented from playing, and be a GUS, i.e., a SG or COTS so that at least one serious component is always explored.

GUS must be played in at least one of the modes, single-player, or multiplayers, in both cases in competitive or cooperative (preferably) form. The GUS are early organized theme game, i.e., sports, action, races, and categorized for tips of serious intervention (motor, cognitive or social), game modes and resources needed to play, i.e., AT or tangible interaction resources.

Furthermore, as accessibility and customisation are key factors for the success of a game for the CP, each game is also programmable so that its features are customisable. Thus, an AI (Artificial Intelligent) system, after selection of the game and before the start, promotes an adaptation of the game variables to the user, through the settings initially selected by the associate healthcare professional and through previous results in that same game.

Interaction/Information Unit. As some studies show, young people with CP consider it is important to have elements of social interaction in their activities [11, 26]. Furthermore, allowing access to different information, such as results obtained, advice from health professionals or other grounds for interest, is also important.

Thus, a unit that allows social interaction and access to different forms of information is considered on this platform. Interaction is achieved in two ways, in-game and out-of-game. In-game, as Hernandez et al. demonstrate [10], keeping the sources of communication (oral or written) opened between players is a good way to promote social interaction between them. On another track, social interaction out-of-game, i.e. in or out the virtual world with individual conversations, is a way to extend communication between players or other agents, outside the game element [26]. In this way, within the game, simple and fast communication is promoted to the players, where the aim is to explore cooperation competence and communication with non-selected players. Outside the game elements, it's intended to promote social interaction in a self-directed way, through individual communication (directly or indirectly navigating in the virtual world) with another player or agent (i.e., therapeutic advice) or in chat groups, where it is intended to explore a communication outside the game context.

As well as interaction with other people, for people with CP, is important obtain other information. As Knibbe et al. demonstrate [26], providing access to information regarding their pathology or areas of interest, is considered as a positive and engagement element to contain in platform for people with CP. To provide access to information of interest, through an initial questionnaire of personal interests, a specific section of the platform scans the news associated with the selected areas, thus allowing easy access to update on topics of interest to the client. To obtain personal information, about results obtained, their evolution, reports and therapeutic feedback, the platform enables access to the data collected throughout the use of the platform and the GUS, such as analyses made by their associated health professional.

Engagement Strategy. To try to ensure maximum player's motivation and interest in that platform, and exploit their skills in the best way, other strategies are applied.

To increase players' engagement with the platform the gamification strategy is applied during its utilization [27]. Thus, outside the use of GUS, game elements are available such as i) immediate feedback; ii) indications regarding individual progress or achievement of goals (through scores, badges, competitions); iii) relational support; iv) social feedback; and v) autonomy.

Inside the use of GUS, a handicapping system makes it possible to balance the different abilities of players to make the game as fair as possible for both [28]. Using an individual handicapping algorithm for each game, each player´s profile is analyzed and if necessary, a handicap, i.e. advantage or disadvantage, is given to one or more players.

3.3 Tangible Unit

To extend the platform scope of intervention, tangible components have been incorporated. This unit have two principal systems, a tangible interaction, so that the user interacts with the GUS through physical objects, and the sensor systems for acquisition of his/her biometrics signals, who control the safe utilization of the platform and obtain important data for clinical analysis.

Tangible Interaction. As CP is a neuromotor disorder, promoting motor development in these individuals is important. Thus, creating solutions that allow clients to have physical interaction with the GUS is an added value for the platform.

In this platform it is considered an additional resource to a tangible interaction system, which through a carpet and pre-identified objects, allows the exploration of motor components in the patients. For GUS programmed for this technology, the carpet connected to the PC reads individual Bluetooth signals from each object. Reading a signal triggers functions programmed into the GUS, which trigger actions in the GUS. Thus, this system enables the promotion of motor and sensory actions in players, allowing its extension to digital components. With this unit, a wider spectrum of motor intervention is achieved where different physical abilities can be developed (i.e.: fine and gross motor skills, small and large movements, different textures). Furthermore, users have a greater variety of solutions, which increases their motivation and interest, and cooperation and interaction components can also be explored [29–31].

Biometric Sensors. Due to the multiple impacts that CP disorder can cause in an individual, measuring different biometric indicators can be important for different parameters. Using biometrics sensors makes it possible to measure tangibly or intangibly user parameters that can become the platform safer and of greater clinical utility [29, 30].

The tangible biometric sensors allow obtaining physical data from the client, which can be profited for direct interaction in GUS or clinical indicators. Through pressure sensors, accelerometers and electromyography, physical components of the player can be acquired and have a direct action on the GUS. His/her control of postural elements is aided and at the same time serves as a controlling element of the game. The motor control of this physical component, i.e.: the good or bad placement of the feet on the ground, associated to their control through the biometric sensors, allows that while GUS is played, this component is considered a game variable, and thus promote change in the game, privileging people health and development. Another example is excessive game use and exposure time to technological devices. In this way, a timer system that identifies and alerts to this problem is considered, so that parents and health professionals can configure and control this factor.

3.4 Data Information

People with CP are usually accompanied by a multidisciplinary team of professionals and carers, that regularly perform considerations and evaluations, in an observational and qualitative analyses. Thus, obtaining and analyzing quantitative data on an individual with CP, allow assisting the assessment of competencies and other indicators.

In specific section of this platform, all the individuals can have access to different information. They can analyze data, through data base associated to different variables, about games (i.e.: results, time played, mode game, time of the responded of the stimuli.), about utilization of the platform (i.e.: time in the virtual world, use or not of the hotkeys, type of the assistive technologies, time and type of the communication with other players, use of the information options), about security and personalization parameters in different times, tangible interactions or biometric indicators (i.e.: postural control, epidermis reactions, ECG or muscle stimulation).

This information is available in two ways, isolated from each data source or across two or three different data sources. In this way users can observe and study the evolution of children, make observations over time, and communicate with the user associated with the same child. In this way, health caregivers can create a historical record, indicate information for other players, promote various changes in the use of the platform or games, and profit this information as unofficial support in clinical evaluations.

4 Conclusion and Future Work

In this work we aim to demonstrate a theorical idea about a type of inclusion strategy for creating different forms for stimulating and developing competencies in children with Cerebral Palsy (CP) using Games Used Seriously (GUS). We described the conceptual development of a tool which promote key development skills (motor, cognitive and social) through serious games, specific adapted for population with CP. Our work is inspired in other research in CP and other disabilities, where was made an intervention with commercial-off-the-shelf (COTS) and serious games, and social inclusion strategies, i.e., games for cognitive development or social networking platforms. Infrastructure simplification is achieved in the platform using GUS which allows servers to be used across multiple customers (with or without CP, and caregivers), in different digital tools.

Being the CP a non-progressive and mutable disorder, intervention in CP can help these people to achieve autonomy. How a clinical picture reveal a very wide clinical range and varying degrees of activity limitations, create the appropriate combinations of interventions tailored to the specific needs, it' is important and can be useful.

Thus, the platform allows individuals to have access to activities where they can talk with other ones, play different games, execute interventions for development keys skills, access personal interest media information and personal data. In addition, it allows caregivers to control this application, to ensure the most appropriate use, to be a source of information on key skills and a way to exchange information between interconnected players. So, the first step is to personalise the platform as much as possible for CP patients and create their network of carers. Second step is using the Cloud Unit, where it is possible to explore the platform and games, to communicate with other customers and to access different interests. Third step is to explore as far as possible, the tangible interaction potential and biometrics data. Lastly, this application can increase knowledge about the development of key competencies, transport intervention strategies to settings outside therapy and increase communication between everyone.

In sum, considering the validated potential of serious games, gamification, correct adaptation and use of assistive technology tools, tangible solutions and intervention in CP, this application tries to offer a way for these strategies to be simultaneously available to children with CP.

In the CP area, continuous tools development, personalization and investigation are fundamental. Thus, in the future, it would be relevant to develop, customize and test various tools to understand their different roles in the intervention context and to identify patron between different population specifications and use of tools or strategies.

Firstly it would be important to develop this solution through small segments and testing it singly in different individuals with CP with distinct characteristics. In this level, it would be important to focus on accessibility, personalization, engagement, and content of the game. These four aspects are fundamental to promote the true and efficient implementation and application of this development strategies.

Secondly it would be also important to adapt these solutions in social environments. The population with CP presents social inclusion difficulties and use this solution with social interaction tool is way to children and adults with CP interact with society in different contexts (i.e.: school or home) and save form.

Lastly, analyzing the data obtained in other two phases permitted to create optimization in different tools, and develop guidelines, for creating other tools for using games with strategies for skills development in CP. Furthermore, created guidelines and open codes for adapted other commercial games (i.e.: Fortnite) for accessing people with CP, it's an important future step for this population.

References

1. Bax, M., Goldstein, M., Rosenbaum, P., Leviton, A., Paneth, N.: Proposed definition and classification of cerebral palsy. Dev. Med. Child Neurol. **47**(8), 571–576 (2005)
2. Odding, E., Roebroeck, M.E., Stam, H.J.: The epidemiology of cerebral palsy: incidence, impairments and risk factors. Disabil. Rehabil. **28**(4), 183–191 (2006)
3. Novak, I., Hines, M., Goldsmith, S., Barclay, R.: Clinical prognostic messages from a systematic review on cerebral palsy. Pediatrics **130**(5), 1285–1312 (2012)
4. Surveillance of Cerebral Palsy in Europe: Prevalence and characteristics of children with cerebral palsy in Europe. Dev. Med. Child Neurol. **44**(9), 633–640 (2002)
5. Novak, I., et al.: A systematic review of interventions for children with cerebral palsy: state of the evidence. Dev. Med. Child Neurol. **55**(10), 885–910 (2013)
6. Rosenbaum, P.: Cerebral palsy: what parents, and doctors want to know. Br. Med. J. **326**(7396), 970–974 (2003)
7. Hoon, A.H.: Neuroimaging in cerebral palsy: patterns of brain dysgenesis and injury. J. Child Neurol. **20**(12), 936–939 (2005)
8. Ravi, D.K., Kumar, N., Singhi, P.: Effectiveness of virtual reality rehabilitation for children and adolescents with cerebral palsy: an updated evidence-based systematic review. Physiotherapy **103**(3), 245–258 (2017)
9. Martins, T., Carvalho, V., Soares, F.: Development of a system for monitoring and tracking of physiotherapeutic movements in patients with neurological diseases. In: IEEE 3rd Portuguese BioEngineering Meeting, 20–22 February 2013, Braga, Portugal (2013)

10. Lawrence, H., Hills, S., Kline, N., Weems, K., Doty, A.: Effectiveness of exercise on functional mobility in adults with cerebral palsy: a systematic review. Physiotherapy Can. **68**(4), 398–407 (2016)
11. Hernandez, H. A., et al.: Design and evaluation of a networked game to support social connection of youth with cerebral palsy. In: Proceedings of the 16th International ACM SIGACCESS Conference on Computers & Accessibility, pp. 161–168. Association for Computing Machinery, New York (2014)
12. Squire, K.D.: Video game–based learning: an emerging paradigm for instruction. Perform. Improve. Q. **21**(2), 7–36 (2013)
13. Girard, C., Ecalle, J., Magnan, A.: Serious games as new educational tools: how effective are they? A meta-analysis of recent studies. J. Comp. Assis. Learn. **29**(3), 207–219 (2013)
14. Green, C.S., Bavelier, D.: Effect of action video games on the spatial distribution of visuospatial attention. J. Exp. Psychol. Hum. Percept. Perform. **32**(6), 1465–1478 (2006)
15. Martins, T., Carvalho, V., Soares, F. Moreira, M.: Serious game as a tool to intellectual disabilities therapy: total challenge. In: 2011 IEEE 1st International Conference on Serious Games and Applications for Health (SeGAH), 16–19 November 2011, Braga, Portugal (2011)
16. Colzato, L.S., van den Wildenberg, W.P., Zmigrod, S., Hommel, B.: Action video gaming and cognitive control: playing first person shooter games is associated with improvement in working memory but not action inhibition. Psychol. Res. **77**(2), 234–239 (2013)
17. Martins, T., Carvalho, V., Soares, F.: An overview on the use of serious games in physical therapy and rehabilitation. In: Cruz-Cunha, M.M. (ed.): Handbook of Research on Serious Games as Educational, Business and Research Tools, Vol. I, Chapter 61. IGI-Global, Hershey (2012)
18. Connolly, T.M., Boyle, E.A., MacArthur, E., Hainey, T., Boyle, J.M.: A Systematic literature review of empirical evidence on computer games and serious games. Comp. Educ. **59**(2), 661–686 (2012)
19. Gros, B.: Digital games in education. J. Res. Technol. Educ. **40**(1), 23–38 (2007)
20. Parong, J., Mayer, R.E., Fiorella, L., MacNamara, A., Homer, B.D., Plass, J.L.: Learning executive function skills by playing focused video games. Contemp. Educ. Psychol. J. **51**, 141–151 (2017)
21. Laver, K.E., George, S., Thomas, S., Deutsch, J.E., Crotty, M.: Virtual reality for stroke rehabilitation. Stroke **43**(2), 20–21 (2012)
22. Lopes, S., et al.: Games used with serious purposes: a systematic review of interventions in patients with cerebral palsy. Front. Psychol. **9**, 1712 (2018)
23. Lindsay, S., McPherson, A.: Experiences of social exclusion and bullying at school among children and youth with cerebral palsy. Disabil. Rehabil. **34**(2), 101–109 (2012)
24. Tan, S., et al.: Factors contributing to the longitudinal development of social participation in individuals with cerebral palsy. Res. Dev. Disabl. **57**, 125–135 (2016)
25. W3C WAI. https://www.w3.org/WAI/standards-guidelines/wcag/. Accessed 27 Apr 2021
26. Knibbe, T.J., McPherson, A.C., Gladstone, B., Biddiss, E.: "It's all about incentive": Social technology as a potential facilitator for self-determined physical activity participation for young people with physical disabilities. Develop. Neurorehabil. **21**(8), 521–530 (2018)
27. Deterding, S., Dixon, D., Khaled, R., Nacke. L.: From game design elements to gamefulness: defining gamification. In: Proceedings of the 15th International Academic MindTrek Conference: Envisioning Future Media Environments, pp. 9–15. Association for Computing Machinery, New York, NY, USA (2011)
28. Rego, P.A., Rocha, R., Faria, B.M., Reis, L.P., Moreira, P.M.: A serious games platform for cognitive rehabilitation with preliminary evaluation. J. Med. Syst. **41**(1), 1–15 (2017)
29. Barros, C., Leão, C.P., Soares, F., Minas, G., Machado, J.: RePhyS: a multidisciplinary experience in remote physiological systems laboratory. Int. J. Online Eng. **9** (Spl.Issue5). 21–24 (2013)

30. Leão, C.P., Soares, F., Rodrigues, H., Seabra, E., Machado, J., Farinha, P., Costa, S.: Web-assisted laboratory for control education: Remote and virtual environments. Commun. Comput. Inf. Sci. **282,** 62–72 (2012)
31. Kalagi, S., Machado, J., Carvalho, V., Soares, F., Matos, D.: Brain computer interface systems using non-invasive electroencephalogram signal: a literature review. In: 2017 International Conference on Engineering, Technology and Innovation: Engineering, Technology and Innovation Management Beyond 2020: New Challenges, New Approaches - ICE/ITMC 2017 – Proceedings, pp. 1578–1583 (2018)

Check for updates

Kinematic Analysis from Solar Panel Tracking System

Andreea Dana Alionte[1], Cristian Canacheu[2], Alexandru Costin Ciobanu[1], Ion Murzac[1], Dana Mirela Văleanu[1], and Liviu Marian Ungureanu[1](✉)

[1] Department of Mechanism and Robot Theory, University Politehnica of Bucharest, Bucharest, Romania
ungureanu.liviu.marian@gmail.com

[2] Department of Mechatronics and Precision Mechanics, University Politehnica of Bucharest, Bucharest, Romania

Abstract. In this article will present you a solar panel tracking system that will be mechanically analyzed. The solar panel can be fixed or mobile using a solar tracker which is an active system that allows the solar panels to move after the sun to capture more sunlight at an angle almost perpendicular to the panel, periodically changing its inclination and axial position. This article presents two different modes of kinematic analysis for the universal joint. The main object of the work is to eliminate the variable angular velocity due to the use of the simple cardan shaft. A technical solution with universal double articulation has been conceived, designed, and analyzed from a kinematic point of view, which, regardless of the 1 or 2 functional motors, will allow the solar system to move at a constant speed. The proposed solar panel solution with the presented adaptable mechanical structure can periodically change its inclination and axial position, regardless of the driven motor, yet the system will move at a constant speed which gives a real advantage from a dynamic point of view.

Keywords: Solar tracker · Mechanical analysis · Solar panel

1 Introduction

The Sun is like a nuclear reactor about 150 million km away from Earth. It emits solar energy that reaches the Earth with a power of 157 TW (157 billion kW) [1]. Though the Sun is an inexhaustible energy source, the total power of the thermal and electrical equipment installed throughout the globe did not exceed 138.2 GW in 2020 [2]. Nowadays, the associated efficiency and costs limit the large-scale expansion of solar energy conversion systems.

The annual amount of solar energy per square meter of horizontal surface varies greatly depending on the weather conditions and astral conditions (latitude, the daytime, and the season) [4]. The latitude influence is due to the curvature of the Earth that made the sunlight touch the ground at a different inclination. Also, the atmosphere thickness is modified and makes the sun radiation to be higher at lower latitudes. Another influence

J. Machado et al. (Eds.): icieng 2022, LNME, pp. 240–249, 2023.
https://doi.org/10.1007/978-3-031-09382-1_21

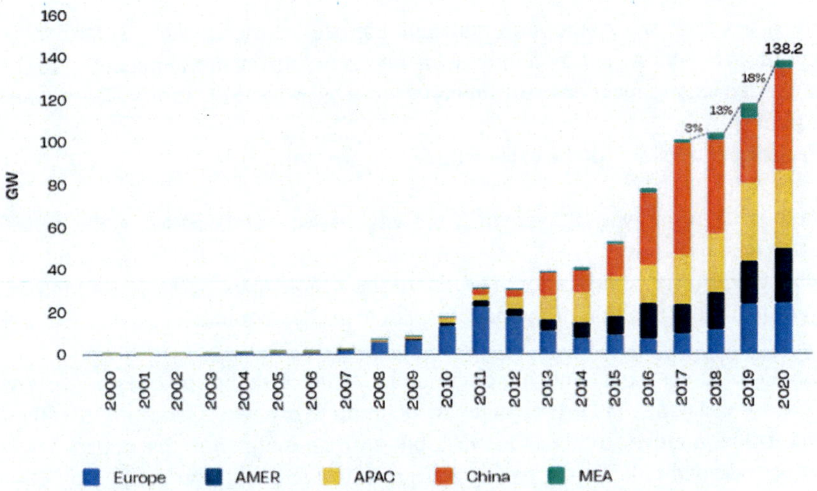

Fig. 1. Figure annual solar PV installed capacity 2000–2020 [2].

related to the sun is the trajectory, that is defined by azimuth angle and elevation angle, and is modified according to astral conditions [5] (Fig. 1).

In the literature, several authors [12–16] present solutions that use equipment that drives solar panels toward the sun, namely solar tracking. For example, in the study [12], it was proposed development a mechanical prototype with a linear actuator and rotator with the move of two-axis solar tracking using photodiodes.

1.1 Solar Panels

A solar panel consists of different modules of photovoltaic cells assembled and mounted in a frame [6]. The photovoltaic cell represents the main part that uses the sunlight as an energy source and generates dc electric current by the photovoltaic effect. Because the alternative current is commonly used for home appliances, an inverter must be used. Excess electricity can be stored separately in a battery. Most of these modules use semi-conductive semi-crystalline silicon (c-Si), polysilicon (poly-Si), or crystalline thin-layer wafers [3] as cells. These solar cells are covered with protective glass (monolayer tempered glass) on the surface exposed to the sun.

The main issue is that these cells have to be protected from potential blows and moisture [7]. Most modules are rigid but semi-flexible in the case of thin-layered ones. The cells are electrically connected in series, relative to each other to reach the required voltage, and then in parallel to increase the amperage, therefore the power of a module is a mathematical product of its voltage and amperage [8].

A photovoltaic panel requires a distribution box that is attached behind the panel, its functions representing the exit interface of the panel at the same time, a digital interface can be used to allow monitoring and sensors [9]. The connection of the modules is serial to achieve the required voltage or in parallel to increase the current, the conductive wires that are used to meet these objectives are sized in accordance with the capacity and may contain silver, copper, or other non-magnetic conductive metal. Some solar panels use concentrators in which light is focused using lenses or mirrors in smaller cells [10]. Also, the frame of a solar panel consists of metal frames to provide the most effective support of the panel.

These solar trackers can be categorized into two types [13]:

- simple, with a one-way movement from east to west and following the rays from sunrise to sunset.
- doubles, with a movement similar to that of the simple ones, and a second movement for adjusting the position of the panels relative to the azimuth.

The double ones have a much higher efficiency because the sun does not follow the same course every day [11]. It changes its position every day. Therefore, double solar trackers follow a move from east to west, but with an inclination that causes the sun's rays to act perpendicular to the panels and increase the energy production by 40–45% [15].

The trajectory of single or double solar panels can have two types of movements[16]:

- active movement orientation using light sensors and trajectory algorithms that adjust the system in real-time according to their position.
- passive/predefined movement according to a trajectory established in advance.

In the following subsection, it will present a tracker with double movement presented in Fig. 2 with two joints (A and B), three motors (1, 2, and 3) a solar panel support (C) and a system support (D).

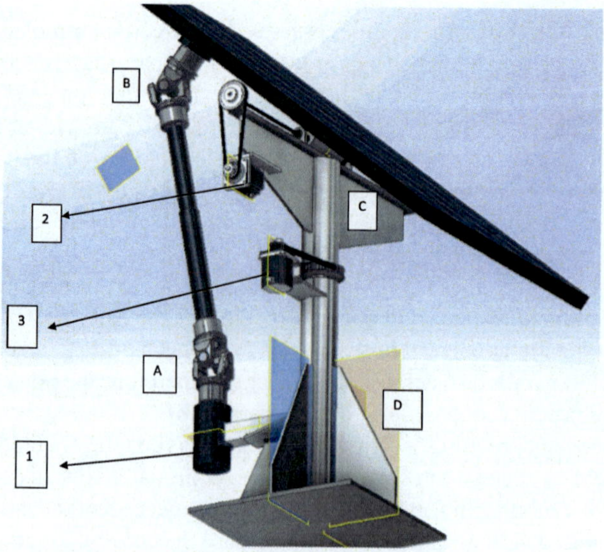

Fig. 2. Positioning mechanism of the solar tracker.

2 Kinematic Analysis

2.1 Analysis of a Joint

The positioning mechanism of the solar panel consists of an ABCD space quadrilateral mechanism with 3 operating possibilities indicated in the figure through the step-by-step 1, 2, 3 m. In the A coupling, it has a universal joint detailed in Fig. 3.

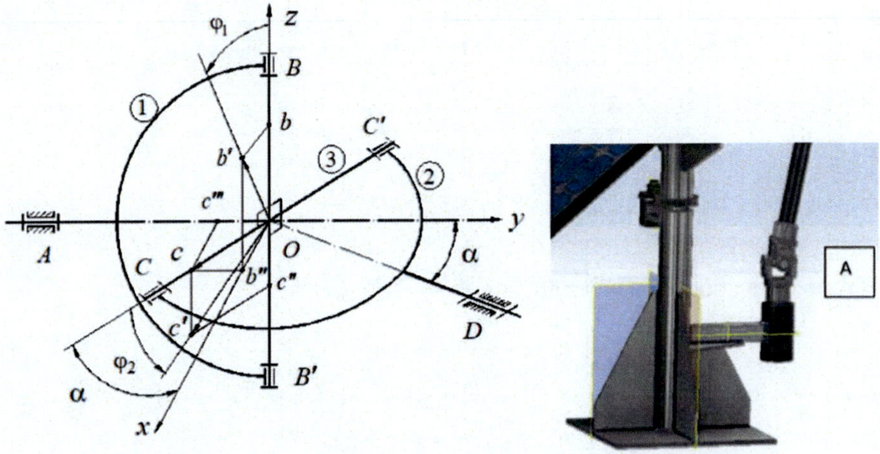

Fig. 3. Universal joint: a) kinematic scheme, b) CAD A joint.

The universal joint can transmit the rotational motion between two shafts whose axes intersect at the *O* point, which it is the origin of the *Oxyz* system of axes and make the α angle between them. The universal joint also bears the name of Cardan cross, due to an element, 3, in the form of a cross with equal arms and perpendicular to each other, element that connects the two forks 1 and 2 of the mechanism. As seen in Fig. 3, the axis of rotation of element 1 overlaps with the *Oy* axis of the coordinate system. The axis of rotation of element 2 is in the *xOy* plane and makes the angle α with the *Oy* axis.

The plane described by the arm of the *BB'* element 3 overlaps with plane *xOz*, and the plane described by the arm of the *CC'* element 3 is plane *cOz*.

If element 1 of the mechanism has a uniform rotational motion around the *Oy* axis, element 2 will have a variable rotational motion around the *OD* axis for a α angle other than zero ($\alpha \neq 0$). Next, the connection relationship between the kinematic parameters of elements 1 and 2 is determined, considering the constant size of the α angle of rotation of the motor 1, imposed ($\alpha = 30°$) which can be determined from the kinematic point of view in two variants depends by plans of movement:

Case 1
For the initial position of the mechanism, the *BB's* arm of the cross is considered superimposed with the *Oz* axis, and the *CC'* arm in the horizontal plane, wound up from the Ox axis with α the angle. This means that fork 1 is for the initial position in the vertical plane, and fork 2 in the horizontal plane (Fig. 3).

Suppose some position of the mechanism in which the vector $\overline{Ob'}$ makes the angle φ_1 with the Oz axis (in the vertical plane xOz), and the vector $\overline{Oc'}$ makes the angle φ_2 with the axis Oc (in the vertical plane cOz). As the vectors $\overline{Ob'}$ and $\overline{Oc'}$ are perpendicular, their inner product is equal to zero, i.e. (Table 1).

$$\overline{Ob'} \cdot \overline{Oc'} = \sin\varphi_1 \cos\varphi_2 \cos\alpha - \cos\varphi_1 \sin\varphi_2 = 0 \tag{1}$$

Table 1. The relation of the $\overline{Ob'}$ and $\overline{Oc'}$ other the three axes.

	$\overline{Ox}$	$\overline{Oy}$	$\overline{Oz}$
$\overline{Ob'}$	$\sin\varphi_1$	0	$\cos\varphi_1$
$\overline{Oc'}$	$\cos\varphi_2 \cdot \cos\alpha$	$-\cos\varphi_2 \cdot \sin\alpha$	$-\sin\varphi_2$

From the relationship (1) results

$$\tan\varphi_2 = \cos\alpha \cdot \tan\varphi_1 \tag{2}$$

or

$$\varphi_2 = \arctan(\cos\alpha \cdot \tan\varphi_1) \tag{3}$$

By derivation of the relation (3) with time, the angular velocity of element 2 is obtained, namely

$$\omega_2 = \frac{\frac{\cos\alpha}{\cos^2\varphi_1}\omega_1}{1 + (\cos\alpha \cdot \tan\varphi_1)^2} = \frac{\omega_1 \cos\alpha}{1 + \left(\cos\alpha \cdot \tan\varphi_1)^2\right]\cos^2\varphi_1} \tag{4}$$

The i_{12} transmission ratio is

$$i_{21} = \frac{\omega_2}{\omega_1} = \frac{\cos\alpha}{\cos^2\varphi_1 + \cos^2\alpha \cdot \sin^2\varphi_1} \tag{5}$$

From the relation (5) it is observed that the ratio of transmission, for a certain angle α, varies between $\cos(\alpha)$ and $1/\cos(\alpha)$.

Figure 4 shows the variation diagram of the transmission ratio i_{21} as a function of the angle φ_1, highlighting the range of variation.

Figure 5 shows the variation diagram of the angular velocity ω_2 as a function of the angle $\varphi 1$. The angular velocity $\omega_1 = 2$ rad/s was considered.

Case 2

The other possibility of positioning is obtained by operating motor 2 (Fig. 6) and it also have two possible variants for kinematic calculation.

For the initial position of the mechanism, the *BB's* arm of the cross is considered superimposed with the *Ox* axis, in horizontal plane, and the *CC'* arm in the vertical

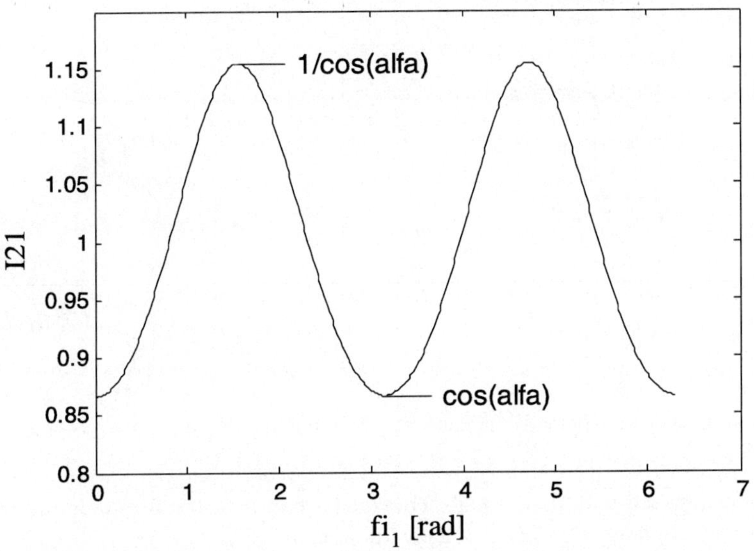

Fig. 4. Variation diagram of the transmission ratio i12; highlighting the range of variation of the transmission ratio

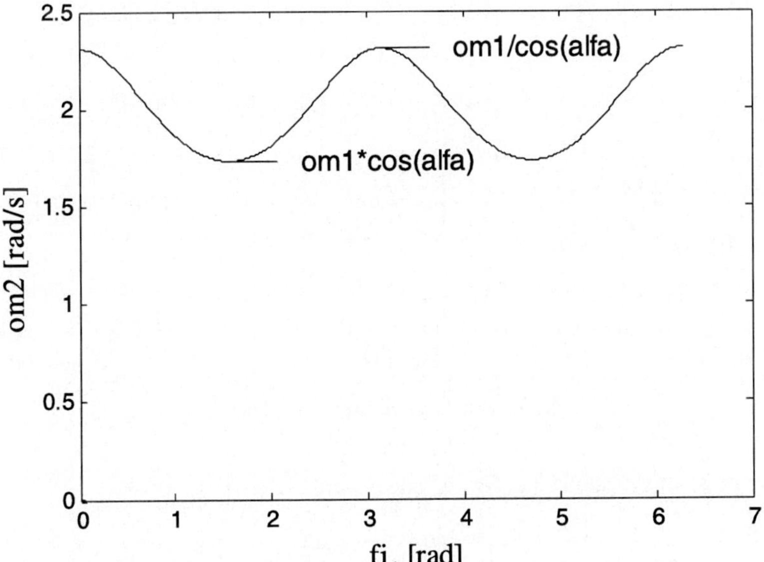

Fig. 5. Angular velocity variation diagram ω_2.

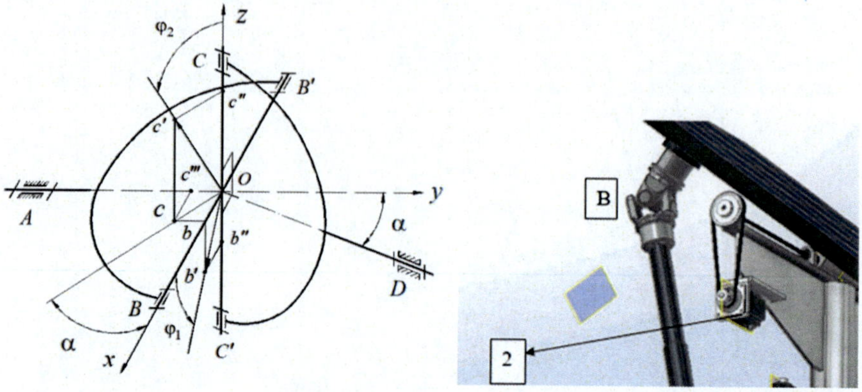

Fig. 6. Kinematic scheme of the universal joint; highlighting of the planes of motion

plane, superimposed with the Oz axis. This means that fork 1 is for the initial position in the horizontal plane, and fork 2 in the vertical plane (Fig. 6).

Suppose some position of the mechanism in which the vector $\overline{Ob'}$ makes the angle φ_1 with the Ox axis (in the vertical plane xOz), and the vector $\overline{Oc'}$ makes the angle φ_2 with the axis Oz (in the vertical plane cOz). As the vectors $\overline{Ob'}$ and $\overline{Oc'}$ are perpendicular, their inner product is equal to zero, i.e.

In Fig. 7 are highlighted the vectors $\overline{Ob'}$, $\overline{Oc'}$, and $\overline{Ob'}$, in the afferent motion planes. The slopes corresponding to these vectors are $\overline{u}_{b'}$, $\overline{u}_c$ and $\overline{u}_{c'}$.

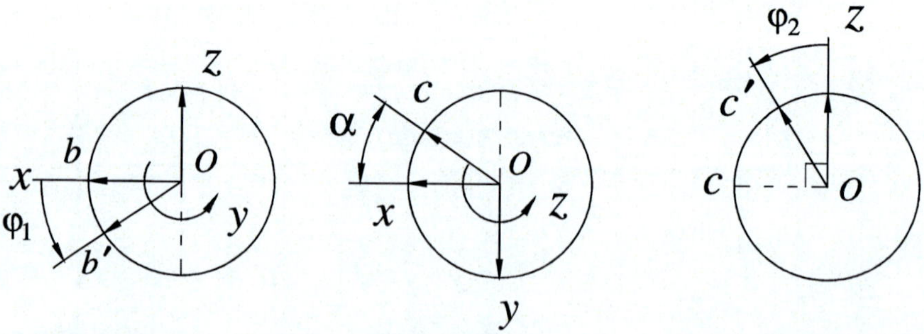

a) Plane xOz b) Plane xOy c) Plane cOc'

Fig. 7. Rotation related to the planes

The expressions of the slopes $\overline{u}_{b'}$, $\overline{u}_c$ and $\overline{u}_{c'}$ are:

$$\overline{u}_{b'} = \overline{i}\cos\varphi_1 - \overline{k}\sin\varphi_1 \tag{6}$$

$$\overline{u}_c = \overline{i}\cos\alpha - \overline{j}\sin\alpha \tag{7}$$

$$\overline{u}_{c'} = \overline{u}_c\cos\varphi_2 + \overline{k}\sin\varphi_2 \tag{8}$$

From the inner product of the slopes $\overline{u}_{b'}$ and $\overline{u}_{c'}$' results

$$\bar{u}_{b'} \cdot \bar{u}_{c'} = \left(\bar{i}\cos\varphi_1 - \bar{k}\sin\varphi_1\right) \cdot \left[\left(\bar{i}\cos\alpha - \bar{j}\sin\alpha\right)\sin\varphi_2 + \bar{k}\cos\varphi_2\right]$$
$$= \cos\varphi_1 \cdot \cos\alpha \cdot \sin\varphi_2 - \sin\varphi_1 \cdot \cos\varphi_2 = 0 \tag{9}$$

in other words

$$\tan\varphi_2 = \frac{\tan\varphi_1}{\cos\alpha} \tag{10}$$

or

$$\varphi_2 = \arctan(\tan\varphi_1 / \cos\alpha) \tag{11}$$

By derivation the relation (11) with time, the angular velocity of element 2 is obtained, namely

$$\omega_2 = \frac{\omega_1 \cdot \cos\alpha}{\cos^2\varphi_1 \cdot \cos^2\alpha + \sin^2\varphi_1} \tag{12}$$

The transmission ratio is

$$i_{21} = \frac{\omega_2}{\omega_1} = \frac{\cos\alpha}{\cos^2\varphi_1 \cdot \cos^2\alpha + \sin^2\varphi_1} \tag{13}$$

2.2 Kinematic Analysis from Solar Panel Tracking System of Movement by Operating Both Motors

The other possibility of positioning is obtained by operating both motors, 1 and 2 (Fig. 8) shows the double universal joint, which consists of two simple universal joints. The peculiarity of this universal double joint is the following: the forks 1 and 5 are in the same plane, the *yOz* plane, and the forks 2 and 2 'are in the horizontal plane *xOy*.

It is considered the simple joint consisting of elements 1, 2 and 3 case 1, also it is considered the simple joint consisting of elements 2', 4 and 5, case 2. It obtained:

$$\tan\varphi_5 = \frac{\tan\varphi_2}{\cos\alpha} \tag{14}$$

In relation (14) it is replaced by the expression in relation (3) and is obtained

$$\tan\varphi_5 = \tan\varphi_1 \tag{15}$$

From relation (9) it is observed that, if it uses two universal joints in series, under the mentioned conditions (forks 1 and 5 in the same plane), it remove that disadvantage obtained in the case of a simple joint (variable angular velocity).

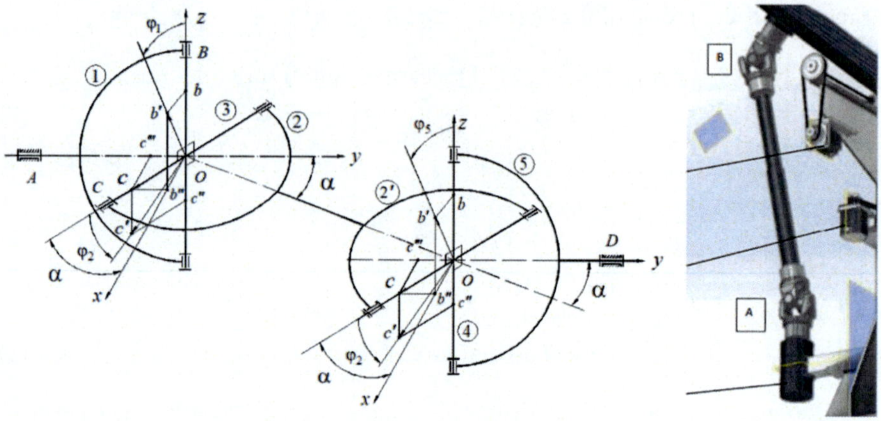

Fig. 8. Kinematic scheme of mechanical system

3 Conclusion

In the paper, it was presented some basic knowledge regarding solar panels and solar trackers. After the bibliographic study, it was proposed a solar racking was analyzed through the mechanical kinematic points of view. The proposed solar panel solution with the presented adaptable mechanical structure can periodically change its inclination and axial position, regardless of the driven motor, the system will move at a constant speed which gives a real advantage from a dynamic point of view. A technical solution with universal double articulation has been conceived, designed, and analyzed from a kinematic point of view, which, regardless of the functional motor, 1 or 2, or simultaneously, will allow the solar system to move at a constant speed $\varphi_5 = \varphi_1$.

References

1. Visa, I., Duta, A.: On the page. https://www.lumeasatului.ro/articole-revista/2248-energia-solara-o-resursa-inepuizabila-pen. Accessed 18 June 2021
2. Solar Power Europe page. https://www.solarpowereurope.org/. Accessed 18 June 2021
3. Einfuhrung in die Material Wissen schaft II page. https://www.tf.uni-kiel.de/matwis/amat/semitech_en/kap_3/backbone/r3_4_1.html. Accessed 14 Jan 2021
4. Jayarama Reddy, P.: Solar Power Generation, CRC Press (2019). ISBN: 9780367381608
5. Deutsche Gesellschaft für Sonnenenergie (DGS): Planning and Installing Photovoltaic Systems. A Guide for Installers, Architects and Motorers, Routledge (2013). ISBN:9780367381608
6. Cotfas, D.T., Cotfas, P.A., Kaplanis, S.: Methods and techniques to determine the dynamic parameters of solar cells: review. Renew. Sustain. Energy Rev. **61**, 213–221 (2016)
7. Charalambous, P.G., Maidment, G.G., Kalogirou, S.A., Yiakoumetti, K.: Photovoltaic thermal (PV/T) collectors: a review. Appl. Thermal Motor. **27**, 275–286 (2007)
8. Awang, J., Rozana, A.: Modified perturb and observe (P&O) with checking algorithm under various solar irradiation. Solar Energy **148**, 128–139 (2017)
9. Mohammed, A.: A way to improve maximum power point (MPP) tracking for PV systems. Department of Electrical Motorering, Aligarh Muslim University (2017)

10. Osamede, A., Ruaan, S., Christo, P.: Simulation and analysis of maximum power point tracking in a standalone PV system: a case study using regression analysis and pulse width modulation, In: International Conference on the Domestic Use of Energy (DUE), IEEE, Cape Town (2017)
11. Oana, O., Babescu, M.: Photovoltaic panels modeling and determination of the maximum power point (MPP). In: ICAS Proceedings, Hunedoara (2019)
12. Haryanti, M., Halim, A., Yusuf, A.: Development of two axis solar tracking using five photodiodes, In: Engineering, Physics, Electrical Power, Electronics, Communicatons, Control and Informatics Seminar (2014)
13. Alexandru, C., Claudi, P.: Simulation of a dual-axis solar tracker for improving the performance of a photovoltaic panel. Proc. Inst. Mech. Eng. Part A J. Power Energy **224**(6), 797–811 (2010)
14. Robles, C., Ospino, A., Naranjo, J.C.: Dual-axis solar tracker for using in photovoltaic systems. Int. J. Renew. Energy Res. **7**(1) (2017)
15. Maharjan, S.: Dual Axis Solar Tracker: Design and Implementation, Project Report, Tribhuvan (2016)
16. Agarwal, R.: Concept of mechanical solar tracking system. IOSR J. Mech. Civil Eng. **1**, 24–27 (2014). (IOSR-JMCE) e-ISSN: 2278-1684, p-ISSN: 2320-334X

Continuous Improvement of Logistics Processes in a Production Line of Electric Buses Chassis

Beatriz Especial Gomes[1], Maria Teresa Pereira[1,2(✉)], Marisa Oliveira[1,2], and Fernanda A. Ferreira[3]

[1] ISEP - School of Engineering, Polytechnic of Porto, 4200-072 Porto, Portugal
mtp@isep.ipp.pt

[2] INEGI - Instituto de Ciência e Inovação em Engenharia Mecânica e Engenharia Industrial, 4200-465 Porto, Portugal

[3] School of Hospitality and Tourism of Polytechnic Institute of Porto, Applied Management Research Unit (UNIAG), Rua D. Sancho I, 981, 4480-876 Vila do Conde, Portugal

Abstract. Nowadays, environmental concerns are deeply rooted in our society. This growing concern is reflected in the strategies implemented by organizations, which are increasingly focused on eco-friendly strategies, not only for its strategic aspect but also for its social responsibility. Therefore, a bus manufacturing company and an automotive company, both located in the north of Portugal, followed this trend by acting in an extremely important area, electric mobility. Since the market is becoming increasingly competitive in these areas it is necessary to ensure the ability of the organization to adapt to its volatility and the efficiency of its processes, in which logistics flows play an important role. It is within the scope of internal logistical flows and processes that this work, developed at the automotive company, in the chassis production line for electric buses, is inserted. The main objective of this project is the continuous improvement of logistics processes, using Lean tools. Firstly, it was necessary to perform a diagnosis and to map the current state of the logistics processes of the electric chassis line, to identify the existing problems and prioritize them in terms of urgency. Based on the Lean philosophy and its inherent tools, such as Standard Work, 5S, Kaizen philosophy and Visual Management, several improvements were presented and implemented, to mitigate the problems that were identified. With the implementation of these improvements in the picking operations, more specifically the picking of the material for the picking boxes, a 44.6% reduction of the time spent on the operation was obtained. Regarding the supply of material to the production line, the duration of this logistic process was reduced by 18.4%.

Keywords: Lean · Picking · Supply · Layout · Warehousing · Electric mobility

1 Introduction

Currently, more than half of the world's population resides in urban areas, which results in numerous challenges regarding urban mobility. These challenges mainly involve air quality due to pollutant gas emissions, high noise levels in urban areas and the clear

J. Machado et al. (Eds.): icieng 2022, LNME, pp. 250–258, 2023.
https://doi.org/10.1007/978-3-031-09382-1_22

dominance of individual transport and the consequent congestion problems caused by the high number of vehicles [1]. Thus, it is possible to state that the transport sector is responsible for a large share of the emission of pollutant gases, such as carbon monoxide. Air pollution becomes a major concern for those who live or work in these urban areas, due to the harmful effects caused by it on human health [2].

The knowledge about the harmful effects of air pollution and noise pollution caused by the transport sector in urbanizations, led to the implementation, in the European Union, of stricter guidelines related to the reduction and limits of pollutant gas emissions. Thus, due to the growing concern with urban mobility, several alternatives focused on electric mobility started to emerge and to be developed, to mitigate the problems and challenges caused by combustion vehicles [3].

One of the organizations that followed this growing trend of electric mobility was the bus manufacturing company. This bus manufacturing company, located in the north of Portugal, was a pioneer in electric mobility worldwide, when it launched the 100% electric airport bus. Later, electric mobility technology was applied to urban transport, with the model named e.CG. This model, both the chassis and the body, were produced at the bus manufacturing company in Gaia. About 4 years ago, the electric chassis line was transferred to the automotive company factory, in Ovar. After this transfer, the material necessary for the manufacture of the chassis was sent, in its totality, by the bus manufacturing company to the automotive company factory, that is, the material was received in the Ovar factory in the right quantities for the production plan.

More recently, about a year ago, material deliveries started to be made directly to the automotive company, leading to the need for several changes regarding the internal logistics of the electric chassis line. Additionally, the production of the second generation of e.CG chassis, the hydrogen buses, began and this second generation has several differences when compared to the first, differences reflected in the material used and quantities.

These changes led to the need to improve and adapt the internal logistics processes of the chassis line, to be able to respond efficiently and effectively to its needs.

This paper focuses on the continuous improvement of the internal logistics processes of the production line. In order to achieve this goal, it was necessary to define which approach to use in order to carry out an adequate analysis of the problem in question. Thus, tools considered relevant for analysis and subsequent decision-making were used, and the Plan-Do-Check-Act (PDCA) management cycle was used as the basis for the research methodology, as well as several Lean tools, such as the Kaizen philosophy, Visual Management, Standard Work and 5S were used to design the improvement opportunities implemented.

2 Literature Review

Currently, and over the last few years, internal logistics has become increasingly important and pertinent, especially for organizations operating in the industrial sector. Intralogistics involves the coordination of all activities that occur within the manufacturing facility, from the receipt of raw materials to the dispatch of the finished product. It is also important to note that it is crucial that internal logistics is coordinated with external logistics. An adequate and flexible management of internal logistics is of utmost importance for organizations that adopt the philosophies inherent to Lean Production [6].

Internal logistics encompasses all material movement and support operations that occur within the organization. It involves several processes, such as storage, stock control, storage systems and handling of material and equipment, as well as all the surrounding information systems [7]. According to [8] the absence of an adequate management of internal logistics processes generates problems in the movement of material and in its supply to the line, which may lead to production stoppages. Hence Lean and Kaizen philosophies and its inherent tools, such as Standard Work, 5S, and Visual Management, etc., can be applied and present several improvements in a short time at an operational level to mitigate the problems that can be identified.

The origins of the Lean philosophy, and consequently the concept of Lean Production, come from the Toyota company, more specifically from the Toyota Production System. The roots of TPS go back to the beginning of the 20th century and its implementation started after the end of the Second World War.

One of the main objectives of TPS is the reduction and, if possible, elimination of waste [9]. According to Taiichi Ohno, all activities should generate value to the final product, and all activities in which this does not happen should be eliminated. In 2001, James P. Womack and Daniel T. Jones defined Lean Manufacturing as the successor to the TPS system, since it is based on the application of the tools first developed by Toyota. In addition to the TPS principles, 5 further principles were also established:

- Define the value of the product from the customer's perspective.
- Map the product value chain.
- Create the flow along the value chain.
- Establish a pull production system.
- Aim for perfection.

The Kaizen philosophy translates into continuous improvement in personal, social, and professional life and is most often applied in the workplace. Its application in an organization refers to continuous improvement with everyone involved, management and employees. This is the place where workers' suggestions are implemented to improve the work process, with a focus on increasing the quality and efficiency of the production line [10]. According to the Kaizen Institute, the Kaizen philosophy is based on principles and tools whose ultimate goal is customer value creation, flow efficiency, *gemba* (shop floor) effectiveness, people involvement and visual management [11].

Standard Work is about the act of identifying, documenting and, in a broad way, improving the performance of processes and tasks. According to Liker and Morgan (2006), it is necessary to stabilize and standardize processes so that it is possible to successfully implement improvement [12]. That is, where standards are not defined, there can be no improvement [13]. This Lean tool is defined as a set of instructions and work procedures in which are mirrored the best methods and sequences for the execution of each task, by each employee. The correct application of this tool leads to high levels of productivity, quality, and safety [13].

Visual Management, or Visual Control, is a display tool whose purpose is to assist employees with visual cues. The use of standard graphics, color systems, markings and space delimitations serve the purpose of facilitating the operator's interpretation of possible problems. One of the clearest benefits of this tool focuses on the fact that it allows managers to have greater control over the process [14].

A warehouse encompasses several characteristics and parameters, such as: the operations performed within the warehouse, the type of warehouse, what resources are available, including human resources, the storage units and the means of handling them, the layout and organization [15].

In relation to the organization and layout of a warehouse, one of the main objectives is almost always to minimize the movements of logistics operators and ensure the flow of materials. Thus, the efficient use of available space, as well as its organization, is crucial for the proper functioning of a warehouse [16].

The available space is one of the most important factors to consider, as it can become a limiting factor when sizing the warehouse. The storage strategy (random or defined locations), the size of the aisles, the material handling and storage systems, the equipment available and necessary for the storage of material are all factors to be considered when designing the layout and organization of a warehouse [17, 18]. It is also important to note that picking operations may represent up to 55% of the costs related to storage, becoming a main target when it comes to continuous improvement, minimizing the time spent on tasks performed and increasing efficiency [19].

The PDCA cycle is a control framework for the execution of a series of continuous improvement activities, applied very frequently in an industrial and production context [4]. Several authors define the PDCA cycle as more than a Lean tool; instead, they defend that this cycle is a philosophy of continuous improvement of processes focused on continuous learning and knowledge creation [5].

3 Case Study Development

Firstly, it was necessary to observe in person the logistics processes of the assembly line so that it was possible to diagnose which problems existed and, after this diagnosis, prioritize them as to their urgency and impact on logistics flows.

3.1 Current Status Mapping

The analysis and description of all the tasks inherent to logistics is fundamental for the mitigation of various problems, as well as for balancing the human resources allocated to it. This analysis led to the detection of various problems to be mitigated. In the table below, Table 1, the problems detected are described as well as the area in which they fall: productivity, quality, safety, ergonomics, or costs.

3.2 Solutions Implemented

After detecting and prioritizing the existing problems, it is necessary to act on them. It is important to consider the resources available for use in this phase, as well as other limitations that may exist.

Table 1. Problems detected

Problem Detected	Productivity	Quality	Safety	Ergonomics	Costs
Picking boxes without photos and quantities and not adapted for 2nd generation chassis	✓	✓			
Lack of location/organization of parts in warehouse (parts on pallets and bulk parts)	✓	✓	✓		✓
Lack of means of supply to the line (dollies and racks) and means of supplying large parts	✓	✓	✓	✓	
No defined layout for pallet racks	✓	✓	✓		
Production line layout hinders logistic supply	✓		✓	✓	
Logistical layout without guaranteeing FIFO		✓			
Start of production of the 2nd generation of chassis: entry of new materials and lack of location for them	✓	✓			

Fig. 1. Picking boxes before and after the implementation of the improvements

Picking boxes without photos and quantities and not adapted for 2nd generation chassis

Through the Bill of Materials (BOM), technical lists, technical drawings, and the production line balancing (the allocation of tasks per workstation) it was possible to remake the picking boxes. This reorganization facilitated the process of picking the material for the boxes since, previously, these, in addition to their non-adaptation to the 2nd generation, only had the material code. In this reorganization (Fig. 1), a photograph of the material was added to the material code, together with the quantity of the material required in that box. After these boxes were designed, they were replicated, so that there is always a set of boxes on the production line, as well as another set of boxes in the warehouse. The before and after the implementation of the picking boxes is represented in the Fig. 1.

Lack of means of supply to the production line and means of supplying large parts

Fig. 2. Dolly created to supply a workstation

The procedure used to make the supply dollies was similar to that of the picking boxes, where the locations of each material and the respective quantity are identified. In order to guarantee product quality when these can be easily damaged, Styrofoam partitions were made to place the refrigerant tubes, in order to avoid possible damage. In Fig. 2 it's possible to see an example of a supply dolly created using visual management tools and 5S's, where the locations of each material and the respective quantity are identified.

Lack of location/organization of parts in warehouse (parts on pallets and bulk parts)/No defined layout for pallet racks/Logistical layout without guaranteeing FIFO

The warehouse is divided into two parts: the racks for the larger pieces and the racks for smaller pieces, both with 9 modules. In a first phase, it was necessary to survey and consequently analyze all the material that is used in the 2nd generation chassis and is stored on pallets. To do this, the BOM and technical lists were used, to analyze the respective quantities used per chassis. Two premises were taken into account when designing the logistics layout: the existence of mixed storage, i.e. fixed storage and random storage; designing the layout based on the ABC analysis, i.e. around 80% of the components applied to the chassis in the first 20% of the warehouse racks. After the implementation of this new solution, the natural procedure would be to identify the racks with the materials and their locations. The solution chosen was to acquire a magnetic strip to place the identifying labels, thus eliminating the whole process of plasticizing and subsequently cutting them out. All modules and respective levels of the warehouse were then identified using an alphanumeric system, where the modules are identified by letters and the levels by numbers. In Fig. 3, represented by the red circle is the alphabetic system used to identify each module; represented by the yellow circle is the numeric system used to identify the levels of each rack.

Production line layout hinders logistic supply

The two major changes made were: the elimination of the pre-assembly cell, which is now located on the line side; the inclusion of a supply aisle. The inclusion of the supply aisle, in addition to facilitating the passage of the logistics operator with the conveyors, reduces the level of risk associated with the tasks of supplying the line, since there is no interference between the logistics operator and the other line employees. Through the supply aisle, the need to supply the material through the aisle where forklifts pass, a task which involved various safety risks, was eliminated. In Fig. 4, on the left side, it´s possible to observe the supply aisle created, whereas on the right side the aisle for the forklift to enter the production line is represented.

Fig. 3. Identification of the materials and their location in the warehouse

Fig. 4. Improvements implemented in the production line layout

4 Conclusions

The design of picking boxes and means of supplying the line suitable for the material were the first tasks to be completed. The study, and consequent implementation, of these solutions became crucial and urgent, due to the unsuitability of the existing boxes and means of supply for the new model produced. The impact of the solutions created for the processes of picking the material and supplying it to the line was very noticeable in the duration of these tasks, with its reduction in the time required being 44.6% and 18.4%, respectively. With the implementation of these solutions, it was possible to obtain an impacting reduction in the duration of the two most time-consuming tasks of the logistics process. When it comes to safety and ergonomic issues related to supplies to the line, with the implementation of new supply means, it was possible to mitigate several risks related to safety and ergonomics. The implementation of the proper boxes for the harnesses resulted in a 19.35% reduction in the duration of the picking operation of this material. Regarding the logistics layout, it was non-existent in the pallet storage area of the warehouse, and its study and implementation became an asset in reducing the duration of tasks related to batch supply. In other words, it was possible to make the task of picking material from the general warehouse a fortnightly task, reducing the impact of its duration on the production Takt Time of a chassis by 50%. Changing the layout of the production line led to the implementation of two very important improvements regarding logistical processes: the inclusion of the supply lane and the safety lanes at the entry points of the stacker onto the line. These improvements have led to a reduction in the level of risk of all tasks associated with supplies to the line as well as in the task of helping to assemble the wheels on the chassis.

Acknowledgments. Fernanda A. Ferreira is grateful to the FCT Portugal for financial support by national funds FCT/MCTES to UNIAG, under Project no. UIDB/04752/2020.

References

1. Ala, G., Castiglia, V., Di Filippo, G., Miceli, R., Romano, P., Viola, F.: From electric mobility to hydrogen mobility: current state and possible future expansions. In: 20th IEEE Mediterranean Eletrotechnical Conference, MELECON 2018 - Proceedings. 2020, pp. 1–6 (2020)
2. Indumathi, R.: Electric vehicles and environmental sustainability. ANWESH Int. J. Manag. Inf. Technol. **5**(2), 5–8 (2020)
3. Mielzarek, J.: Mobility turnaround, air pollution control and secure the budget - electric mobility in the urban area of tension. In: International Conference on Electric Vehicle and Vehicle Engineering, pp. 1–5 (2019)
4. Song, M.H., Fischer, M.: Daily plan-do-check-act (PDCA) cycles with level of development (LOD) 400 objects for foremen. Adv. Eng. Inform. **44**, 101091 (2020)
5. Realyvásquez-Vargas, A., Arredondo-Soto, K.C., Carrillo-Gutiérrez, T., Ravelo, G.: Applying the plan-do-check-act (PDCA) cycle to reduce the defects in the manufacturing industry a case study. Appl. Sci. **8**(11), 1–17 (2018)
6. Ucar, C., Bayrak, T.: Improving in-plant logistics: a case study of a washing machine manufacturing facility. Int. J. Ind. Eng. Theory. Appl. Pract. **22**(2), 195–212 (2015)
7. Burganova, N., Grznar, P., Gregor, M., Mozol, Š: Optimalisation of internal logistics transport time through warehouse management: case study. Transp. Res. Procedia **55**(7), 553–560 (2021)
8. Emmett, S.: Excellence in Warehouse Management: How to Minimise Costs and Maximise Value. John Wiley & Sons, New York (2005)
9. Ohno, T.: Das Toyota-Produktionssystem, 3rd edn. Campus Verlag, Frankfurt/Main (2013)
10. Imai, M.: Kaizen: The key to Japan's Competitive Success. Random House, New York (1986)
11. Institute K. Kaizen Institute: Glossary (2021)
12. Liker, J.K., Morgan, J.M.: The Toyota way in services: the case of lean product development. Acad. Manag. Perspect. **20**(2), 5–20 (2006l)
13. Pereira, A., et al.: Reconfigurable standardized work in a lean company - a case study. Procedia CIRP **52**, 239–244 (2016)
14. Silva, F., Ferreira, L.C.P.: Lean Manufacturing: Implementation, Opportunities and Challenges. Nova Cience Publishers (2019)
15. Rouwenhorst, B., Reuter, B., Stockrahm, V., Houtum, G.J., Mantel R.J., Zijm W.H.M. Warehouse design and control: Framework and literature review. Top. Cogn. Sci. **122**l(122), 515–533 (2000)
16. Rebelo, C.G.S., Pereira, M.T., Silva, F.J.G., Ferreira, L.P., Sá, J.C.: The relevance of space analysis in warehouse management. Procedia Manuf. **55**(C), 471–478 (2021)
17. Martins, R., Pereira, M.T., Ferreira, L.P., Sá, J.C., Silva, F.J.G.: Warehouse operations logistics improvement in a cork stopper factory. Procedia Manuf. **51**, 1723–1729 (2020)
18. Gu, J., Goetschalckx, M., McGinnis, L.F.: Research on warehouse design and performance evaluation: a comprehensive review. Eur. J. Oper. Res. **203**(3), 539–549 (2010)
19. De Koster, R., Le-Duc, T., Roodbergen, K.J.: Design and control of warehouse order picking: a literature review. Eur. J. Oper. Res. **182**, 481–501 (2007)

3D Print Technology Used in High Technical Education

Jan Novotný(✉), Martin Jaskevič, and Tomáš Vysloužil

Faculty of Mechanical Engineering, Jan Evangelista Purkyně University in Ústí nad Labem, Pasteurova 7, 400 96 Ústí nad Labem, Czech Republic
jan.novotny@ujep.cz

Abstract. The paper deals with technical project-based as an alternative form of schooling. Its focus is on problematic of 3D printings, their essence and their practical application. This article treats these topics as a necessity for former technical experts studying at technical faculty at University. It presents the application of 3D printing as a total challenge task. The task was to create own 3D printer and on the printer to create and print in 3D model some actual object. Described technical project is an actual for students of technical studies, but is suitable to build cross disciplinary cooperation. It was realized in cooperation to Nature sciences faculty and with a Historical department of Faculty of Philosophy. This real problematic of finding a real shape and appearance of real historical building, exactly of destroyed church.

Keywords: 3D print · Technical education

1 Introduction

Investigative student activity leads to a very satisfactory result of their work. As was mentioned, students technical project was interconnect to other branches as a History and Geography. Anyway the work and result is a clear technical work with a 3D printer and adequate software. Student project fits into curricular unit of CAD/CAM technology and technology of 3D print, which is realized at our University.

Student skills to already have before start of the project are not necessary to be at advanced level. All missing skills can students learn during project process and teaching lessons.

2 Preparation of a Project

The project was realized on the basis of the invitation for Collaboration of the Faculty of Philosophy, Department of History.

Collaboration consisted in making 3D visualization of old building monuments destroyed during the post-war history.

The students of the Faculty of Mechanical Engineering have been provided with a number of historical photographs and drawings. According to these materials they have been tasked with creating a 3D model of a church printed on a rapid prototyping printer.

J. Machado et al. (Eds.): icieng 2022, LNME, pp. 259–265, 2023.
https://doi.org/10.1007/978-3-031-09382-1_23

2.1 Partial Steps of Implementation

The individual steps of implementation could be characterized as:

- study of available information sources - chronicles, drawings, photographs, discussions with witnesses,
- build up own 3D printer,
- create a drawing based on these materials,
- creating an accurate 2D image with recorded construction dimensions,
- converting to a 3D model,
- print a 3D model,
- arrangement of the exhibition at the place where the church originally stood.

3 Design and Realization of 3D Printer Construction

The first step of the construction was a design of 3D printer created in the modelling program (CAD), in which the dimensions of all necessary components were determined. A 3D model was created with given dimensions of guide rods and other components necessary for the construction. On the basis of these was chosen metallurgical and connecting material and especially electrical components [1].

The printer has been designed to be self-replicating, it means to be capable of printing all plastic parts onto its clone, a replica. The main task is physical construction of CNC machine, more specifically 3D printer. This printer will be controlled by firmware [2].

Figure 1 shows the design, modelling and assembly of the printer. And the following paragraphs describe the connection of each component for printer functionality. The design is intended for the Cartesian coordinate system, with the pad having freedom of movement in the Y-axis direction, an extruder in the X and Z-axis directions. As shown in Fig. 2, the printer can also be built from flat parts. However, due to the availability and price for piece production of printers, It was determined to build the printer on modular aluminium profiles with dimensions 30 × 30 mm [2].

4 Realization of a Project

The implementation of the project consisted of observing the principles described above. Project was implemented in conjunction with four basic rules. Project objectives (drawing up a project plan) consisted of specific ideas about a historical content of a need of historical human memory and its conservation.

Students spoke to the buildings still living contemporary witnesses, ensured historical pictures and photographs (Figs. 3, 4). Than mapped the place where the church stood and found its basement. On this foundation determined its original dimensions of the ground plan [3].

Formulation of the project basis - the goal of the project is to preserve the historical awareness of the buildings of our ancestors. This awareness coupled with our ability to use modern technologies can provide the ability to create a real 3D model that can perfectly document and remind you of the already abolished construction.

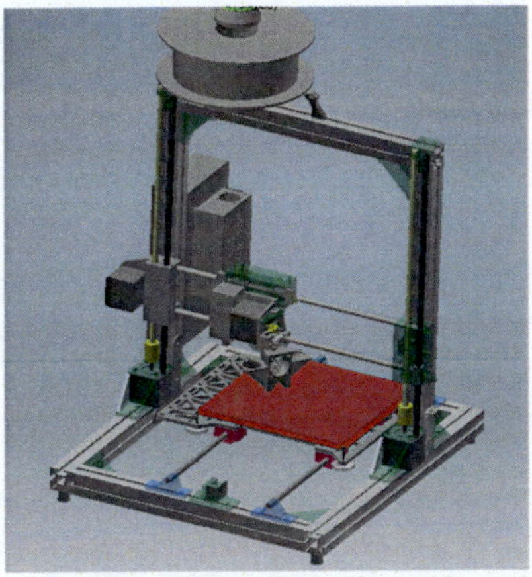

Fig. 1. Custom printer design

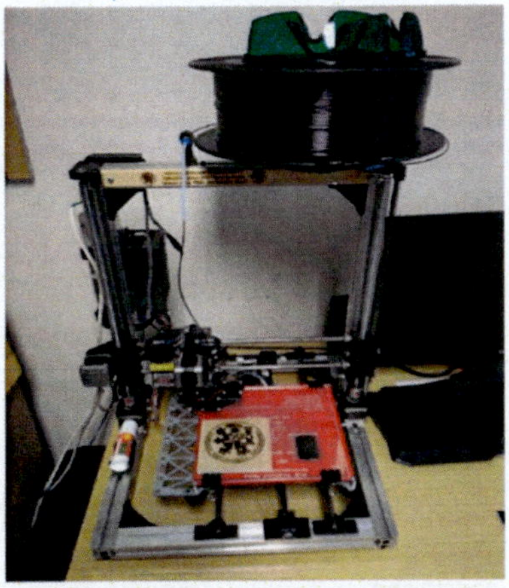

Fig. 2. The resulting functioning, real-life machine

Fig. 3. Preserved photos of the church

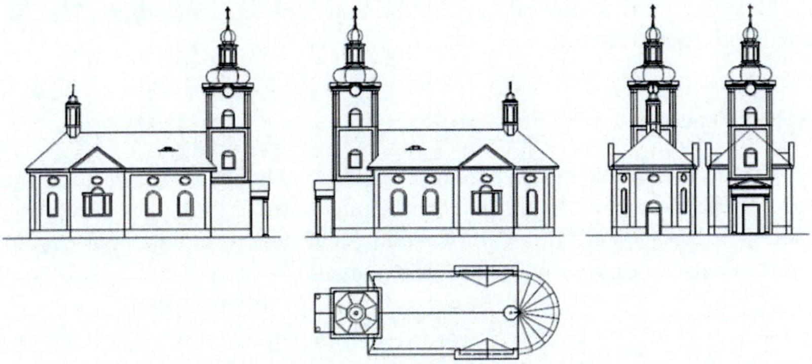

Fig. 4. Basic sketch for creating of technical drawings in correct scale

Processing the plan - the implementation of the project has already been outlined above.

It was a complete construction of a detailed layout of the building. The drawing (Fig. 5) has to be processed, including all the details, and it is created to the exact scale 1:100. In this scale, the resulting 3D model will also be printed.

Based on this information and drawings, the final part of the project was implemented. It means to create the documentation, a 3D model and print it on a fast prototyping printer. Creating a 3D model and its printing is described in detail in the following chapter.

4.1 Creating a 3D Model

All PC modelling was implemented in software Autodesk Inventor in school laboratories at Faculty of Mechanical Engineering of the Jan Evangelista Purkyně University. All 3D models were created by students on the basis of their previous collected documents [4].

The whole construction was built gradually, which ensured team collaboration. Each part of the building was developed and designed by one student (Figs. 5, 6, 7).

Fig. 5. Software model of the church base

The process followed continued by the printing process on a 3D printer. The process is described in the next chapter.

4.2 Printing Process on a 3D Printer

Print runs on a 3D printer using ABS plastic. This material has been chosen as best suited with respect to the number of details and overlapping parts. Printing has also taken into account the possibility of printing complications, and therefore the whole church has been printed in parts as originally divided into software.

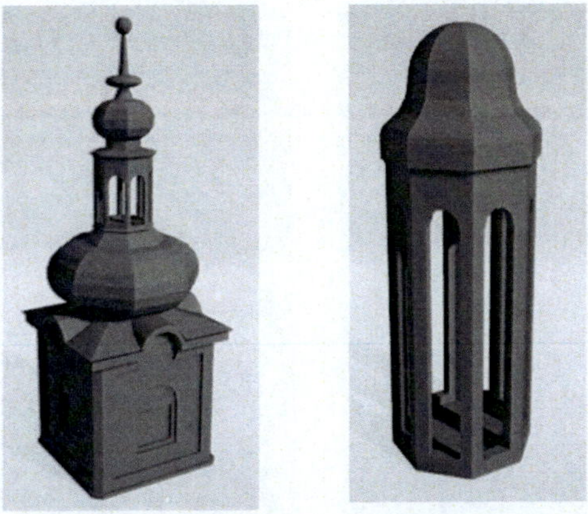

Fig. 6. Models of the church towers

Fig. 7. Software model of the assembled church

In the overall construction of the resulting church model the individual parts of the residual resin were removed. Then they were glued up with adhesive as a whole.

5 Results and Discussion

We can positively evaluate the whole project not only in the frame of the capabilities of the students of the Faculty of Mechanical Engineering. They created the perfect 3D model design and its printing. We can positively assess the overlap in other personal sectors as well. Resulting 3D model is visible at Fig. 8 and Fig. 9.

Multidisciplinary relations are:

- the project didn't concern only manual training, but penetrated through other subjects as well,

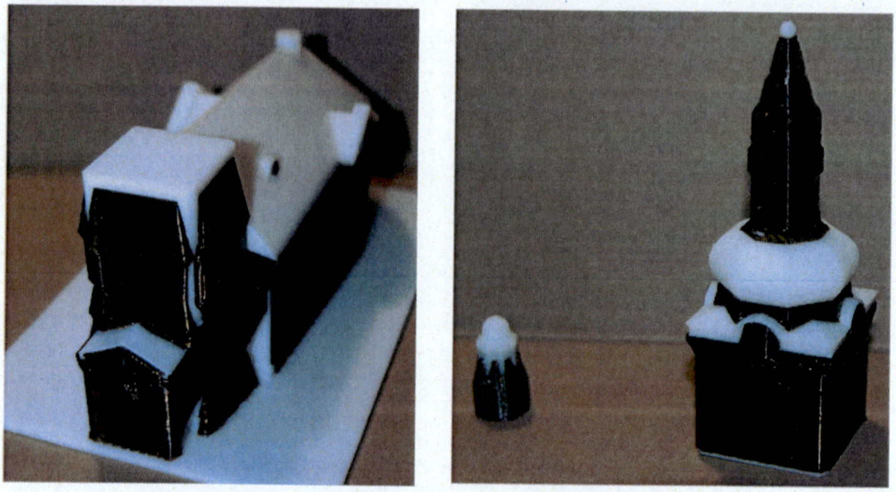

Fig. 8. Printed parts of the model - still with a supporting resin

Fig. 9. The resulting model

- history and geography – the history of buildings, examination of a map,
- sociology – living with German minority in the Czech borderlands,
- arts and crafts - was organized a final exhibition in cooperation with the local cultural centre (Fig. 10).

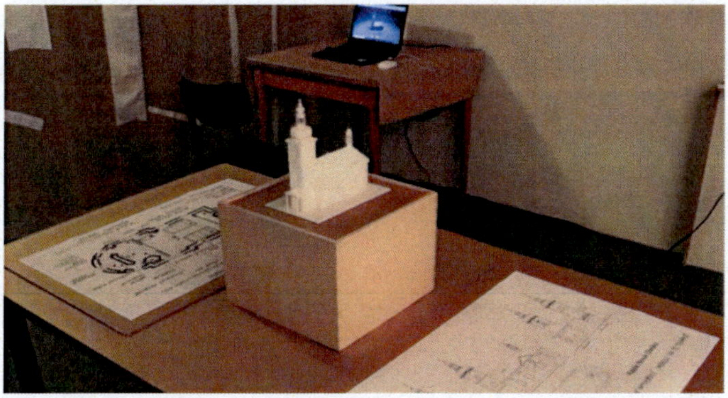

Fig. 10. Final exhibition in cooperation with the local cultural centre

6 Conclusions

Suppose, that introducing of 3D modelling in technical subjects is useful for our students and gives more technical skills and understandings, like:

- technical arrangement, production technology knowledge, which are necessary in working progress assessment, work organisation, material determination, instruments, tools and set out,
- graphic communication understanding, important in simple engineering drawing and working instruction orientation,
- designing knowledge, which are needed accumulate and disassembly work,
- elementary knowledge of information systems – communication technology areas, which are necessary in data obtaining, and particular effect simulation.

References

1. Markers Lab Homepage. https://makerslab.cz. Accessed 10 May 2019
2. Svoboda, P.: Návrh a sestavení 3D tiskárny. Univerzita J. E. Purkyně v Ústí nad Labem, Ústí nad Labem (2018)
3. Novotny, J.: Technical education and conditions of technical education as a means to develop creativity. J. Technol. Inf. Educ. **7**(1), (2015). ISSN:1803-537X-1 (2015)
4. 3D print Encyclopedia Homepage: https://www.3d-tisk.cz. Accessed 1 Jan 2022
5. 3D Printing Homepage: https://all3dp.com. Accessed 1 Jan 2022
6. History of 3D print. Homepage: http://www.o3d.cz. Accessed 10 May 2019
7. Prusa - 3D print Homepage: https://josefprusa.cz. Accessed 28 Mar 2019

Is Critical Thinking a Skill or a Way to Develop Skills? An Overview in Engineering Education

Celina P. Leão[1(✉)] and Ana C. Ferreira[1,2,3]

[1] ALGORITMI Centre, University of Minho, Guimarães, Portugal
cpl@dps.uminho.pt

[2] Mechanical Engineering and Resource Sustainability Center (MEtRICs), University of Minho, Guimarães, Portugal

[3] Centro de Investigação em Organizações, Mercados e Gestão Industrial (COMEGI), Universidade Lusíada Norte, Vila Nova de Famalicão, Portugal

Abstract. One of the most highly valued skills for higher education in Engineering is the development of critical thinking. Critical thinking skills have great importance in cultivating personal and professional competences for the transition to the industrial context. Also, critical thinking achieves a vital importance when students start their professional engineering activity. Engineers are challenged to analyze and think creatively through problems from multiple perspectives in order to reach an appropriate solution. Some studies suggest campus culture and out-of-class experiences may play a role in the improvement of critical thinking. So, this paper aims to identify and to understand how this topic is handled in engineering education, based on scientific literature. Despite the acknowledgement of critical thinking importance in engineering education, there is no practical evidence of its inclusion in the university curricula.

Keywords: Critical thinking · Engineering education · Skills · Competences

1 Introduction

Generally speaking, everyone thinks. The author Tom Chatfield, in his book "How to Think" [1] refers that critical thinking has been known as a discipline in the beginning of the 20th century, in 1910. However, critical thinking can be considered or compared to que method of questioning of Socrates, known as Socratic questioning and as the best known critical thinking teaching approach [2]. As mentioned in [2], critical thinking has been diversely defined, from traditional definitions to more contemporary, according to the area of study that can require different approaches of critical thinking.

According to Elder (2007) [2], and using her own definition "*Critical thinking is the intellectually disciplined process of actively and skillfully conceptualizing, applying, analyzing, synthesizing, and/or evaluating information gathered from, or generated by, observation, experience, reflection, reasoning, or communication, as a guide to belief and action*". This comprehensive definition shows that critical thinking is an important

J. Machado et al. (Eds.): icieng 2022, LNME, pp. 266–278, 2023.
https://doi.org/10.1007/978-3-031-09382-1_24

skill for knowledge development at higher education level, across a range of disciplines and scientific domains [3–7].

The need to include critical thinking in engineering studies has drawn the attention of academics and professionals who claim for restructuring of the curricula of engineering schools and also recommend different pedagogical postures to be followed by those who have the task of transmitting knowledge in this field [3, 6].

Current studies support that engineering education should have a global focus on the application technology and pedagogical tools [8]. Students must be prepared to retain information and produce knowledge through the synthesis of their analyses. These requirements must be sufficient to cope with complex problems and scenarios, where creativity must be stimulated. In addition to technological tools, the engineer must take into account when thinking critically, his individual experiences, cognitive intuition, logical reasoning and the ability to ask the right questions to reach the appropriate and expected result [9].

Taking as a starting point the research question that gives title to the paper:

"Is Critical thinking a skill or a way to develop skills?"

it will help to gain a more sensitivity and knowledge concerning how critical thinking is applied, what type of critical thinking approaches are used, and if critical thinking itself is a skill or it involves several skills in its use, if yes, identify which are the most representative and associated to critical thinking. To meet these aspects, two sub research questions raised: SRQ1: Are critical thinking skills included in Engineering education as part of its curricula?; and SRQ2: How critical thinking importance is perceived in Engineering when compared with other scientific fields?

This paper is divided into four sections. Following the first section, introduction, Sect. 2 presents the methodology. In Sect. 3, findings and discussion are presented. Final remarks and suggestions for future research are depicted in the last section.

2 Methodology

A Systematic Literature Review (SLR) was led based in published studies, in the Scopus database related to critical thinking and how it is used in the area of engineering education [10]. The SLR comprises a set of three main searches done sequentially, in order to refine the search based on exclusion and inclusion criteria, Fig. 1.

The selected documents were analyzed through a descriptive analysis, allowing a ´big picture´. And, based on a selected set of documents, a critical analysis was made. Also, where applicable, for viewing bibliometric maps, helping the analysis and discussion, the free software VOSviewer was used [11, 12], for the construction of the co-authorship and co-citation networks maps.

3 Findings and Discussion

In the following section, the findings are presented and discussed. The analysis is conducted in order to understand the links between ideas discussed within the literature documents, reflect and appraise arguments.

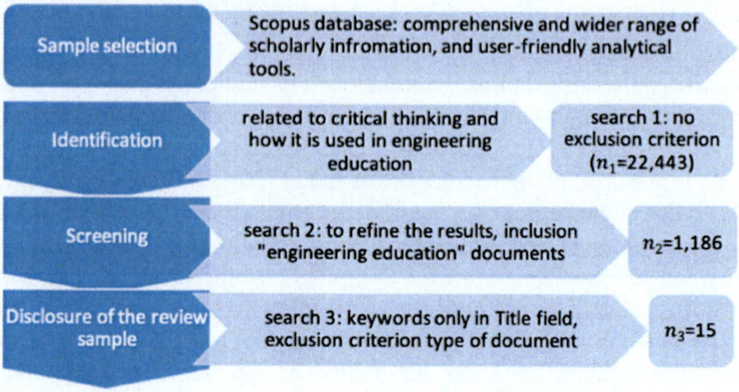

Fig. 1. Steps of documents' sample selection used in the SLR.

3.1 Search 1

The first search (Search 1) considered the set of the two words: "critical thinking", used in the Article title, Abstract and Keywords, with no restriction in the time frame, allowing to verify how this topic has evolved over the years, and in the type of documents published. The search, conducted on February 1st 2022, yielded 22,443 documents results. Figure 2 illustrate the evolutions of the total published documents by year across 1928 and 2021. In this figure the documents published in 2022 where not considered (152 documents). The majority of the documents were articles type (63.4%), followed by conference paper (15.4%) and by Review and Book Chapter with 6.4% and 6.3%, respectively (Fig. 3). Regarding the subject area, Social Sciences with the highest percentage (37%), followed by Medicine with 8% (Fig. 4). Sixthly, the Engineering subject area with 6% of documents.

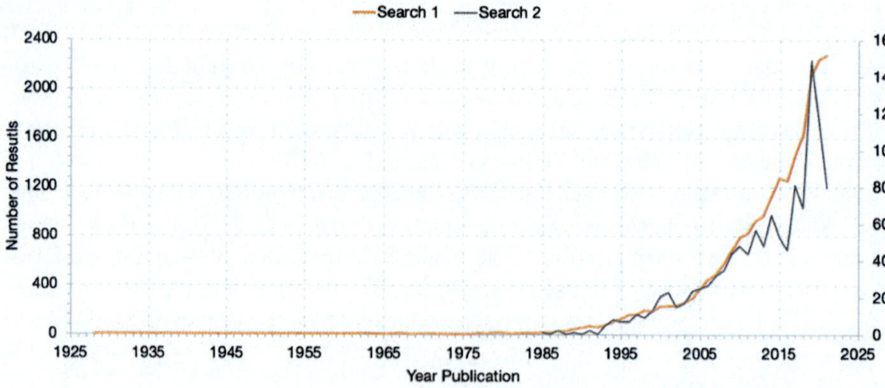

Fig. 2. Evolution of the number of published documents by year (from 1928 to 2021).

Being Engineering the study area that this article analyzes the application and development of critical thinking, a second search was done considering only this area, refining also the search process.

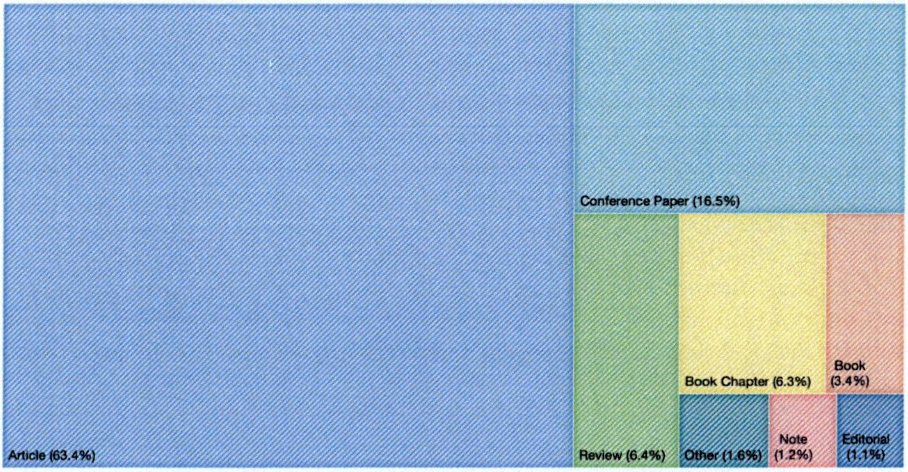

Fig. 3. Published documents by type (from 1928 to 2021).

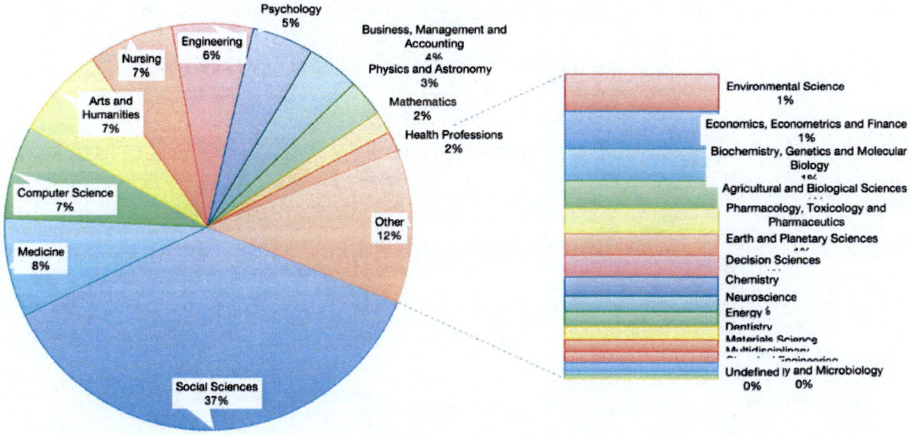

Fig. 4. Published documents by subject area (from 1928 to 2021).

3.2 Search 2

Refining the first search, Search 2, was performed considering ("Critical Thinking" AND "Engineering Education"), all other conditions remaining equal. This search yielded to a total of 1,186 documents results. The documents published in 2022 where not considered (3 documents). Figure 2 illustrate the evolutions of the total published documents by year across 1985 and 2021 (values on secondary axes). Also presents a trend of exponential growth nevertheless the ups and downs. The first document was published in 1985, and 2019, corresponds to the year with the highest number (149 documents). Regarding the type of documents, articles represent 17.5% being the majority conference paper (79.5%), showing a different behavior regarding the type of documents obtained in Search 1. Regarding the subject area, though Engineering was one of the keywords used in the search, Social Sciences represents the second documents' subject area with 23.7%, followed by Computer Science with 15.9%. Engineering as subject area embodies 42.2% of the documents. The remaining subject areas are Physics and Astronomy (3,9%), Decision Sciences (2.6%), Business, Management and Accounting (2.1%), Mathematics (2.0%), and Other (6.2%) including Arts and Humanities, Chemical Engineering, Energy among other areas.

Noteworthy that the spectrum of authors' countries has broadened, in recent years, not only European countries (like Norway, Netherlands) but also South America countries (like Columbia) and Asia countries (like Indonesia), start publishing information related this topic (Fig. 5), being the USA the central with the highest total link strength.

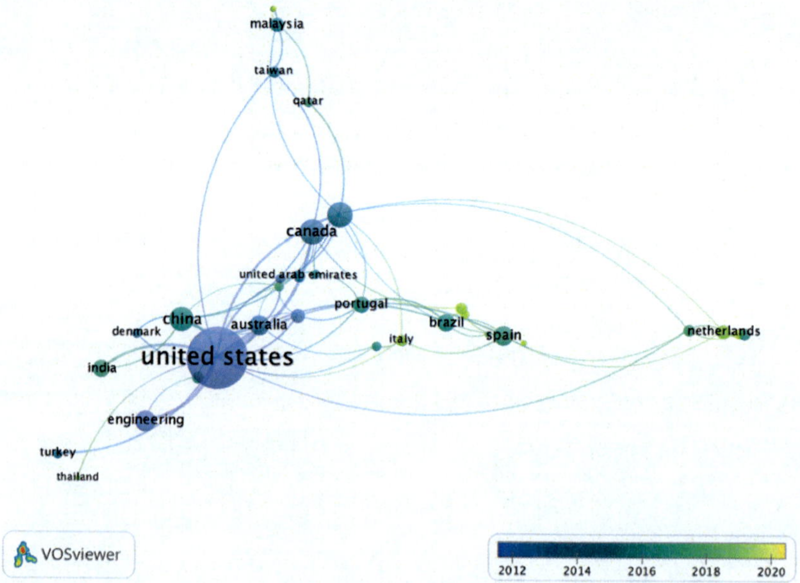

Fig. 5. Overlay visualization for authors' countries over the years considering documents' total links strength.

Regarding how the selection of keywords, by average occurrence, evolved over the publication year, "stem", "project-based learning", "educational innovation" started to being used, together with "active learning", "engineering education", "critical thinking" and "creativity" (Fig. 6). Regarding "Critical thinking", it can be said that is the earliest keyword, and then became associate to "engineering education".

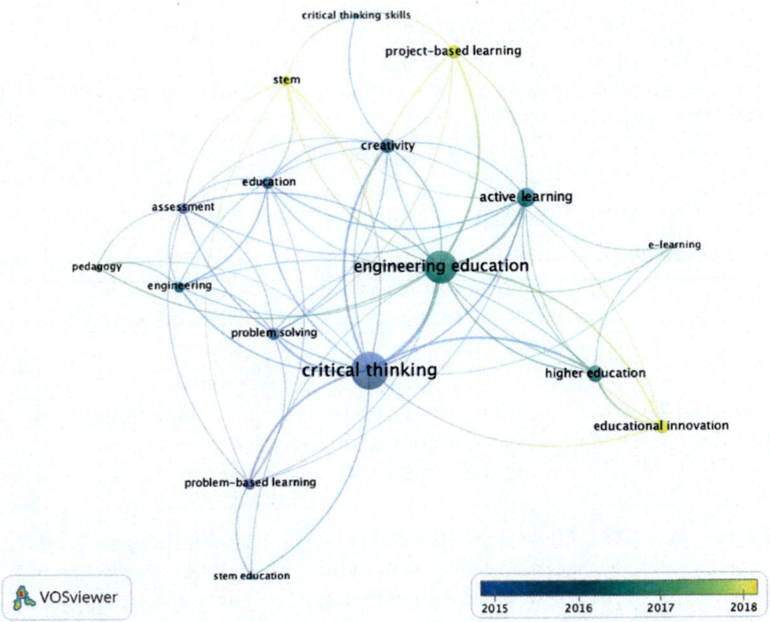

Fig. 6. Overlay visualization of keywords over the years.

3.3 Search 3

The next stage and in order to better specify the study and the number of published documents to perform a content analysis, a new search was made considering the same set of words used in Search 2, however only in the "Article title" field. It was authors believe that considered these two set of words in the title could point out the intrinsic relationship and weighting between them.

"Critical Thinking" AND "Engineering Education" search in the title filed led to 16 documents. From these documents, one was excluded since it was career advisor document, and the remaining 15 considered for analysis: 9 were published in conferences proceedings and 6 as articles, ranged between 2006 and 2020 (Table 1).

Based on the authors' keywords, a word cloud was constructed, Fig. 7. Notice that the "thinking", "education" and "engineering" are the most frequent words as would be expected. However, words like "judgment", "learning", "reflexivity", "questioning", "skills", "assessment", are also be selected.

Papadopoulos *et al.* [4] developed an assessment tool that considers examples in mechanics education where students fail in applying basic principles of mechanics. By doing this, students are encouraged to think critically, grounding their problem-solving skills in the mechanics knowledge. Also, authors emphasize the need of a mutually engaged relationship between the two actors: student and teacher, where teacher keeps in challenging the student to understand the subject matter critically.

Table 1. Studies included in chronological order.

No	Author(s) and publication year	Title	Keywords	Type
[4]	Papadopoulos *et al.* (2006)	Assessing **Critical Thinking** in Mechanics in **Engineering Education**	---	CP
[13]	Ahern *et al.* (2012)	**Critical thinking** in the university curriculum – the impact on **engineering education**	Creativity · curriculum design · science education · university education · critical thinking	A
[7]	Claris and Riley (2012)	Situation critical: critical theory and **critical thinking** in **engineering education**	power/knowledge · critical pedagogy · reflexivity · reflective judgment	A
[14]	Pierce *et al.* (2013)	Environments for fostering effective **critical thinking** in geotechnical **engineering education** (Geo-EFFECTs)	active learning · critical thinking · deep learning · engineering judgement · estimation · geotechnical education · reflection	A
[6]	Chouseinoglou and Bilgen (2014)	Introducing **critical thinking** to software **engineering education**	Software engineering education · critical thinking · practicum · SQ4R	CP
[15]	Rocke *et al.* (2014)	Use of Argument Maps to Promote **Critical Thinking** in **Engineering Education**	---	CP
[16]	Adair and Jaeger (2015)	Cultivation of **critical thinking** in undergraduate **engineering education**	critical thinking · undergraduate	CP
[17]	Osman *et al.* (2016)	Identifying Pertinent Elements of **Critical Thinking** and Mathematical Thinking Used in Civil Engineering Practice in Relation to **Engineering Education**	Critical Thinking · Engineering Education · Mathematical Thinking · Modified Grounded Theory	A

(*continued*)

Table 1. (*continued*)

No	Author(s) and publication year	Title	Keywords	Type
[18]	Luksanasakul and Chanthong (2017)	Questioning techniques promote the **critical thinking** in **engineering education**: Case study in microcontroller course of electrical engineering	Questioning Technique · Critical Thinking · Instructional Design	CP
[19]	Endut *et al.* (2018)	Reformulation of **critical thinking** in the Malaysian tertiary **engineering education**: An Islamic approach	Critical thinking · Engineering education · Islamic education	A
[5]	Sirinterlikci and Mativo (2019)	**Critical Thinking** in Manufacturing **Engineering Education**	---	CP
[20]	Ahern *et al.* (2019)	A literature review of **critical thinking** in **engineering education**	Critical thinking skills · dispositions · engineering · intervention · assessment	A
[21]	Baghdadchi *et al.* (2019)	Creating Environments for **Critical Thinking**: Building Upon Multiple Choice Problems in Electrical **Engineering Education**	---	CP
[22]	Ismail *et al.* (2019)	The essential of **engineering education** involving **critical thinking** and problems solving skills among	---	CP
[23]	Castro and Sancristobal (2020)	From Technology Enhanced Learning to Ethics and **Critical Thinking** as part of the **Engineering Education**: Skill Driven with Humanities Comprehension Editorial	---	CP

CP: Conference Paper; A: Article.

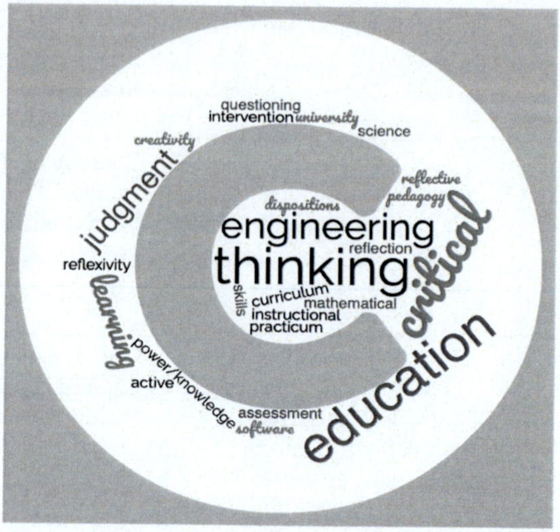

Fig. 7. Word cloud created based on the authors' keywords of the 15 documents in analysis.

Ahern *et al.* [13] based on in-depth, semi-structured interviews with academics involved in teaching and learning in different fields with the objective to understand how the different areas define critical thinking and how it is taught in their courses, and how the activities developed, based on student's work led to the development of a model of critical thinking. Thirteen academics from different fields, from Chemistry to Law, including Mechanical Engineering, were selected for the study. As results, critical thinking definition across fields of knowledge are similar, however they identified differences in the formulation and how the definitions have been reached, in line with Elder work [2]. Authors concluded that in engineering, there is a need to understand better critical thinking and that can be learn from the humanities, law and architecture.

Claris and Riley [7] defended in their study that there has been an increasing focus on developing critical thinking within engineering curriculum. They have implemented an integrative approach in engineering education, considering multiple curricula contexts. They have concluded that critical thinking is imperative in the learning process, unifying the technical, personal and social skills.

Pierce *et al.* [14] showed that in geotechnical engineering courses, innovative strategies seek to improve teaching and learning process, and that most students develop critical thinking if they can relate or visualize the problems. Active learning provides skills in discipline-specific engineering problems.

Chouseinoglou and Bilgen [6] developed a model for assessing organizational learning skills in software development. The Survey, Question, Read, Recite, Review and wRite (SQ4R) technique is proposed as a critical thinking method, with positive results where students, independently of their personality types and learning styles, performed successfully their tasks with knowledge and judgment capabilities vitals to a software engineering.

Rocke *et al.* [15] develop an interesting study through the use of argument maps to promote critical thinking in Engineering curricula. They considered critical thinking as a skill that is difficult to cultivate and to assess. Based on the use of the argument map in a course in electromagnetic field theory (EFT) within an undergraduate Electrical and Computer Engineering programme. Firstly, authors introduced a module with 4 h, introducing critical thinking topic, with practical activities. The results show that the using graphic organizers helps to promote critical discourse in a collaborative setting. Also, authors address critical thinking as a competence/skill that enables students to analyze data and information and to evaluate proposed engineering designs.

Adair and Jaeger [16] emphasize the importance of the quality of both students' and engineers' thinking, since how they think determines the quality of what they design, produce or make. So, authors explore several methods to develop critical thinking in engineering education, for both students and academics. For students, an initial course about the definition of critical thinking, why it is important to an engineer has such a skill and how to incorporate this skill into everyday engineering practice. For academics, seminar/discussion group discussing the integration of critical thinking into their respective engineering modules.

Osman *et al.* [17], as in [13], use conducted semi-structured interviews and writing memos to obtain data. The difference was regarding the participants in the study. In this study, 8 engineers from consultancy firms participated, aiming to identify pertinent elements of critical thinking and mathematical thinking that can be used in real-world engineering practice. The information gathered helps, from one side, students to understand the importance of thinking skills in solving engineering problem with regard to engineering design. And from the other side, the engineering educators strengthen the need of using more closely representation of real-world engineering practice. From the literature it is difficult to understand if critical thinking is, in fact, a skill for itself or a way/path to develop critical thinking.

Sawhney and Anda [3] developed a model, comprising six steps in three different levels of the Critical thinking process applied to industrial engineers. The steps are (1) define, (2) recognize, (3) identify, (4) visualize, (5) execute and (6) sustain. These are divided by: (*i*) the first phase – the ability to define the right problem to solve; (*ii*) second phase – the ability to solve the problem; and (*iii*) third level – the ability to sustain the solution. Despite it is not directly stated, the study implies that the critical thinking is a path to develop skills, mostly in the transition to the industry because it improves decision-making, helps engineers to form their own opinion and develops their personal ethics and contributes to improve quality and efficiency.

Lukasnasakul and Chanthong [18] by a process of questioning, students developed thinking including critical thinking skills. Authors emphasizes that since "*questioning practices in classrooms have been the same for hundred years*". So, a questioning technique was developed to promote critical thinking based on the instructional model and questioning pattern. The students of electrical engineering course in the topic of microcontroller performed a mini project. At the end, students express a positive attitude towards questioning, claiming that with this technique they review prior knowledge, create new knowledge, develop understanding, develop thinking skill, motivate them to higher thinking.

Endut *et al.* [19] adopts and reformulate the conceptual approach of critical thinking at higher learning based on Islamic education. Authors based their approach on the fact that in engineering education values of humanities in the curriculum are needed to complement the technical competencies.

Sirinterlikci and Mativo [5] with students of a Manufacturing Engineering course, authors developed several activities, integrated in a project-based approach, with the application of several tools that allows student to acquire competences on how to gather information, organize and present it, that is, develop analysis, synthesis, and evaluation skills.

Baghdadchi *et al.* [21] presents a study based on the impact on student learning when integrating several tasks, like take-home tests, peer instruction, pre-lecture videos, and guided quizzes, in a lower division analog circuits course of an Electrical and Computer Engineering programme. The majority of students refers the positive impact that the development of these activities has in developing a better understanding of the course topics. Also, by the comparative analysis of students' exams grades, strong evidence of the implementation of these tasks through the semester, have supported the development of problem-solving and critical thinking skills among students.

Ismail *et al.* [22] emphasizes that soft skills are related to critical thinking and to the ability in solving problems. Authors defend that there is a relationship between creativity and critical thinking, essential for professional endeavors.

Castro and Sancristobal [23] as guest editorial article, emphasize an important concern by including ethics and critical thinking as the core part of engineering education: "*.. our ethical and sustainable vision of engineering education must continue to penetrate inside our practitioner and academic communities, to reinforce our actions and commitment to the global society.*".

Ahern *et al.* [20] with a literature review work of critical thinking in engineering education, somehow, underline that critical thinking to be embedded in university curricula, "*there needs to be well-funded research programmes that allow different methods to be developed and trialed over extended periods in higher education engineering programmes*".

In sum, critical thinking must always supplement logical and rational thinking for problem solving and decision making. Collaborative/cooperative learning should be encouraged, based on cognitive flexibility. Thus, critical thinking is seen both as a skill and a path to Engineering students to develop skills and competencies that they will have to put into practice in the transition to their professional activity.

4 Final Remarks

Taking as a starting point the research question that gives title to the paper:

"Is Critical thinking a skill or a way to develop skills?",

and based on a SLR, it was possible to identify that critical thinking can be both considered as a skill or as an approach that need to incorporate skills to be applied.

Most of the works are unanimous on the importance that the exercises/examples need to became as close as possible to the real-world engineering problems where students develop ability in solving problems. To do so, critical thinking must be considered as a continuous process and students must be encouraged continuously. However, and as pointed out by Ahern *et al.* [13], there is a need of clarification on the definition of critical thinking in engineering, and that can be learn from the humanities, and health sciences on the basis of long-term use and experience. With the change of teaching/learning methodologies based on integrated project [18, 21], efforts have been made to include critical thinking in Engineering Curricula, answering to question "Are critical thinking skills included in Engineering education as part of its curricula?" (SQR1).

Considering the large number of documents obtained in search 1, ranged between 1928 and 2021, where 37% of documents comes from social sciences area. Engineering represents 6% of the total however with an exponential growth on the use or consideration of critical thinking in their practices. This evidence answers to the question "How critical thinking importance is perceived in Engineering when compared with other scientific fields?" (SRQ2).

Acknowledgments. The authors acknowledge support from FCT, through the project UIDB/00319/2020.

References

1. Chatfield, T.: How to Think: Your Essential Guide to Clear Critical Thought. SAGE Publications Ltd, London (2021)
2. Elder, L.: Defining critical thinking. https://www.criticalthinking.org/pages/defining-critical-thinking/766
3. Sawhney, R., de Anda, E.M.: Industrial engineers articulate their critical thinking for problem solving via the D.R.I.V.E.S. model. In: Engineering Leaders World Congress on Engineering Education 2013. Hamad bin Khalifa University Press (HBKU Press) (2014)
4. Papadopoulos, C., Rahman, A., Bostwick, J.: Assessing Critical Thinking in Mechanics in Engineering Education. American Society for Engineering Education (2006)
5. Sirinterlikci, A., Mativo, J.: Critical thinking in manufacturing engineering education. In: 2019 ASEE Annual Conference & Exposition Proceedings. ASEE Conferences (2019)
6. Chouseinoglou, O., Bilgen, S.: Introducing critical thinking to software engineering education. In: Lee, R. (ed.) Software Engineering Research, Management and Applications. Studies in Computational Intelligence, vol. 496, pp. 183–195. Springer, Heidelberg (2014). https://doi.org/10.1007/978-3-319-00948-3_12
7. Claris, L., Riley, D.: Situation critical: critical theory and critical thinking in engineering education. Eng. Stud. **4**, 101–120 (2012)
8. Sola, E., Hoekstra, R., Fiore, S., McCauley, P.: An investigation of the state of creativity and critical thinking in engineering undergraduates. Creat. Educ. **08**, 1495–1522 (2017)
9. Luksanasakul, A., Chanthong, A.: Questioning techniques promote the critical thinking in engineering education: case study in microcontroller course of electrical engineering. In: 2017 IEEE Global Engineering Education Conference (EDUCON), pp. 1054–1057. IEEE (2017)
10. Briner, R.B., Denyer, D.: Systematic review and evidence synthesis as a practice and scholarship tool. In: Handbook of Evidence-Based Management: Companies, Classrooms and Research. Oxford University Press, New York (2012)
11. van Eck, N.J., Waltman, L.: Manual for VOSviewer (2020)
12. van Eck, N.J., Waltman, L.: Software survey: VOSviewer, a computer program for bibliometric mapping. Scientometrics **84**, 523–538 (2010)
13. Ahern, A., O'Connor, T., McRuairc, G., McNamara, M., O'Donnell, D.: Critical thinking in the university curriculum - the impact on engineering education. Eur. J. Eng. Educ. **37**, 125–132 (2012)

14. Pierce, C.E., Gassman, S.L., Huffman, J.T.: Environments for fostering effective critical thinking in geotechnical engineering education (Geo-EFFECTs). Eur. J. Eng. Educ. **38**, 281–299 (2013)
15. Rocke, S., Radix, C., Persad, J., Ringis, D.: Use of argument maps to promote critical thinking in engineering education. In: Proceedings of the 2014 IEEE Frontiers in Education Conference (FIE), pp. 1–4. IEEE (2014)
16. Adair, D., Jaeger, M.: Cultivation of critical thinking in undergraduate engineering education. In: 43rd Annual SEFI Conference, pp. 1–8. , Orléans, France (2015)
17. Osman, S., Abu, M.S., Mohammad, S., Mokhtar, M.: Identifying pertinent elements of critical thinking and mathematical thinking used in civil engineering practice in relation to engineering education. Qual. Rep. **21**, 212–227 (2016)
18. Luksanasakul, A., Chanthong, A.: Questioning techniques promote the critical thinking in engineering education: case study in microcontroller course of electrical engineering. In: IEEE Global Engineering Education Conference (EDUCON), pp. 1054–1057 (2017)
19. Endut, M.N.A.A., Abdullah, M.R.T.L., Yaacob, R.A.I.R.: Reformulation of critical thinking in the Malaysian tertiary engineering education: an Islamic approach. J. Soc. Sci. Humanit. **26**, 121–136 (2018)
20. Ahern, A., Dominguez, C., McNally, C., O'Sullivan, J.J., Pedrosa, D.: A literature review of critical thinking in engineering education. Stud. High. Educ. **44**, 816–828 (2019)
21. Baghdadchi, S., Nemerever, Z., Hadjipieris, P., Serslev, S., Sandoval, C.: Creating Environments for Critical Thinking: Building Upon Multiple Choice Problems in Electrical Engineering Education, pp. 1–18. American Society for Engineering Education (2019)
22. Ismail, W.O.A.S.W., Hamzah, N., Fatah, I.Y.A., Muhammad, A.K.: The essential of engineering education involving critical thinking and problems solving skills among mechanical engineer employees. IOP Conf. Ser. Mater. Sci. Eng. **697** (2019)
23. Castro, M., Sancristobal, E.: From Technology enhanced learning to ethics and critical thinking as part of the engineering education: skill driven with humanities comprehension editorial. Int. J. Eng. Pedagog. **10**, 4 (2020)

Solving Manufacturing Orders Scheduling Problem Using Annealing Simulation

Kamil Musiał, Artem Balashov, Anna Burduk(✉),
and Dominika Rysińska-Wojtasik

Faculty of Mechanical Engineering, Wroclaw University of Science and Technology, 27 Wybrzeże Wyspiańskiego street, 50-370 Wrocław, Poland
anna.burduk@pwr.edu.pl

Abstract. For many years, the decision-making process has been based on a human factor and the foreman's know-how, what was error prone. Nowadays production processes are more and more complex. The company that wants to be competitive in the market must take under consideration many factors and basing on human factor only is not efficient at all. The paper presents the production scheduling optimization problem in the furniture company. Until now, company engineers used to schedule production based on customers' orders and their know-how. The problem was that customers make orders at various times with fixed deadlines, and for the company, it is profitable to group customers' orders together. Wise order scheduling makes a significant difference in total production time and cost. In the paper, the use of a greedy algorithm, Simulated Annealing, and Tabu Search algorithms to verify the current method of production process scheduling and improve the process has been proposed.

Keywords: Production process · Scheduling · Tabu search · Simulated annealing · Greedy algorithm · Heuristic methods · Intelligent methods in manufacturing · Sustainable technology

1 Introduction

Nowadays, manufacturing companies are required to develop good quality products that are both at a competitive price and delivered on time. To keep up and be competitive with other producers, companies need to find new organization methods, for example, intelligent solutions [1, 2]. The research was carried-out on the example of a large manufacturing company that grows rapidly last years. The management of the company after investigation noticed decline in competitiveness comparing with their competitors. Additionally, new orders are in the process of being tendered and company decided to deploy smart solutions to make production more effective. In 3 years the company wants to deploy industry 4.0 solutions starting with automation of at least 3 production processes using smart scheduling.

The paper examines the case where production engineers create production schedules based on customers' orders. Process of decision making about production schedule based

J. Machado et al. (Eds.): icieng 2022, LNME, pp. 279–292, 2023.
https://doi.org/10.1007/978-3-031-09382-1_25

only on human conclusions always means the possibility of making mistakes. This is one of the aspects that does not appear in computer methods that analyze only the given data. The study aimed to verify if the methods of scheduling used in a company can be improved by using intelligent methods and releasing human resources. There was a need to compare various aspects of the production process and orders in different options. The greedy algorithm, Simulated Annealing, and Tabu Search was used.

2 Research Methodology

The research was performed in a company producing components for the furniture industry. Because of the components' character, most of the operations performed in the company are cutting, grinding, and painting.

The company producing furniture offers 42 types of products realizes orders for customers. The production consists of 3 different manufacturing cells, and the specification assumes that each product can be produced in at least 2 of them. (if any of the manufacturing cells is disabled, all the products can be produced. In practice, there are two manufacturing cells (MC1 and MC2) for various products, and 3rd, the most advanced manufacturing cell (MC3) is capable of proceeding with all the products. Manufacturing cells MC1, MC2, and MC3 are consecutively equipped in 8, 9, and 14 workstations. Each product has production time per piece, extra time (to start and finish production process) and profit per piece. Company makes customers' orders. Each order has:

- product (1 of 42 offered products),
- date of placing the order,
- volume: number of pieces of the order,
- deadline – given by company,
- weekly penalty if delayed.

Each order contains only one product and the maximum number of pieces is realized within 2 days according to the product parameters. If the customer orders more pieces, the order is automatically divided into 2 orders. Each manufacturing cell has:

- Matrix of retooling time between products;
- Matrix of retooling cost between products.

The production process is about producing orders in 3 manufacturing cells. The manufacturing cell must be retooled between the production process of different products, which takes specific time and cost. Orders can be grouped by products and be produced together. In such a situation, there is no retooling and "extra time" is counted just once.

The company works with week plans and orders with various priorities can be placed anytime. It means that sometimes it is more profitable to wait with a specific product to reach other orders of the same product and group them. To estimate the probability of receiving the order of a specific product, "the table of the popularity of products" is used. On the other hand, orders received may have high priority, which causes its realization nearest week and – very often – additional retooling. The second common scenario is about waiting and grouping process and receiving additional urgent orders that mast be finished first what causes a penalty for delay of other orders of these urgent orders are not accepted what money lost is.

The company works in 2-shift mode in 5 days a week. If necessary, there is a possibility of running a weekend shift that is more expensive for the company. In the following example human resources are not taken under consideration.

Based on accepted orders, the planning department elaborates the production plan of each shift for the following week and when the plan is done should not be changed.

Company planning department tries to calculate and estimate the following variables:

- how to schedule (group) the orders to avoid unnecessary retooling,
- how big time buffer should be kept for additional urgent orders,
- to run weekend shift to speed up or not,
- how many orders to accept to maximize the usage of manufacturing cells and not to pay the penalty.

2.1 Case Study

Formally, this task belongs to the class of dynamic job shop scheduling problem (DJSSP); however, inherently, this task is somewhere between DJSSP and the one-dimensional packing problem of the finite number of figures into an infinite strip. The workload of each workstation can be represented as a container with a single width and an infinite length (Fig. 1).

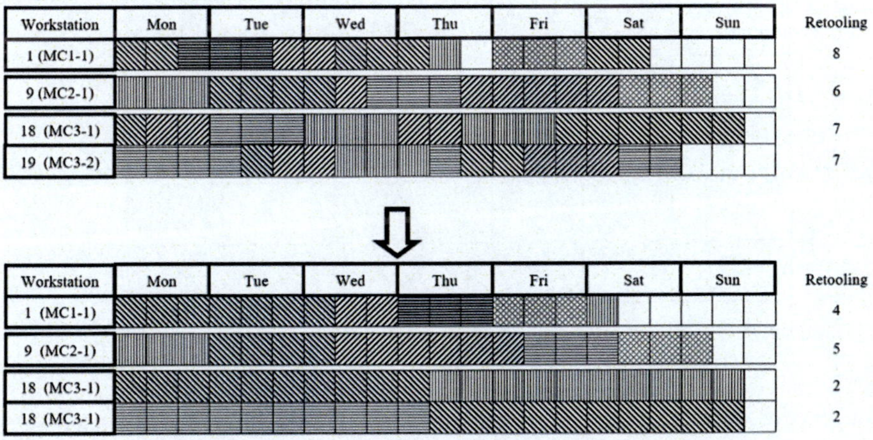

Fig. 1. Simplified graphical representation of the algorithms' operation

Figure 2 shows the workload schedule of several workstations. For the compactness representation, tasks production schedules are presented in shifts. In reality, of course, this is not the case, and several retooling for the manufacture of different products may occur in one shift. In addition, the figure graphically displays the current solution of algorithm X, which is—the schedule of task production for each workstation per each product. The objective function of this case is set profit maximization $f(x) \rightarrow max$.

Algorithm characteristics: products from the set $P = \{1,\ldots, p_i,\ldots, 42\}$ are manufactured on workstation $WS = \{1,\ldots, WS_i,\ldots, 31\}$. Workstation corresponds to Manufacturing Cells as MC1 = [WS1—WS8], MC2 = [WS9—WS17], MC3 = [WS18-WS31], each of which is capable of producing the next set of products: MC1 = [p_1-p_{18}], MC2 = [p_{19}-p_{42}], MC3 = [p_1-p_{42}]—these are the only limitations of the algorithm. In addition, the algorithm does not have binary criteria for feasibility—the algorithm in any case gives a result.

Each order consists of $o = [p, V, d_{out}]$, where p is the product number [1–42], V is the volume (quantity of the product to produce), d_{out} is the estimated date of order fulfillment, that is, the deadline after which the profit from will decrease correspondingly to penalties and delays in realization, din is the date of order acceptance, is not taken into account during optimization. The pool of orders will be denoted by $O = \{o_1,\ldots, o_i,\ldots, o_n\}$. Since these are not optimized parameters, but constants necessary for the correct operation of the algorithm, a vector of time limits $D = \{d_1,\ldots, d_i,\ldots, d_n\}$ is compiled, where n is the current total number of orders.

Two more constants of the algorithm are conversion matrices. The matrices consist of the time and retooling costs for each product correspondingly, shown in Fig. 2.

Product	**1**	**2**	**...**	***i***	**...**	**42**
1	**0**	Rf_{2_1}	**...**	Rf_{i_1}	**...**	Rf_{42_1}
2	Rt_{2_1}	**0**	**...**	Rf_{i_2}	**...**	Rf_{42_2}
...	**...**	**...**	**0**	**...**	**...**	**...**
i	Rt_{i_1}	Rt_{i_2}	**...**	**0**	**...**	Rf_{42_i}
...	**...**	**...**	**...**	**...**	**0**	**...**
42				Rt_{42_i}	**...**	**0**

Fig. 2. Matrices of time and funds of retooling.

Since we assume that the retooling workstation time from product p_i to product p_j is the same as it is from p_j to p_i, the matrix is diagonally symmetric. Since we assume that the cost of retooling workstation from product p_i to product p_j is the same as it is from p_j to p_i, the matrix is diagonally symmetric as well. Therefore, for display simplification, the figure shows both matrices in one table: the upper right area is the Rf_{i_j} cost matrix, the lower left is the Rt_{i_j} time matrix. Dimensions of both matrices are 42×42.

3 Optimization Methods

Literature review and previous research of the authors showed that Tabu Search algorithm is very efficient in production scheduling problem. Authors research also showed that Simulating Annealing - because of its specific parameters can propose better solution then Tabu Search. Greedy - as a very simple algorithm may turn out to be efficient enough while guaranteeing fast implementation and calculation time.

3.1 Greedy Algorithm

The greedy algorithm solves optimization problems by taking locally optimal decisions. It makes the optimal choice at each stage without assessing its effect on further steps [1, 3] – it seems to be the best at the analyzed moment. The search procedures used to find optimal solutions are adopted to generate alternative moves. With the use of all the locally optimal solutions, the algorithm can approximate a global optimal solution in a relatively short time [4]. The greedy algorithm is often used to minimize or maximize the target value. The main advantages of this type algorithms include fast action, moderate memory requirements and easy implementation [3, 5].

In the context of this article, a greedy algorithm is used to construct the initial solution of X0.

3.2 Neighborhood's Alteration and Heuristics

Introduce operations on permutations, with which neighborhoods of solutions are constructed. Selecting of neighborhoods strongly affects algorithms construction. The complexity of one iteration, the total number of iterations, and ultimately, the precision of the resulting solution significantly depends on it. We will use the following opera-tions on neighborhood alteration:

Substitution (*i, j*) – places the task of manufacturing product i in the position of task manufacturing product j. Given the one-dimensionality of the space, the schedule of tasks before the replaced (either replaced) product remains unchanged. Only the tasks that are in the schedule later than the replacement are shifted. Changeover from the process i to j takes the same time and resources as the changeover from the process j to i (Fig. 3).

Shift with displacement *(i, j)* —places the task of manufacturing product *i* in the position of task manufacturing product *j*, while all the tasks in between, including *j*, move further along the schedule. When a task is shifted from the schedule of one workstation to another, it is evident that this operation will create an uneven distribution of tasks in the schedules of different workstations. So after the Shift, task *k*, which appeared to be the last in the schedule of workstation *j* (where the task was added), was displaced to the initial position of task *i*. This operation will allow grouping tasks within the schedule of each workstation in order to avoid retooling (Fig. 4).

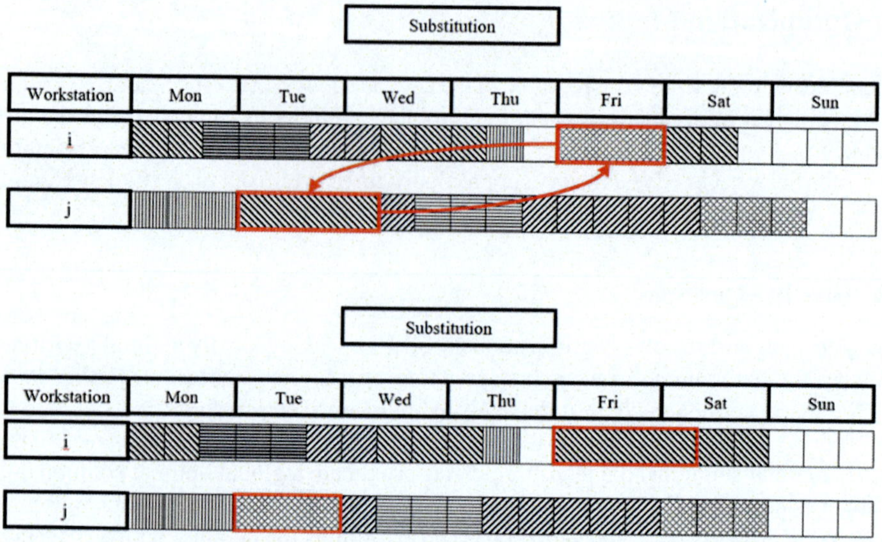

Fig. 3. Visual representation of Substitution neighborhood permutation

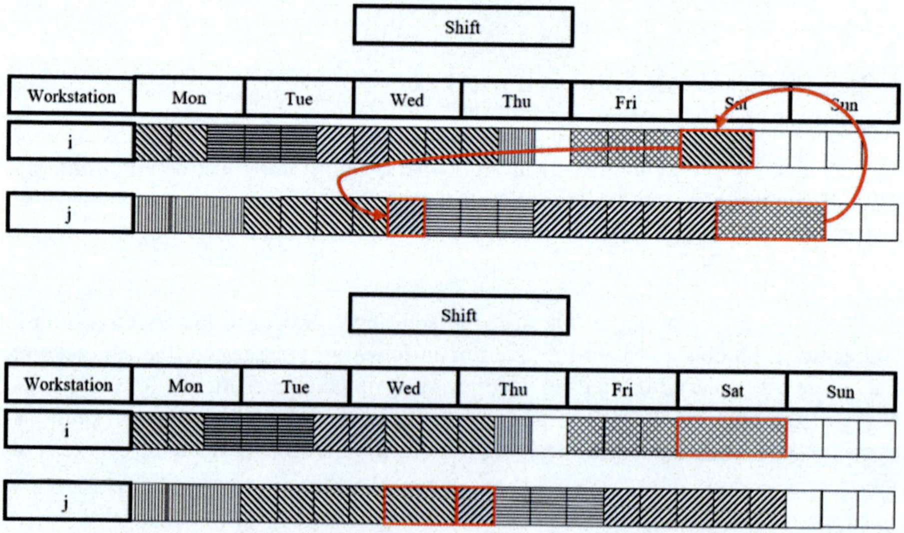

Fig. 4. Visual representation of shift neighborhood permutation

As it was empirically established, using only one operation gives worse results than using both. The operations have different effects on the packing of tasks corresponding to some permutation, and together generate a sufficiently efficient neighborhood with the power of $O\left[n \cdot (n-1) + \frac{n \cdot (n-1)}{2}\right]$, since the order of tasks does not matter for the Substitution operation $(i \to j) = (j \to i)$, but for operations shift, the order does matter.

"Local packaging" are the operations of "packing" tasks in the schedule so that they occupy weekend shifts in order to don't exceed the corresponding deadlines. A single operation adds one weekend shift to the schedule of each workstation, where there are existed production tasks that exceed permissible time limits. The operations are applied after application the current solution with the deadline vector D. "Packing" operations are performed sequentially until one of the following conditions is met:

- all tasks in the schedule are performed within time limits (no deadlines were exceeded),
- the number of free weekend shifts has ended.

"Local packing" operations are performed serially after establishing optimal neighborhood by a series of Substitution and Shift operations. The power of the neighborhood in this case will be $O\left[n \cdot (n-1) + \frac{n \cdot (n-1)}{2} + k\right]$, where k—the number of free weekend shifts, in fact, the number of orders exceeding the time limits should also be considered, whatever happens first, in practice however the number of free weekend shifts is almost always more than orders, we count by a larger value. When perform a series of "local packing" operations after each substitution or Shift permutation, the result should theoretically be more accurate, the power of the neighborhood in this case will be $O\left[k \cdot \left(n \cdot (n-1) + \frac{n \cdot (n-1)}{2}\right)\right]$. In practice, difference in the objective function insignificant, and it is decided to ignore it. The decision was made in favor of the speed of the algorithms.

It should be noted that this operation makes only sense when the company's losses due to penalties for fulfilling orders after the deadline are higher than losses due to working in weekend shifts, which happens in absolutely the vast majority of cases.

3.3 Tabu Search Algorithm

The Tabu search algorithm presented in this section performs a probabilistic local search in the neighborhood, improving the objective function and degrading it, which allows the algorithm not to stop at the local optimum. As prescribed in the local descent algorithm, but to move from one local optimum to another to find the best solution among them [6–8]. The TABU list is the main mechanism that allows the algorithm to leave local optima. It is based on the background of the search, i.e., the last few iterations, and prohibits part of the neighborhood X of the current solution [9–11]. Pseudocode for Tabu Search algorithm:

```
X0 = greedy(O)
Set iniial Tabu_Tenor, Iterations_total
TABU = []
while iterations < Iterations_total:
for i in range(O*(O-1)):
          X" = Substitution(Xi)
          if (F(X") > F(X')) and X" ∉ TABU:
               X' = X"
for i in range(O*(O-1)/2):
          X'' = Swipe(X')
          if (F(X") > F(X')) and X" ∉ TABU:
               X' = X"
X' = local_packing(X')
ΔF = F(X') - F(X)
if ΔF ≥ 0:
          X = X'
else:
         if length(TABU) ≤ Tabu_Tenor:
               Tabu.append(X)
               X = X'
          else:
               Break
iterations += 1
Tabu.append(X)
X = best(TABU)
```

In order to increase algorithm efficiency, the principle of dynamic updating of the taboo list was used. When adding a new element to the list of prohibitions, in case its length becomes longer than the specified one in Tabu tenor, the element with the lowest index of the objective function is removed. In this case, the overflow of the taboo list is no longer a stop criterion, and an additional criterion for exiting the algorithm was introduced—the number of overwriting of the taboo list. Empirically, it was possible to determine the optimal characteristics of the search with prohibitions:

- Iterations_total = 17000
- Tabu_Tenor = 60
- Tabu_Tenor_Rewriting = 120

Typically, Tabu Tenor and Tabu Tenor Rewriting relate to the intensity of one local peak on the iterative search graph and the total number of peaks.

3.4 Simulated Annealing

Algorithms of simulated annealing in finding the optimal solution with some probability allow a transition to a state with a worse value of the objective function. This property allows them to get out of local optima [12, 13]. At the beginning of the algorithm, this probability should be large enough so that it is possible to move from the selected initial solution to any other solution. During the algorithm's operation, the probability of transition decreases by the chosen law [14, 15, 16]. Pseudocode for annealing algorithm:

```
X0 = greedy(O)
Set iniial T0, Set final Tf
X, X', X", T = X0, 0, 0, T0
while T>Tf:
     for i in range(O*(O-1)):
          X" = Substitution(Xi)
          if F(X") > F(X'):
               X' = X"
     for i in range(O*(O-1)/2):
          X'' = Swipe(X')
          if F(X") > F(X')
               X' = X"
     X' = local_packing(X')
     ΔF = F(X') - F(X)
     if ΔF ≥ 0:
        X = X'
     else:
          r=random()
          p = exp(-ΔF/T)
          if r<P:
               X = X'
    T = A
```

To increase the algorithm efficiency, a few different laws A of lowering the temperature were observed (Fig. 5):

- reduction by a static value $T- = k$;
- Boltzmann law $T = \frac{T_0}{ln(1+i)}$;
- Cauchy law $T = \frac{T_0}{1+i}$;
- $T = T_0 \cdot \frac{ln(1+i)}{1+i}$;
- Cauchy law modified $T = \frac{T_0 \cdot k}{(1+i)}$, where k is empirical parameter, in current study set as 20.

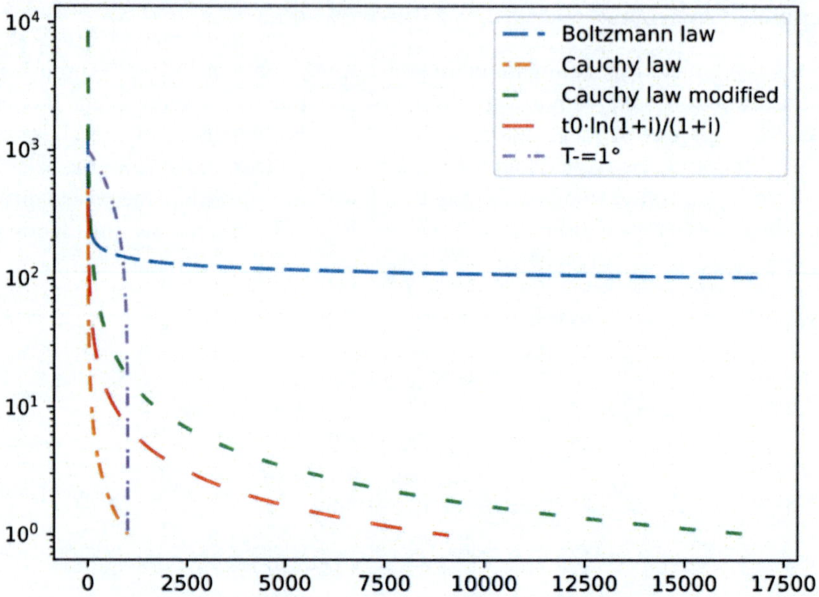

Fig. 5. Temperature drop in accordance with the chosen law.

For better visualization, the ordinate axis on the graph is set to logarithmic. The most gentle law was chosen for this calculation, which corresponds to the modified Cauchy law with a coefficient k = 17. It was empirically determined the optimal characteristics of the annealing simulation:

- T0 = 1000°;
- Tf = 1°;

4 Algorithms' Calculation Results

The tested algorithms do not differ regarding the involvement of employees in the preparation of input data. The following paper has chosen and implemented three algorithms according to company requirements. For verification 10 samples have been taken, where each sample contains a set of orders in a timeline of 12 weeks. As mentioned before, each order's production time is longer than one day is automatically divided, and various deadlines are proposed for various "parts" of the base order (Tables 2, 3 and 4).

Table 1. Company's data

Sample no	No of products	No of orders	No of retooling	No of penalties	No of weekend shifts	Profit [PLN]
1	27	86	62	0	0	436,956
2	12	93	64	0	1	395,462
3	17	72	47	3	1	245,965
4	29	134	55	2	2	319,865
5	36	112	74	4	2	284,545
6	37	99	70	1	1	513,221
7	15	86	42	1	0	344,565
8	18	102	58	5	0	387,756
9	22	126	74	2	3	346,688
10	35	142	63	3	3	295,671

Table 2. Greedy algorithm

Sample no	No of products	No of orders	No of retooling	No of penalties	No of weekend shifts	Profit [PLN]	Gain
1	27	86	77	2	2	312,566	−28.467
2	12	93	69	3	1	374,112	−5.399
3	17	72	55	5	3	211,855	−13.868
4	29	134	83	4	3	259,121	−18.991
5	36	112	92	6	5	221,194	−22.264
6	37	99	82	4	2	501,963	−2.194
7	15	86	57	6	2	295,646	−14.197
8	18	102	71	5	3	341,298	−11.981
9	22	126	98	3	5	306,204	−11.677
10	35	142	86	4	7	254,144	−14.045

In Table 1, company results have been presented. 10 samples meaning 10 weeks of 12 weeks each, contain 86 to 142 orders. Because the number of retooling is lower, one can notice that the company was able to group some of the orders. In 8 samples, the company paid penalties because of exceeding the deadline and in 7 samples weekend shifts were run.

Table 3. Simulation annealing algorithm

Sample no	No of products	No of orders	No of retooling	No of penalties	No of weekend shifts	Profit [PLN]	Gain
1	27	86	59	0	0	451,160	3.251
2	12	93	62	0	0	440,039	11.272
3	17	72	47	3	1	261,040	6.129
4	29	134	54	2	2	341,026	6.616
5	36	112	70	2	1	291,146	2.320
6	37	99	68	1	0	534,122	4.073
7	15	86	38	0	0	356,414	3.439
8	18	102	55	3	0	412,500	6.381
9	22	126	70	0	1	374,521	8.028
10	35	142	62	1	2	326,121	10.299

Table 4. Tabu Search algorithm

Sample no	No of products	No of orders	No of retooling	No of penalties	No of weekend shifts	Profit [PLN]	Gain
1	27	86	61	0	0	443,002	1.384
2	12	93	62	0	1	405,525	2.545
3	17	72	47	3	1	261,040	6.129
4	29	134	53	2	2	330,654	3.373
5	36	112	70	3	1	284,545	0.000
6	37	99	68	1	0	534,122	4.073
7	15	86	38	0	0	356,414	3.439
8	18	102	56	4	0	396,401	2.229
9	22	126	71	2	2	369,956	6.712
10	35	142	62	1	2	326,121	10.299

5 Comparison of Methods

Figure 6 presents the comparison of profits: achieved by the company and calculated by greedy, Simulated Annealing and Tabu Search algorithms.

The greedy algorithm gave worse quality solutions in all 10 samples compared to the company solution. Following this algorithm, the company would have more penalties, more weekend shifts and 14% lower profit in the end. The greedy algorithm did not

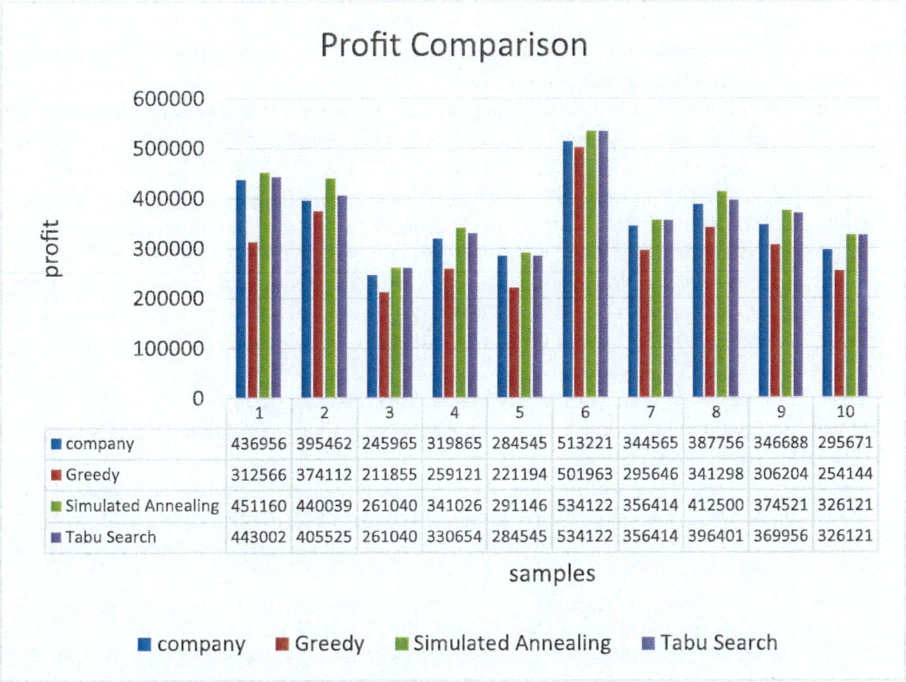

	1	2	3	4	5	6	7	8	9	10
company	436956	395462	245965	319865	284545	513221	344565	387756	346688	295671
Greedy	312566	374112	211855	259121	221194	501963	295646	341298	306204	254144
Simulated Annealing	451160	440039	261040	341026	291146	534122	356414	412500	374521	326121
Tabu Search	443002	405525	261040	330654	284545	534122	356414	396401	369956	326121

Fig. 6. Comparison of chosen algorithms for the company profit

meet the company requirement and is not an algorithm that the company should implement. Simulated Annealing and Tabu Search propositions are promising. The number of retooling, the number of penalties, and the number of weekend shifts are lower than the company solution. Within 10 samples, Simulated Annealing calculated profit higher up to 11% (average 6%) and Tabu Search calculated profit higher up to 10% (average 4%).

6 Conclusion

Presented implementations of greedy, Simulated Annealing and Tabu Search algorithms clearly show that the greedy algorithm is insufficient for this complexity of tasks. Tabu Search and Simulated Annealing algorithms can improve the company production process and in the nearest future, will be verified in the real production process. Also next step in the following investigation is to gather a bigger set of data and apply neural networks. It is supposed that with the use of a combination of neural networks and Simulated Annealing or Tabu Search, it is highly probable to receive a better solution (higher profit) than presented in the following paper.

References

1. Musiał, K., Kotowska, J., Górnicka, D., Burduk, A.: Tabu search and greedy algorithm adaptation to logistic task. In: Saeed, K., Homenda, W., Chaki, R. (eds.) CISIM 2017. LNCS, vol. 10244, pp. 39–49. Springer, Cham (2017). https://doi.org/10.1007/978-3-319-59105-6_4

2. Zwolinska, B., Grzybowska, K., Kubica, L.: Shaping production change variability in relation to the utilized technology. In: DEStech Transactions on Engineering and Technology Research. Destech Publications, Inc (2017)
3. Kahraman, C., Engin, O., Kaya, I., Öztürk, R.E.: Multiprocessor task scheduling in multistage hybrid flow-shops: a parallel greedy algorithm approach. Appl. Soft Comput. **10**, 1293–1300 (2010)
4. Goyal, A., Lu, W., Lakshmanan, L.V.: CELF++: optimizing the greedy algorithm for influence maximization in social networks. In: Proceedings of the 20th International Conference Companion On World Wide Web, pp. 47–48 (2011)
5. Zhang, Z., Schwartz, S., Wagner, L., Miller, W.: A Greedy algorithm for aligning DNA sequences. J. Comput. Biol. **7**, 203–214 (2000)
6. Glover, F.: Future paths for integer programming and links to artificial intelligence. Comput. Oper. Res. **13**(5), 533–549 (1986)
7. Glover, F.: Tabu search part i. ORSA J. Comput. **1**(3), 190–206 (1989)
8. Brandão, J.: A tabu search algorithm for the open vehicle routing problem. Eur. J. Oper. Res. **157**(3), 552–564 (2004)
9. Kotowska, J., Markowski, M., Burduk, A.: Optimization of the supply of components for mass production with the use of the ant colony algorithm. In: Burduk, A., Mazurkiewicz, D. (eds.) ISPEM 2017. AISC, vol. 637, pp. 347–357. Springer, Cham (2018). https://doi.org/10.1007/978-3-319-64465-3_34
10. Cordeau, J.F., Gendreau, M., Laporte, G.: A tabu search heuristic for periodic and multi-depot vehicle routing problems. Networks **30**(2), 105–119 (1997)
11. Bożejko, W., Gnatowski, A., Pempera, J., Wodecki, M.: Parallel tabu search for the cyclic job shop scheduling problem. Comput. Ind. Eng. **113**, 512–524 (2017)
12. Kirkpatrick, S., Gelatt, C.D., Vecchi, M.P.: Optimization by simulated annealing. Science **220**(4598), 671–680 (1983). https://doi.org/10.1126/science.220.4598.671
13. Franzin, A., Stützle, T.: Revisiting simulated annealing: a component-based analysis. Comput. Oper. Res. **104**, 191–206 (2019)
14. Kirkpatrick, S., Gelatt, C., Vecchi, M.: Optimization by simulated anealing. Science **220**, 671–680 (1983)
15. Kłosowski, G., Gola, A., Thibbotuwawa, A.: Computational intelligence in control of AGV multimodal systems. IFAC-PapersOnline **51**(11), 1421–1427 (2018)
16. Antosz, K., Pasko, L., Gola, A.: The use of intelligent systems to support the decision-making process in lean maintenance management. IFAC-PapersOnLine **52**(10), 148–153 (2019)

Dynamic Sectorization - Conceptualization and Application

Filipe Soares de Sousa[1(✉)], Maria Margarida Lima[2], Elif Göksu Öztürk[3,5], Pedro Filipe Rocha[3], Ana Maria Rodrigues[2,3], José Soeiro Ferreira[3,4], Ana Catarina Nunes[1,6], and Cristina Oliveira[2]

[1] ISCTE - University Institute of Lisbon, Lisbon, Portugal
{Filipe.Soares.Sousa,catarina.nunes}@iscte-iul.pt
[2] CEOS.PP, ISCAP, P.PORTO, Porto, Portugal
{mmal,cteles}@iscap.ipp.pt
[3] INESCTEC - Technology and Science, Porto, Portugal
{elif.ozturk,pedro.f.rocha,ana.m.rodrigues,
jose.soeiro.ferreira}@inesctec.pt
[4] FEUP - Faculty of Engineering, University of Porto, Porto, Portugal
[5] FEP.UP - Faculty of Economics, University of Porto, Porto, Portugal
[6] CMAFcIO - Faculty of Sciences, University of Lisbon, Lisbon, Portugal

Abstract. Sectorization is the division of a large area, territory or network into smaller parts considering one or more objectives. Dynamic sectorization deals with situations where it is convenient to discretize the time horizon in a certain number of periods. The decisions will not be isolated, and they will consider the past. The application areas are diverse and increasing due to uncertain times. This work proposes a conceptualization of dynamic sectorization and applies it to a distribution problem with variable demand. Furthermore, Genetic Algorithm is used to obtain solutions for the problem since it has several criteria; Analytical Hierarchy Process is used for the weighting procedure.

Keywords: Sectorization · Dynamic sectorization · Genetic algorithm · Analytical hierarchy process

1 Introduction

Sectorization is the division of a large area, territory or network into smaller parts considering one or more objectives, intending to optimize or simplify a problem. Sectorization problems have a vast field of applications, many of which arise due to real-world dilemmas. Such as designing sales territories, health care services and schools zones.

As uncertainty increases, distinct dynamic attributes emerge in sectorization problems, depending on the application at hand. In the case of a delivery company, their clients will not be the same every day, the demand of a client may differ each time s/he requires service or the number of trucks to make the

J. Machado et al. (Eds.): icieng 2022, LNME, pp. 293–304, 2023.
https://doi.org/10.1007/978-3-031-09382-1_26

distribution may change. Companies must make sectorization considering such fluctuations, to be better prepared without major changes. Then dynamic sectorization arises, although this concept may differ from author to author.

Dynamic sectorization deals with situations where it is convenient to discretize the time horizon in a certain number of periods ($t = 1, 2, ..., T$). A new sectorization is performed in each period, but the decisions will not be isolated since the information from previous periods is considered. Different strategies may be used to deal with the dynamic component, such as fitting a distribution or forecasting techniques.

This paper applies this concept of dynamic sectorization to a delivery company problem, where the demand changes over time. A Genetic Algorithm (GA) and Analytical Hierarchy Process (AHP) are used to solve the problem.

The remainder of the paper is organized as follows. Section 2 presents a review of the literature. Section 3 describes the dynamic sectorization problem. A solution method is proposed in Sect. 4 and, in Sect. 5, the corresponding results and a brief discussion are included.

2 Literature Review

This section briefly summarizes some of the literature relevant to the problem.

Sectorization problems have many real-world applications. Political districting is one of the oldest, aiming to divide the territory neutrally and avoid gerrymandering. Most of the time, political districting is subject to resectorization to adapt the current solution regarding the updates in the territory. Applications of this type of problem consider various criteria, such as equilibrium, compactness and contiguity [1,2]. Bação *et al.* [1] called it a zone design problem and presented a solution method based on a genetic algorithm for which they considered two objective functions and proposed two different types of gene encoding.

Another study field is the design of sales territories. It aims to divide a large commercial territory into certain zones and assign them to salespeople [3,4]. Salazar-Aguilar *et al.* [5] proposed a multi-objective scatter search procedure to a territory design problem, considering equilibrium and compactness as objectives. The results obtained with this metaheuristic are compared to the Scatter Tabu Search Procedure [6] and Non dominated Sorting Genetic Algorithm (NSGA-II) [7]. Lei *et al.* [8] solved a multi-objective dynamic stochastic districting and routing problem, in which the customers of a territory stochastically evolve over several periods of a planning horizon. They developed a preference-inspired co-evolutionary algorithm using mating restriction for the problem; results confirmed the superiority and effectiveness of the algorithm.

Moreover, health care districting is another application of sectorization problems. Home Health Care (HHC) represents an alternative to the traditional hospitalisation [9]; care services are provided to patients at their homes by a team composed of care providers, using a distribution network. Sectorization divides this network in a more optimized way. Gutierrez-Gutierrez *et al.* [10] studied this type of problem in a rapidly growing city and proposed a bi-objective mathematical model, identifying a trade-off that allows reaching better-compromised solutions. Their model is evaluated with real data instances from an HHC institution, which delivers services in the largest cities in Colombia. Lin *et al.* [11] formulated the care service districting as a multi-objective mixed-integer nonlinear programming model, with equilibrium and compactness as objective functions. NSGA-II was used to obtain results for this formulation.

Other attributes can be dynamic in sectorization problems. For instance, in [8,12,13], the clients are seen as the dynamic attribute, since they may leave or enter the territory from one period to another.

In the literature of dynamic sectorization, it is possible to see the usual criteria, compactness, equilibrium and contiguity, being used. However, from one period to another, the similarity of the sectors has a crucial role in dynamic sectorization problems. Yanık *et al.* [14] proposed a multi-period multi-criteria districting problem applied to a primary care scheme, a specific healthcare service in Turkey. Six criteria are considered: workload balance, capacity, accessibility, compactness, income equity, and similarity.

Airspace design appears to be the most studied field in the literature of dynamic sectorization. Airspace design consists of changing the airspace that the air traffic controllers manage to handle modifications in the airspace. The objective is to create sectors with a balanced workload but maintain a certain similarity from one period to another. Several approaches are considered in the literature. Wong *et al.* [15] used a rolling horizon optimization approach, where the shapes are changed based on a finite horizon of the traffic in future intervals, and evaluated the method's performance based on historical data from Singapore. On the other hand, Sergeev *et al.* [16] formulated the problem as a graph partitioning problem and solved it using a Genetic Algorithm, obtaining solutions that outperform the existing ones.

GA has been used to solve sectorization problems, as is the case of [17–20]. Some of these problems have more than one objective. In these cases, a composite weighted single objective function is used, whose weights have to be determined in some manner. Analytical Hierarchy Process (AHP), proposed by Saaty [21–23] is characterized for regarding the preferences of the decision-maker into consideration. For each pair of objectives, the decision-maker evaluates the importance of one objective over the other. This evaluation is expected to be coherent with the comparison scale, which takes integer values from 1 to 9, where 1 shows equal importance, while 9 shows extreme importance. Since unilateral comparisons are enough, $k(k-1)/2$ comparisons are required in AHP when k objectives are in question.

3 Problem Definition and Criteria

This section describes the dynamic sectorization problem of a delivery company and presents the criteria taken into account.

The time horizon is divided into periods, $t \in P = \{0, ..., T\}$ is the set of periods. Let $V = \{1, ..., n\}$ be the set of customers. The location of each customer does not change from one period to another and is known, given by its coordinates. The demands change over time and therefore constitute the dynamic component of the problem. l_i^t represents the demand of customer i in period t. Demands will follow a given probability distribution, as described later in Sect. 4.

Let $D = \{1, ..., k\}$ be the company's set of facilities whose locations o_j, $j \in D$, are known in advance. Moreover, each facility $j \in D$ has a limited capacity, a_j, which means that it can only supply up to this amount on total; the capacity is the same for all periods.

Finally, let us consider the binary decision variables x_{ij}^t, where

$$x_{ij}^t = \begin{cases} 1 & if\ customer\ i\ is\ assigned\ to\ facility\ j\ in\ period\ t \\ 0 & otherwise \end{cases} \tag{1}$$

The number of sectors must be equal to the number of facilities. Furthermore, each facility must have at least one customer, and each customer must be assigned to exactly one of the facilities.

The company's purpose is to design compact, contiguous and balanced sectors for each period, maintaining the sectors similar from one period to another. The details concerning the three considered objective functions and the measure to evaluate the similarity end this section.

Compactness. The compactness of a sectorization improves by minimizing the dispersion within sectors. This way, the Euclidean distance $dist(o_j, p_j)$ between the location o_j of a facility j and the location p_j of the farthest customer assigned to j is considered. The compactness measure d, in Eq. (2), sums up this distance for all facilities. Logically, the smaller the value, the higher the compactness of the sectors.

$$d = \sum_{j=1}^{k} dist(o_j, p_j) \tag{2}$$

Balance. It is desirable to balance the sectors, which means that a facility has to supply more or less the same as the others. The balance measure proposed by Rodrigues and Ferreira (2015) [24] is used. The mean demand of the sectors, $\bar{q}$, is given by Eq. (3).

$$\bar{q} = \frac{\sum_{j=1}^{k} q_j}{k}, \text{ where } q_j = \sum_{i=1}^{n} l_i^t x_{ij}^t \tag{3}$$

The Balance is given by the standard deviation of the demands of the sectors, $s_q^{'}$, which is calculated by Eq. (4). The value of $s_q^{'}$ has to be minimized to obtain a better balance.

$$s_q^{'} = \sqrt{\frac{1}{k-1} \sum_{j=1}^{k} (q_j - \bar{q})^2} \tag{4}$$

Contiguity. The third objective considered is contiguity; it evaluates the connectivity of the sectors. The contiguity measure proposed by Rodrigues and Ferreira (2015) [24] is used. Consider, for two customers w and i ($w \neq i$) of sector j:

$$m_{wi}^{j} = \begin{cases} 1 & if\ there\ is\ a\ path\ between\ customers\ w\ and\ i\ of\ sector\ j \\ 0 & otherwise \end{cases} \tag{5}$$

Equation (6) is used to calculate the contiguity of sector j, given by the total number of paths between its customers divided by the maximum number of possible links in sector j.

$$c_j = \frac{\sum_{i=1}^{n_j} (\sum_{w=1, w \neq i}^{n_j} m_{wi}^{j})}{n_j(n_j - 1)} \tag{6}$$

where n_j represents the number of customers in sector j. Contiguity, $\bar{c}$, is given by the weighted average of the contiguities of the sectors, as shown in expression (7). In this case, the higher the value, the better the contiguity. Thus $(1 - \bar{c})$ is minimized, as it is for the other criteria.

$$\bar{c} = \frac{\sum_{j=1}^{k} c_j n_j}{n} \tag{7}$$

Similarity. The similarity is a measure to evaluate the resemblance, or coincidence, between two solutions. In dynamic sectorization, it is relevant to measure how different is the sectorization of period $t + 1$ from the one of the previous period t. In the delivery company problem, it is important that there are no major changes from a period to the following. The similarity measure used in this work is the percentage of customers that stay unchanged, which is given by:

$$\frac{\sum_{j \in D} \sum_{i \in V} x_{ij}^{t} x_{ij}^{t+1}}{\sum_{j \in D} \sum_{i \in V} x_{ij}^{t}}, \qquad t \in P. \tag{8}$$

This measure evaluates how much the similarity is being ensured by the solution method.

4 Solution Method

In the current work, GA, presented by Goldberg and Holland [25], and AHP are used in the solution method described in this section.

Some GA specific definitions, such as genetic encoding scheme, selection and crossover or mutation, should be decided carefully due to their effect on the final solution/s. This paper considers the matrix form binary grouping (MFBG) genetic encoding system due to its suitable scheme for sectorization, assignment, and related problems. MFBG is a matrix form genetic encoding scheme where rows represent basic units (customers, nodes, etc.) and columns represent sectors.

Binary values show the assignment status of basic units to the sectors. More precisely, when a basic unit is assigned to a sector, the corresponding row takes the value of one in the cell where the column corresponds to that sector and takes the value of zero in every other cell of that row. MFBG has two restrictions: (i) each basic unit can be assigned to only one sector, and (ii) each sector must have at least one basic unit assigned. We assume that solutions that do not respect these restrictions are infeasible.

Tournament selection is used for the selection step where two randomly selected solutions are picked, and the superior one is selected as a parent for the crossover procedure. A multi-point crossover method proper for the chosen genetic encoding system is employed. In this method, several randomly selected rows are switched between two-parent solutions to generate two offspring solutions. The mutation is applied to offspring only every two generations. We use a very low mutation rate that randomly changes a basic unit's assignment to a sector if the mutation rate holds. Finally, the number of generations, G, is used as the stopping criteria.

As mentioned earlier, we consider a dynamic problem where changes in demand may occur from one period to the next. Over time, the sectorization is updated according to these changes, using the sectorization found in period t to create the initial population in period $t+1$ so that similarity between sectorizations is promoted. A whole GA procedure is completed independently from the previous or the next period. In other words, the two consecutive periods are only connected through the best solution found in the first of the two periods. This is evidenced in the pseudocode of Algorithm 1.

Algorithm 1 starts with a random initial population only in period $t = 0$. In all subsequent periods, the best solution of the previous period is used to create the initial population of the current period. Moreover, the demands start to change only after the initial period, and according to a uniform or a normal distribution with the selected percentage. These changes in the demands are reflected in Expression (9).

$$L_i^{t+1} \sim U\left(l_i^t - \frac{\Delta\%}{100} l_i^t,\quad l_i^t + \frac{\Delta\%}{100} l_i^t\right) \quad or \quad L_i^{t+1} \sim N\left(l_i^t,\quad \frac{\frac{\Delta\%}{100} l_i^t}{3}\right) \tag{9}$$

Here l_i^t is the demand of the basic unit i at period t, and $\Delta\%$ is the deviation allowed to the demands, in percentage. Finally, L_i^{t+1} is the random variable representing the demand in period $t+1$, and U and N stand for the uniform and the normal probability distributions, respectively. The expressions in U are respectively the minimum and maximum parameters required in the uniform probability distribution. Regarding N, the expressions are respectively the mean and variance parameters needed in the normal probability distribution. If the demand decreases below zero, the basic unit is assumed to disappear from the instance.

Algorithm 1. GA for Dynamic Sectorization

```
Generate N random feasible solutions and insert into Population (Pop_size = N)
Evaluation of the solutions according to the fitness function
Period := 0
while Period < T do
   if Period > 0 then
      Consider the demands of period Period
      Generate N feasible solutions based on the best solution found in the previous
      period and insert them into Population (Pop_size = N)
      Evaluation of the solutions according to the fitness function
   end if
   Generation := 0
   while Generation < G do
      while Pop_size < N × 2 do
         Select parents through tournament selection
         Create two offsprings using two-point crossover in each turn
         Pop_size := Pop_size + 2
      end while
      Mutate off-springs (for selected P_mut)
      Merge offsprings into the population
      Evaluation of the solutions according to the fitness function
      while Pop_size > N do
         Refinement: Delete the worst solution from the population
         Pop_size := Pop_size − 1
      end while
      Generation := Generation + 1
   end while
   Keep the best solution
   Calculate the similarity
   Period := Period + 1
end while
```

The evaluation step takes place by analysing the solutions regarding their performance according to the fitness function. Whether a solution is adequate for the problem depends on its fitness. Given h objectives, the composite weighted single objective function in Eq. (10) is used to measure the fitness of the solutions.

$$F(x) = w_1 f_1(x) + w_2 f_2(x) + ... + w_h f_h(x), \ \ where \ \sum_{i=1}^{h} w_i = 1 \tag{10}$$

$F(x)$ shows the fitness value of a solution x, and $f_i(x)$ is the performance of solution x concerning objective i. Finally, w_i is the weight assigned to objective i using the AHP procedure.

When weighted composite single objective functions are used, there are two necessary operations: normalisation and weighting. Normalisation is necessary when the objective functions have different measurement units. In the current paper, min-max normalisation was used to standardise Balance and Compactness factors to set their values between 0 and 1, where zero is the best, and one is the worst. No normalisation was necessary for contiguity since its values already occur between 0 and 1.

5 Data and Results

This section describes the data and presents the main results. Additionally, results are briefly discussed.

The instance used in this paper was generated by us based on real data from a delivery company which included the coordinates of 350 customers and 10 facilities, the demand of each client, and the capacity of each facility. To simulate the dynamic demand, we considered the normal and uniform distributions, with the percentage of the demand to change equal to 10%. In the genetic algorithm, we used a population size of 50, a rate of mutation of 10% and the number of generations $G = 250$.

Table 1 presents the results obtained in each period for both distributions. The first thing to notice is how much the solutions worsen while moving in time. In the case of the normal distribution, the solution fitness gets worst by approximately 118%, and 98% in the uniform. Additionally, both compactness and contiguity also have increased in their values. The differences are not significant in terms of balance. The similarity values between solutions are above 85%, which is a high similarity value.

Table 1. GA results

Distribution	Period	Compactness	Balance	Contiguity	Fitness	Similarity	Time (min)
Normal	0	0.703	0.054	0.023	0.260	–	38.65
	1	0.456	0.067	0.171	0.231	90.29%	
	2	0.798	0.068	0.315	0.394	85.88%	
	3	0.805	0.059	0.416	0.427	85.00%	
	4	0.860	0.101	0.579	0.513	87.35%	
	5	0.898	0.042	0.680	0.540	87.65%	
	6	0.951	0.084	0.667	0.567	86.76%	
Uniform	0	0.643	0.148	0.032	0.274	–	37.87
	1	0.411	0.059	0.114	0.195	84.71%	
	2	0.778	0.048	0.380	0.402	88.53%	
	3	0.891	0.084	0.450	0.475	85.88%	
	4	0.891	0.111	0.534	0.512	90.00%	
	5	0.916	0.070	0.592	0.526	86.76%	
	6	0.904	0.069	0.652	0.542	86.76%	

Table 2 presents the fitness values for totally random populations in period $t+1$. It shows much lower similarity values than the ones obtained when the population in period $t+1$ is generated from the best solution of the previous period

(Table 1). However, fitness values tend to be smaller. These results emphasize the effectiveness of the proposed method in maintaining the similarity of the sectors for later periods.

Table 2. GA Results for totally random initial populations

Period	0	1	2	3	4	5	6
Fitness	0.260	0.201	0.352	0.362	0.293	0.302	0.322
Similarity	–	5%	19%	21%	32%	19%	17%

Figure 1, Fig. 2 and Fig. 3 respectively show the visual representation of the sectors for periods $t = 0$, $t = 3$, and $t = 6$ when the normal distribution is used. Each colour is a sector associated with a facility, whose location is represented by a cross. The other geometric shapes are the clients.

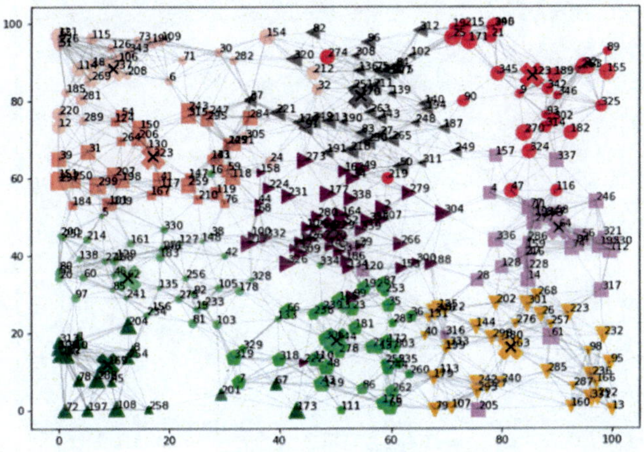

Fig. 1. Period = 0

It is possible to see that the points are more scattered when the period increases; this happens because of the higher compactness values. Moreover, most of the points stay in the same sector, from one period to another, maintaining the sectors similar, as we intended.

Finally, we decided to see what happens when changing the percentage that the demand can deviate from. We observed that the fitness values remained within the same ranges.

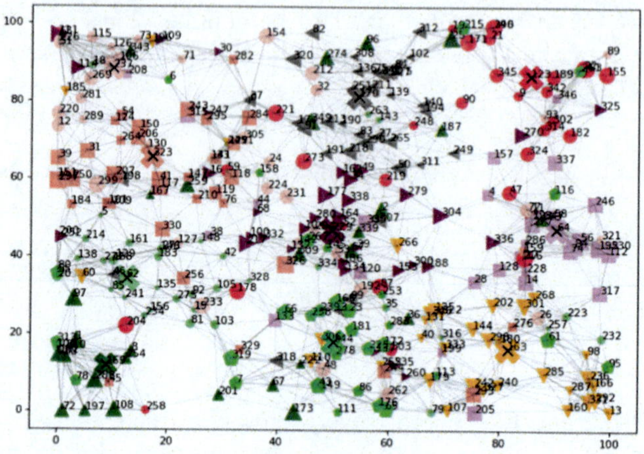

Fig. 2. Period = 3

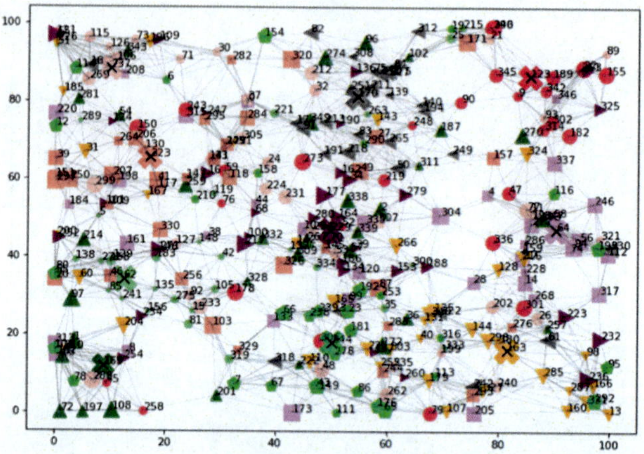

Fig. 3. Period = 6

6 Conclusions and Future Work

Sectorization divides a set or region into smaller parts to help solve complex situations. Decision-making in sectorization problems can also be affected by growing uncertainty about the future. Thus, a dynamic perspective to handle such problems may be relevant. It is convenient to discretize the time horizon in a certain number of periods to deal with a dynamic situation, ensuring a sectorization with no major changes.

The main contributions of this work were the proposal of a conceptualization of dynamic sectorization and an application to a distribution problem, where the demand changes over time. The goal was to assign customers to a facility and create compact, contiguous, and balanced sectors. Moreover, we used a genetic algorithm to obtain a solution to the problem. Given that we considered three objectives, we used a composite weighted single objective function to calculate the solution's fitness. AHP was used to obtain the weights. The solution method was tested in two different demand cases, following a normal or uniform distribution. In both cases, the similarity values were above 85%, which means no major changes from one solution to another. The balance of the sectors stayed in the same range. However, compactness, contiguity and the solution's fitness were worse when the period increased. On the other hand, when similarity is not priority the solution's fitness are better.

In the future, the solution method should be tested with the real data from the company. In addition, we will address the case where the clients' locations change over time and integrate it with the dynamic demand. Another development will consist of proposing measures to evaluate the entire time horizon, particularly concerning the similarity of the solutions.

Acknowledgments. This work is financed by the ERDF - European Regional Development Fund through the Operational Programme for Competitiveness and Internationalisation - COMPETE 2020 Programme and by National Funds through the Portuguese funding agency, FCT - Fundação para a Ciência e a Tecnologia within project 'POCI-01-0145-FEDER-031671'.

References

1. Bação, F., Lobo, V., Painho, M.: Applying genetic algorithms to zone design. Soft. Comput. **9**, 341–348 (2005)
2. Bozkaya, B., Erkut, E., Laporte, G.: A tabu search heuristic and adaptive memory procedure for political districting. Eur. J. Oper. Res. **144**, 12–26 (2003)
3. Kalcsics, J., Nickel, S., Schröder, M.: Towards a unified territorial design approach - applications, algorithms and GIS integration. TOP **13**, 1–56 (2005). https://doi.org/10.1007/BF02578982
4. Lodish, L.: Sales territory alignment to maximize profit. J. Marketing Res. **12**, 30–36 (1975)
5. Salazar-Aguilar, A., Ríos-Mercado, R., Gonzalez-Velarde, J., Molina, J.: Multiobjective scatter search for a commercial territory design problem. Ann. Oper. Res. **199**, 343–360 (2012)
6. Molina, J., Laguna, M., Marti, R., Caballero, R.: SSPMO: a scatter tabu search procedure for non-linear multiobjective optimization. INFORMS J. Comput. **19**, 91–100 (2007)
7. Deb, K., Pratap, A., Agarwal, S., Meyarivan, T.: A fast and elitist multiobjective genetic algorithm: NSGA-II. IEEE Trans. Evol. Comput. **6**(2), 182–197 (2002)
8. Lei, H., Wang, R., Laporte, G.: Solving a multi-objective dynamic stochastic districting and routing problem with a co-evolutionary algorithm. Comput. Oper. Res. **67**, 12–24 (2016)
9. Benzarti, E., Sahin, E., Dallery, Y.: Modelling approaches for the home health care districting problem. In: 8th International Conference of Modeling and Simulation-MOSIM, pp. 10–12 (2010)
10. Gutierrez-Gutierrez, E.V., Vidal, C.J.: A home health care districting problem in a rapid-growing city. Ingenieria y Universidad **19**, 87–113 (2015)

11. Lin, M., Chin, K.S., Ma, L., Tsui, K.L.: A comprehensive multi-objective mixed integer nonlinear programming model for an integrated elderly care service districting problem. Ann. Oper. Res. **291**(1), 499–529 (2020)
12. Lei, H., Laporte, G., Liu, Y., Zhang, T.: Dynamic design of sales territories. Comput. Oper. Res. **56**, 84–92 (2015)
13. Xue, G., Wang, Z., Wang, G.: Optimization of rider scheduling for a food delivery service in O2O business. J. Adv. Transp. **2021**, 1–15 (2021)
14. Yanık, S., Kalcsics, J., Nickel, S., Bozkaya, B.: A multi-period multi-criteria districting problem applied to primary care scheme with gradual assignment. Int. Trans. Oper. Res. **26**, 1676–1697 (2019)
15. Wong, C., Suresh, S., Sundararajan, N.: A rolling horizon optimization approach for dynamic airspace sectorization. IFAC J. Syst. Control **11**, 100076 (2020)
16. Sergeeva, M., Delahaye, D., Mancel, C., Vidosavljevic, A.: Dynamic airspace configuration by genetic algorithm. J. Traffic Transp. Eng. (English Edition) **4**(3), 300–314 (2017). https://www.sciencedirect.com/science/article/pii/S2095756417301927
17. Agustín-Blas, L.E., Salcedo-Sanz, S., Ortiz-García, E.G., Portilla-Figueras, A., Pérez-Bellido, Á.M.: A hybrid grouping genetic algorithm for assigning students to preferred laboratory groups. Expert Syst. Appl. **36**(3), 7234–7241 (2009)
18. Konak, A.: Network design problem with relays: a genetic algorithm with a path-based crossover and a set covering formulation. Eur. J. Oper. Res. **218**(3), 829–837 (2012)
19. Di Nardo, A., Di Natale, M., Santonastaso, G.F., Tzatchkov, V.G., Alcocer-Yamanaka, V.H.: Water network sectorization based on a genetic algorithm and minimum dissipated power paths. Water Sci. Technol. Water Supply **13**(4), 951–957 (2013)
20. Noorian, S.S., Murphy, C.E.: Balanced allocation of multi-criteria geographic areas by a genetic algorithm. In: Peterson, M.P. (ed.) ICACI 2017. LNGC, pp. 417–433. Springer, Cham (2017). https://doi.org/10.1007/978-3-319-57336-6_29
21. Saaty, T.L.: A scaling method for priorities in hierarchical structures. J. Math. Psychol. **15**(3), 234–281 (1977)
22. Saaty, T.L.: How to make a decision: the analytic hierarchy process. Eur. J. Oper. Res. **48**(1), 9–26 (1990)
23. Saaty, T.L.: Decision making with the analytic hierarchy process. Int. J. Serv. Sci. **1**(1), 83–98 (2008)
24. Rodrigues, A.M., Ferreira, J.S.: Measures in sectorization problems. In: Barbosa Póvoa, A.P.F.D., de Miranda, J.L. (eds.) Operations Research and Big Data. SBD, vol. 15, pp. 203–211. Springer, Cham (2015). https://doi.org/10.1007/978-3-319-24154-8_24
25. Goldberg, D.E., Holland, J.H.: Genetic algorithms and machine learning. Mach. Learn. **3**(2), 95–99 (1988). https://doi.org/10.1023/A:1022602019183

Research of Features of Oxide Coatings Deposition by D–gun Spraying

Oleksiy Romanchenko, Olga Porkuian, Yuriy Kharlamov, Volodymyr Sokolov, and Oleg Krol(✉)

Volodymyr Dahl East Ukrainian National University, 59-a Tsentralnyi Pr., Severodonetsk 93400, Ukraine
krolos@i.ua

Abstract. Protective and functional coatings based on oxides are of considerable interest for high-tech industries. Of nparticular interest is alumina, which is cheap and accessible, the production of powders of which has been mastered on an industrial scale. The features of oxide coatings deposition by D – gun spraying method are considered in the article. When developing technological processes for D – gun spraying of oxide coatings, it is necessary to take into account and provide optimal conditions for the occurrence of leading transformations and phenomena that have a decisive effect on the structure and properties of the resulting coatings. For this, the flow rate and composition of the combustible mixture, the geometry and dimensions of the barrel, the conditions for introducing the powder into the barrel, a single dose of powder, the distance of spraying, the cyclogram of the working cycle and other technological methods can be used. The possibility of controlling the phase composition and properties of the coatings is shown on the example of applying coatings from aluminum oxide powder. A significant effect of thickness of single coating layers on the kinetics of polymorphic transformations and the phase composition of aluminum oxide coatings has been established. The method of D – gun spraying makes it possible to obtain coatings of aluminum oxide with an increased content of the α phase. Also, D – gun spraying allows the application of durable coatings based on titanium, iron and cobalt oxides.

Keywords: D-gun sprayed coatings · Functional coatings · Aluminum oxide · Titanium oxide · Iron oxide · Cobalt oxide

1 Introduction

An important reserve for increasing the operational characteristics of technological equipment is the application of thermal spraying in the manufacturing technology of their parts and units [1–5]. The application of protective coatings makes it possible to solve various problems that reduce the material and energy consumption of production, repair and operation of machines and mechanisms, increase their durability and reliability, create and master modern technology [6–10].

Among the methods of coatings thermal spraying from powder materials the D – gun spraying is one of the most prospective in mechanical engineering [11–15]. Among the

J. Machado et al. (Eds.): icieng 2022, LNME, pp. 305–316, 2023.
https://doi.org/10.1007/978-3-031-09382-1_27

main advantages of D – gun spraying of coatings are the lower requirements for spraying surface quality; wider possibilities for regulating the thermal cycle of the formed coatings; high velocity of coating thickness formation; relative simplicity of designs and high reliability of technological equipment; high density of the resulting coatings, etc. The application and development of D-gun sprayed coatings is associated with the requirements of dense and strong coatings. In most publications on these questions, much attention is paid to such important property of coatings as the adhesion strength. However, to date, the mechanisms of strong coatings formation for D-Gun spraying have not yet been sufficiently researched. This complicates the development of technological processes and designs of coated products, as well as corresponding technological equipment.

Creation and introduction into production of effective technological processes and equipment [16–20] for D – gun spraying of oxide coatings requires studying the features of their deposition.

2 Literature Review

Among the materials used for the deposition of coatings, refractory oxides belong to a group of materials that are of great interest [21–25]. They are characterized by stability in an oxidizing environment, high melting point, high compressive strength. Coatings on aluminium base have good tribological properties, corrosion resistance, high dielectric strength, etc. Aluminium oxide powder is also attracted by its low cost in comparison with special powders for thermal spraying.

The mechanical properties, including the hardness of coating, can be influenced by a number of factors [26–30]:

- granulometric composition of the initial powder material, on which depends the energy state of particles at the exit from the barrel – their temperature and velocity;
- the quantitative ratio of Al_2O_3 modifications in a certain part of the coating layer, which, in turn, depends on the cooling conditions of sprayed material during formation of the coating;
- the distribution of powder particles by size when they approaching the sprayed part, due to the fact that smaller particles, due to low inertia, are carried away by the gas flow to periphery of spraying spot. This can lead to a significant decrease of particle velocity and an increase of coating porosity.

As a result, aluminium oxide coatings are characterized by an inhomogeneous structure due to different conditions of heat removal from the coating formation zone, which affect the process of polymorphic transformations in it. The quantitative ratio of α-, γ-Al_2O_3 and the glass phase varies both by thickness of sprayed layer and diameter of the spray spot, which leads to unequal levels of mechanical properties of the coating in its various areas. In connection with the dependence of the process of polymorphic transformations on the intensity of heat removal from the zone of formation of the coating, its mechanical properties are influenced by such parameters of the coating process as thickness of the sprayed layer, rate of fire of detonation machine, velocity of movement of sprayed surface, as well as size and weight of the part.

Thus, the information about the effect of the technological parameters of D – gun spraying on the structure and properties of coatings from aluminum oxide is contradictory, which complicates the development and optimization of technology for obtaining coatings with the required properties.

For a number of technical solutions, a significant interest has the oxides of zirconium, titanium, iron, cobalt, etc. However, there is no information in literature by the application and technological features of D – gun spraying to obtain coatings from these oxides.

The objective of work is to study the possibility of controlling the mechanisms of structure and phase formation during obtaining of coating by D – gun spraying with powders of aluminum, zirconium, titanium, iron, and cobalt oxides.

3 Research Methodology

The formation of coatings during D – gun spraying depends on a large number of factors that determine the features of used equipment, initial powders, design and preparation of sprayed parts, technological modes of spraying, etc. [1–3]. These features of D – gun spraying process have a noticeable effect on the process of polymorphic transformations during obtaining of coatings associated with effect of particle composition and type of powders and other parameters on the state of particles upon collision with surface of a part, the rate of cooling of particles during formation of a coating, etc. Depending on the physicochemical properties of the initial powder material and the technological parameters of D – gun spraying, the structure and phase formation of coatings can occur under the conditions of formation: by liquid particles with cooling rates that ensure their crystallization, including in metastable states; liquid particles with velocities providing an amorphous and microcrystalline structure; solid particles in a plastic state, including those with fixation of metastable states; in the mode of flowing or ending the combustion of metal particles to obtain oxymetallic coatings; a combination of two or more of the above modes. By changing the technological modes of process of D – gun spraying, it is possible to control mechanisms of structure and phase formation during the formation of various layers of coating, for example, adjacent to the base, intermediate and outer layers of the coating. It is possible to provide conditions of coatings formation when one of mechanisms of structure and phase formation is the leading one, and the others are concomitant.

The coatings were deposited on a DNP5M machine with replaceable barrels of variable cross-section, tapering from a diameter of 50 mm to 10… 20 mm to the open end, with a volume of 0,6 × 10^{-3} m^3 at a rate of fire of 2… 4 Hz. The machine operated in the modes of cyclic gas supply or with continuous oxygen supply with a flow rate in the range from 0,6 to 1,1 m^3/h. The method of jet-axial powder supply from the side of the closed end of the barrel was used. The coatings were sprayed on polished steel plates with 4… 5 mm thickness made of Steel 3. The hardness of coatings was determined by the Vickers method. By measurements of the distribution of coating thickness in the spot, obtained by repeating the cycles of D – gun spraying on stationary samples, the cross-sectional area of the «metallization» figure was determined. This gives a more complete and qualitative information in comparison with the traditional assessment of thickness of a single layer of coatings and the achieved coefficient of powder use.

The determination of qualitative and quantitative phase composition of coatings was carried out by the X-ray diffraction method on a DRON-3M installation [1, 3]. The recording of samples for X-ray structural analysis was carried out at room temperature in monochromatic FeK_{α} radiation. When determining the relative content of α-, γ-Al_2O_3, the reflections (400) and (113) of these phases were analyzed. The presence of an intermediate δ-phase was determined by the characteristic splitting of the (400) and (440) reflections.

4 Results

4.1 Features of Obtaining of Coatings Based on Aluminum Oxide

The patterns of coatings formation from white electrocorundum powder in accordance with GOST 21445-84 with a fineness of 10–14 microns have been studied. The coating was deposited using a DNP-5M machine with barrel that have variable cross section tapering towards the open end, the length of the barresl is 0,8 m, volume $0{,}6 \cdot 10^{-3}$ m^3 and rate of fire 2,1 Hz. With an increase in continuous oxygen consumption from 0,6 to 1,1 m^3/h and other unchanged process parameters, the patterns of change of the cross-sectional area of the coating (Fig. 1) and Vickers hardness have an extremum and at different flow rate of acetylene (0,47 and 0,68 m^3/h) correspond to the same composition of combustible mixture O_2: $C_2H_2 = 1{,}5$. The oxygen consumption values for the maximum of the cross-sectional area and coating hardness at the same acetylene consumption are coincided. Moreover, with a lower value of acetylene consumption (0,47 m^3/h), these dependences have a gentler character, that is, they are less sensitive to oxygen consumption. The lines 7 and 8 show the change in the filling coefficient of the barrel by combustible mixture.

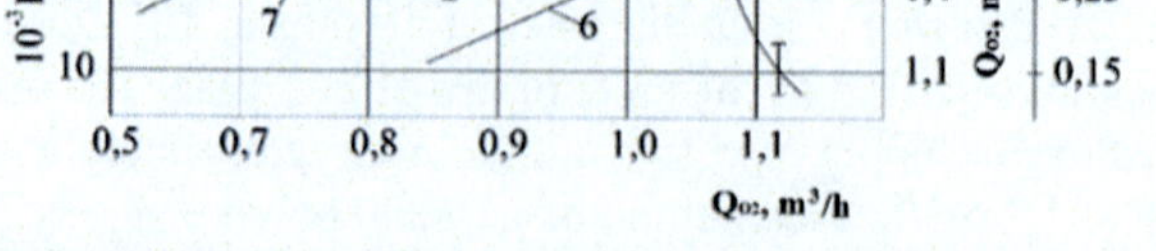

Fig. 1. Dependence for the cross-sectional area of a single spot (1, 2) and the hardness (3, 4) of coatings from alumina on the oxygen consumption, the volumetric ratio of gases $\beta = O_2/C_2H_2$ (5, 6) and the filling coefficient of the barrel by combustible mixture (7, eight). Spraying distance is 150 mm. Acetylene consumption, m^3/h: 0,47 (1, 3, 5, 7); 0,68 (2, 4, 6, 8).

The hardness of Al_2O_3 plasma coatings is usually in the range of 5440... 7300 MPa, since they consist mainly from γ- Al_2O_3. The observed increase in coatings hardness can be explained by the change in the relative content of the α- and γ - phases. A certain influence is also exerted by porosity of coatings and strength of bonds (adhesion) along the boundaries of contact between individual particles. The retention of a significant amount of α-Al_2O_3 in D – gun coatings are associated with a reduced cooling rate of a single coating layer, as well as with the possibility of forming of coatings with completely melted particles at high velocities of collision with the surface of the part. This is also evidenced by the monotonic decrease of the hardness and cross-sectional area of coatings with an increase in spraying distance (Fig. 2). With an increase in spraying distance, not only velocity and temperature of particles decrease, but the degree of opening of the detonation jet also increases, and thereby the thermal effect on the surface of the part weakens. The dependences of the coating's hardness on the consumption of transported gas have an extremum, and the values of the consumption of transporting gas corresponding to the maximum hardness of the coatings decreases with the increase in spraying distance (Fig. 3). Experimental verification showed that the flow rate of transported gas practically does not affect the flow rate of powder and mainly determines the intensity of powder spraying in the barrel, the length of the powder cloud and its initial position in the barrel. The extrema of the experimental dependences of hardness correspond to the optimal energy state of Al_2O_3 particles at the moment of coating formation. As the spraying distance increases, the powder particles undergo more intense cooling.

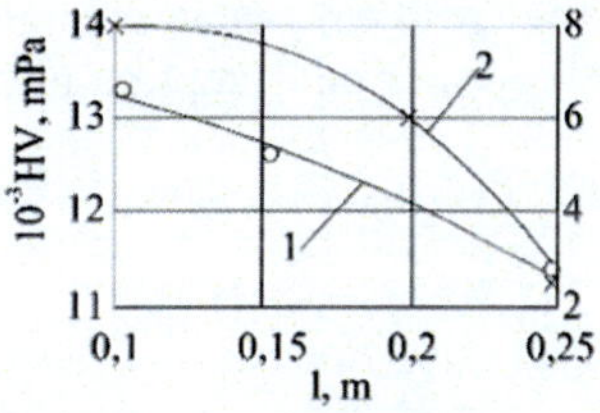

Fig. 2. Dependence of hardness (1) and cross-sectional area for the single spot (2) of coatings from aluminum oxide on spraying distance. Gas consumption, m^3/h: O_2 – 0,83; C_2H_2 – 0,68.

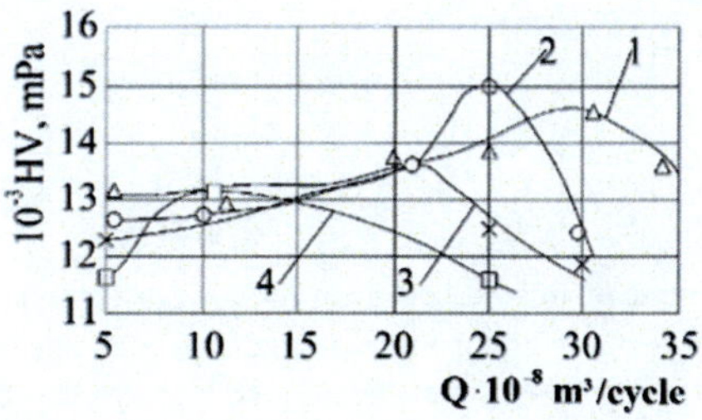

Fig. 3. Dependence of the hardness for coatings from the aluminum oxide on the flow rate of transported gas. Gas consumption, m^3/h: O_2 – 0,93; C_2H_2 – 0,68. Spraying distance, mm: 100 (1), 150 (2), 200 (3), 250 (4).

An increase in the content of corundum in coatings during D – gun spraying is obtained with an increase in the thickness of a single layer to 8... 18 microns. However, the formation of α-Al_2O_3 in the coating is also influenced by other factors of D – gun spraying: an increase in the contact temperature by increasing the frequency of collision of powder particles with surface of part during the formation of a single coating layer, additional heat release by converting the kinetic energy of a particle into thermal energy upon impact. Even without taking these factors into account, at collision velocities above 500 m/s, the crystallization temperature of materials increases significantly, the contact temperature increases by hundreds of degrees. As the temperature rises, the time required for the transition γ-$Al_2O_3 \rightarrow \alpha$- Al_2O_3 decreases.

The influence of technological factors on the structure and properties of D – gun coatings from a mechanical mixture of powders of oxides $Al_2O_3 + Cr_2O_3$ (5%) with a particle size of ~50 μm has been studied. For the deposition of coatings, a small-sized D – gun machine that operating on a propane-butane-oxygen mixture was used.

The possibility of obtaining high-quality coatings from aluminum oxide using gases that replace acetylene has been established. The bond strength, determined by the pin method on steel samples is 40–50 MPa. The decisive influence on the phase composition of coatings from powder is exerted by the thickness of the applied layer per single cycle. With an increase in the thickness of a single layer, the cooling rate of molten particles decreases and polymorphic transformations intensify: $L \rightarrow \gamma \rightarrow \delta \rightarrow \theta \rightarrow \alpha$-$Al_2O_3$. For coatings formed from single layers of thinner thickness, an increased content of γ-Al_2O_3 is characteristic (Fig. 4 and 5). A clearly expressed regularity of an increase in the content of the α-phase with an increase in the thickness of a single coating layer is observed. The use of a short barrel with a variable cross-section makes it possible to obtain high-quality coatings with a lower consumption of propane-butane and oxygen (Fig. 6).

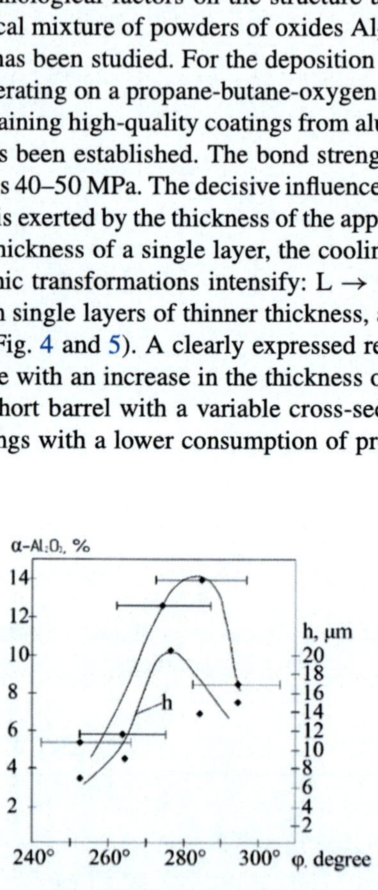

Fig. 4. Influence of the moment of pulsed injection of powder (mechanical mixture Al_2O_3 + Cr_2O_3 (5%)) on the content of α-phase and the thickness of a single layer of coating. Barrel length is 0,8 m. Gas flow rates, m^3/s: propane-butane – $0{,}145 \cdot 10^{-3}$; oxygen – $0{,}7 \cdot 10^{-3}$. Rate of fire is 4 Hz, spray distance is 90 mm.

The microhardness of the coatings in the cross section is nonuniform. In layers directly adjacent to the substrate (steel base), it is lowest and then increases as it approaches the outer surface of the coating. When using technological modes corresponding to the upper point of the curve for a barrel 0,8 m long, the microhardness of the layer adjacent to the base is 10130–14150 MPa, at a distance of 0,15 mm from the base it is 14150–21140 MPa, for the outer layer (at a distance of 0, 25 mm from the base) microhardness is 18020–23200 MPa. This is explained by a decrease in the cooling rate of single layers of the coating as the thickness of the formed coating increases, and thereby an increase in its thermal insulation properties. Therefore, as the coating layer grows, the $L \rightarrow \gamma \rightarrow \delta \rightarrow \theta \rightarrow \alpha\text{-}Al_2O_3$ transformation intensifies. To obtain coatings with a high content of $\alpha\text{-}Al_2O_3$, technological modes of spraying with an increased thickness of a single layer are recommended, as well as a decrease in the time of formation of a single layer by creating more concentrated flows of powder particles in the barrel. It is also recommended to overheat the molten particles and additional heating of the sprayed parts.

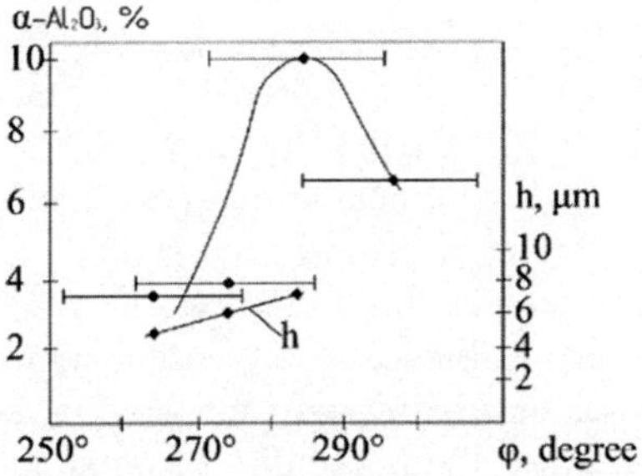

Fig. 5. Influence of the moment of pulsed injection of powder (mechanical mixture Al_2O_3 + Cr_2O_3 (5%)) on the content of the α-phase and the thickness of a single layer of coating. Barrel length is 0,435 m. Gas flow rates, m^3/s: propane-butane – $0{,}145 \cdot 10^{-3}$; oxygen – $0{,}7 \cdot 10^{-3}$. Rate of fire is 4 Hz, spray distance is 90 mm.

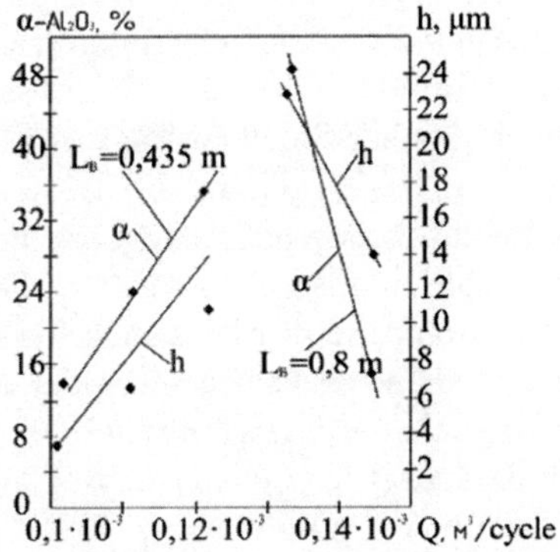

Fig. 6. Influence of the consumption of propane-butane on the content of the α-phase and the thickness of a single layer of aluminum oxide coating. Oxygen consumption, m^3/s: barrel 0,435 m – $0{,}6 \cdot 10^{-3}$; barrel 0,8 m – $0{,}7 \cdot 10^{-3}$.

4.2 Features of Obtaining of Coatings Based on Zirconium Dioxide

Zirconium dioxide ZrO_2 is a chemically resistant refractory compound with a melting point of 2688 °C. Among all oxides, it is characterized by the lowest thermal conductivity and is widely used in plasma spraying of heat-stressed products. When heated above 1182 °C, the monoclinic modification of ZrO_2 with a density of 5,56 g/cm^3 turns into a tetragonal one with a density of 6,1 g/cm^3, which leads to the destruction of the coatings upon cooling. Therefore, stabilization of ZrO_2 with oxides of calcium, cerium, yttrium or manganese is used, which leads to an increase in the linear coefficient of thermal expansion and resistance to thermal shock.

The peculiarities of depositing of coatings from ZrO_2 are associated with its low thermal conductivity, high density, and low coefficient of heat accumulation. The heating time of ZrO_2 particles to melting is 1,5... 2 times longer than for Al_2O_3 particles. Therefore, when depositing coatings from ZrO_2, it is necessary to increase the residence time of the powder in the flow of detonation products by a factor of 1,5... 2, as well as to provide more intensive heating of the powder particles.

For ZrO_2, the existence of two modifications was established: stable monoclinic and metastable tetragonal. Transformation temperature ≈1000 °C. In a number of works, the existence of a cubic high-temperature modification of ZrO_2 was established. The temperature of transformation of cubic ZrO_2 into tetragonal for pure ZrO_2 is 2285 °C.

Experiments carried out using a barrel with a tapered outlet section and a propane-butane-oxygen mixture confirmed the possibility of obtaining durable coatings from ZrO_2 powder with a particle size of up to 200 μm. On the X-ray diffraction patterns of coatings from highly dispersed stabilized ZrO_2 powder (up to 200 μm), it has a cubic modification in the initial state, and only lines of the cubic ZrO_2 modification are stored. By some distortion of the lattice, all the peaks of the lines of cubic ZrO_2 are shifted towards smaller angles. When using the output fine powder ZrO_2 (50 μm) as coating, the output powder has a monoclinic lattice with an insignificant content of cubic modification. In the coating, the content of the cubic modification slightly increased, as evidenced by higher peaks of cubic ZrO_2.

A study was carried out on the application of coatings from aluminized ZrO_2 powder containing 20% (wt.) of aluminum, has a cubic modification and stabilized CaO (5%). The phase line peaks in the X-ray diffraction patterns are slightly shifted towards smaller angles. The coating is highly durable compared to pure ZrO_2. However, there are no lines of aluminum in the X-ray diffraction patterns of coatings. There are only lines of cubic modification of ZrO_2. However, the line peaks became wider, asymmetrical and blurred towards larger angles. It is possible that Al interacts with ZrO_2, Al does not interact with ZrO_2 at 1600 °C. Aluminum oxides, as in plasma spraying, are not detected by X-ray diffraction. Probably, the interaction of aluminum with ZrO_2 occurs in case of D – gun spraying. Moreover, it is possible to reduce zirconium oxide and appearance of particles of Zr – Al solid solutions in the surface layer.

4.3 Features of Obtaining of Coatings Based on Titanium Oxide

Titanium oxide TiO_2 is a chemically stable crystalline compound with a melting point of 1840 °C. It is used to obtain coatings such as dielectric, wear-resistant, etc. A study by the application of coatings from a rutile concentrate that contains $TiO_2 = 93{,}5\ldots 95{,}5\%$; $Fe_2O_3 = 1{,}4\ldots 2{,}5\%$; $ZrO_2 = 0{,}5\ldots 1{,}2\%$; $SiO_2 = 0{,}3\ldots 1{,}7\%$; $Al_2O_3 = 0{,}2\ldots 0{,}4\%$ was carried out. The oxides accompanying titanium dioxide in the rutile concentrate carry out its natural alloying, which leads to a decrease in the melting point of the system and the formation of solid solutions with increased hardness compared to TiO_2. Rutile concentrate is a mixture of two forms of TiO_2 – rutile and anatase. The density of rutile is 4240 kg/m^3, anatase is 3830 kg/m^3. The melting parameter is lower than in Al_2O_3 and ZrO_2 and is equal to $D = 3{,}85 \cdot 10^{10}\ kj^2 \cdot kg^{-1} \cdot m^{-3}$, the heat accumulation coefficient is $36\ W \cdot m^{-2} \cdot K \cdot s^{-0{,}5}$. Liquid titanium dioxide has surface tension and viscosity values about half as compared to Al_2O_3. The coating was carried out on a DNP3 machine with a barrel length of 800 mm of variable cross-section with a volume of 0,6 L at a rate of fire of 1,5 Hz. The continuous oxygen consumption was 10^4 cm^3/cycle. Rutile concentrate powder with a particle size of up to 40 μm was fed with nitrogen at a flow rate of 30 cm^3/cycle through a nozzle coaxial with the barrel from the side of the closed end.

With a decrease in acetylene consumption and an increase in the spraying distance, the thickness of a single layer of coating decreases (Fig. 7), which is explained by a decrease in the degree of particle heating. In all cases, the formation of sufficiently strong coatings was ensured. With help of X-ray diffraction, the presence of various modifications of TiO_2, as well as other titanium oxides are revealed in coatings. Titanium oxide reduction takes place in stages: $TiO_2 \rightarrow Ti_3O_5 \rightarrow Ti_2O_3 \rightarrow TiO \rightarrow Ti$.

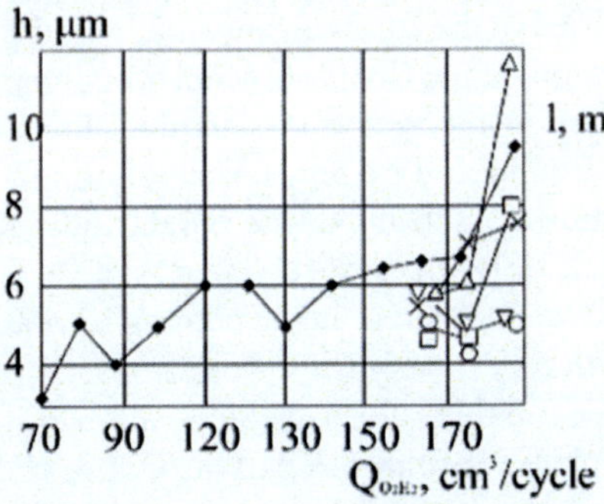

Fig. 7. Dependence of the thickness for the single layer of coating from the natural rutile powder on the consumption of acetylene and spraying distance.

In the investigated range of modes, the reduction of TiO_2 to Ti_2O_3 was observed. With a decrease in the acetylene content in the combustible mixture and an increase in the spraying distance, which is equivalent to an increase in the oxygen content in the combustible mixture and the residence time of TiO_2 powder particles in the flow of detonation products, the degree of completeness of the TiO_2 reduction in higher oxides increases. Simultaneously, when TiO_2 particles are heated, polymorphic transformations take place.

4.4 Features of Obtaining of Coatings Based on Iron and Cobalt Oxides

Aluminum oxide, zirconium dioxide, titanium dioxide belongs to the category of thermodynamically strong. When developing a technology for deposition of coatings from such oxides, usually try to prevent possible crystal-chemical transformations and to preserve dense and durable output structures in the coating. However, there are a number of oxides with low thermodynamic strength and high dissociation capacity. Typical representatives of this group are oxides of iron and cobalt. These materials can be used as electrode materials for electrochemical devices to replace electrodes with an active coating from expensive ruthenium oxide. The possibility of D – gun sprying of coatings from oxides Fe_3O_4 and Co_3O_4 were studied. Magnetite Fe_3O_4 melts congruently at 1597 °C, transforming into a melt of oxides (Fe_2O_3, Fe_3O_4, FeO). At room temperature, there are only two stable Fe oxides; hematite Fe_2O_3 and magnetite Fe_3O_4. At temperatures below 570 °C, wustite FeO (or FeO_{1-k}) decomposes according to the disproportionation reaction $4FeO = Fe_3O_4 + Fe$. With an increase in the volumetric content of nitrogen or oxygen in the combustible mixture, the content of Fe_3O_4 in the coating increases, and the effect of both components is approximately the same. With an increase in the degree of dilution of the combustible mixture with oxygen and nitrogen, the coefficient of use of powder decreases.

Obtaining of coatings from Co_3O_4 oxide is a difficult task, since it decomposes at 967 °C. For the deposition of coatings, Co_3O_4 powder with a particle size of ≤ 30 μm was used. A barrel with a diameter of 28 mm and a length of 850 mm was used. The rate of fire of machine is 2 Hz. The coating was applied to aluminum plates. At a spraying distance of 150 mm, the composition of the combustible mixture $C_2H_2{:}\,O_2 = 1{:}\,3$, and the nitrogen flow rate from 1,1 to 1,5 m^3/h, the Co_3O_4 powder was completely decomposed into CoO. With a further increase in nitrogen consumption, the content of Co_3O_4 in the coating increases. After improving the system for supplying of gases and powder to the barrel, changing the ignition conditions of the combustible mixture, increasing the oxygen content in the combustible mixture (C_2H_2: O_2 = 1: 3) and reducing the spraying distance to 35 mm, the Co_3O_4 content in the coating increases, reaching 100% at a nitrogen flow rate of 1,6 m^3/h.

When using acetylene-oxygen mixtures close in composition to stoichiometric without adding nitrogen, almost complete decomposition of the Co_3O_4 powder occurred, and metallic Co was predominantly present in the coating. A decrease in the distance of spraying and the distance between point of introduction of powder and barrel cut decreased the rate of decomposition of Co_3O_4 by decreasing the time of its stay in a high-temperature heterogeneous flow. However, under optimal conditions with a phase composition, it was not possible to obtain a coating with a thickness of more than 0,1 mm.

5 Conclusions

1. D – gun spraying allows the deposition of durable coatings based on titanium, iron and cobalt oxides.
2. When developing technological processes for D – gun spraying of the oxide coatings, it is necessary to take into account and ensure optimal conditions for the flow of leading transformations and phenomena that have a decisive effect on the structure

and properties of the resulting coatings. For this, the flow rate and composition of the combustible mixture, the geometry and dimensions of the barrel, the conditions for introducing the powder into the barrel, the single dose of powder, the spraying distance, the cyclorama of the working cycle, and other technological methods can be used.

3. On the example of depositing coatings from the aluminum oxide powder, the possibility of controlling the phase composition and properties of the coatings is shown. A significant effect of the thickness of single coating layers on the kinetics of polymorphic transformations and the phase composition of the aluminum oxide coatings has been established. The method of D – gun spraying makes it possible to obtain coatings with an increased content of α-Al_2O_3.
4. To obtain durable coatings from the aluminum oxide with a high α-phase content, D – gun spraying modes are recommended, which provide heating of the powder particles to 2000...2300 °C, the collision velocity with substrate 300...400 m/s, temperature of contact between the particles and the substrate during the formation of single microlayers is less than 2000 °C and the cooling rate is not more than $10^3 \ldots 10^4$ °C/s.

References

1. Udalov, Yu., German, A., Zhabrev, V.: Technology of Inorganic Powder Materials and Functional Coatings. Janus, Saint Petersburg (2001)
2. Bartenev, S., Fedko, Yu., Grigorov, A.: Detonation coatings in mechanical engineering. Mashinostroenie, pp. 58–60 (1982)
3. Budag'janc, N.: D – Gun Processes in Industry. Publishing house of VUGU, Luhansk (1998)
4. Baldaev, L., Khamitsev, B., Prokofiev, V.: Features of polymorphic transformations of detonation coatings from aluminum oxide. Strength. Technol. Coat. **4**(124), 25–33 (2015)
5. Psyllaki, P., Jeandin, M., Pantelis, D.: Microstructure and wear mechanisms of thermal-sprayed alumina coatings. Mater. Lett. **47**, 77–82 (2001)
6. Fesenko, A., Basova, Y., Ivanov, V., Ivanova, M., Yevsiukova, F., Gasanov, M.: Increasing of equipment efficiency by intensification of technological processes. Periodica Polytech. Mech, Eng, **63**(1), 67–73 (2019)
7. Ivanov, V., Pavlenko, I., Liaposhchenko, O., Gusak, O., Pavlenko, V.: Determination of contact points between workpiece and fixture elements as a tool for augmented reality in fixture design. Wireless Netw. **27**(3), 1657–1664 (2019). https://doi.org/10.1007/s11276-019-02026-2
8. Shevchenko, S., Mukhovaty, A., Krol, O.: Gear transmission with conic axoid on parallel axes. In: Radionov, A.A., Kravchenko, O.A., Guzeev, V.I., Rozhdestvenskiy, Y.V. (eds.) ICIE 2019. LNME, pp. 1–10. Springer, Cham (2020). https://doi.org/10.1007/978-3-030-22041-9_1
9. Tsankov, P.: Modeling of vertical spindle head for machining center. J. Phys: Conf. Ser. **1553**, 012012 (2020)
10. Krol, O., Sokolov, V.: Research of modified gear drive for multioperational machine with increased load capacity. Diagnostyka **21**(3), 87–93 (2020)
11. Saravanan, P., Selvarajan, V., Srivastava, P.: Study of plasma and d–gun sprayed alumina coatings using taguchi experimental design. J. Therm. Spray Technol. **9**(4), 505–512 (2000)
12. Semenov, S., Cetegen, B.: Experiments and modeling of the deposition of nano–structured alumina–titania coatings by detonation waves. Mater. Sci. Eng. **335**, 67–81 (2002)

13. Kundrák, J., Mitsyk, A., Fedorovich, V., Markopoulos, A., Grabchenko, A.: Simulation of the circulating motion of the working medium and metal removal during multi-energy processing under the action of vibration and centrifugal forces. Machines **9**(6), 1–22 (2021)
14. Kotliar, A., et al.: Ensuring the economic efficiency of enterprises by multi-criteria selection of the optimal manufacturing process. Manage. Product. Eng. Rev. **11**(1), 52–61 (2020)
15. Li, C.-J., Ohmori, A.: The lamellar structure of a detonation gun sprayed $A1_20_3$ coating. Surf. Coat. Technol. **82**, 254–258 (1996)
16. Sokolov, V., Porkuian, O., Krol, O., Baturin, Y.: design calculation of electrohydraulic servo drive for technological equipment. In: Ivanov, V., Trojanowska, J., Pavlenko, I., Zajac, J., Peraković, D. (eds.) DSMIE 2020. LNME, pp. 75–84. Springer, Cham (2020). https://doi.org/10.1007/978-3-030-50794-7_8
17. Sokolov, V., Porkuian, O., Krol, O., Stepanova, O.: Design calculation of automatic rotary motion electrohydraulic drive for technological equipment. In: Ivanov, V., Trojanowska, J., Pavlenko, I., Zajac, J., Peraković, D. (eds.) DSMIE 2021. LNME, pp. 133–142. Springer, Cham (2021). https://doi.org/10.1007/978-3-030-77719-7_14
18. Krol, O., Sokolov, V.: Research of toothed belt transmission with arched teeth. Diagnostyka **21**(4), 15–22 (2020)
19. Krol, O., Sokolov, V.: Modeling of spindle node dynamics using the spectral analysis method. In: Ivanov, V., Trojanowska, J., Pavlenko, I., Zajac, J., Peraković, D. (eds.) DSMIE 2020. LNME, pp. 35–44. Springer, Cham (2020). https://doi.org/10.1007/978-3-030-50794-7_4
20. Krol, O., Porkuian, O., Sokolov, V., Tsankov, P.: Vibration stability of spindle nodes in the zone of tool equipment optimal parameters. Comptes rendus de l'Acade'mie bulgare des Sciences **72**(11), 1546–1556 (2019)
21. Sobiecki, J., Ewertowshib, J., Babulc, T., Wierzchona, T.: Properties of alumina coatings produced by gas-detonation method. Surf. Coat. Technol. **180–181**, 556–560 (2004)
22. Venkataraman, R., Ravikumar, B., Krishnamurthy, R., Das, D.: A study on phase stability observed in as sprayed Alumina-13 wt. % Titania coatings grown by detonation gun and plasma spraying on low alloy steel substrates. Surface Coat. Technol. **201**(6), 3087–3095 (2006)
23. Andrenko, P., Rogovyi, A., Hrechka, I., Khovanskyi, S., Svynarenko, M.: Characteristics improvement of labyrinth screw pump using design modification in screw. J. Phys: Conf. Ser. **1741**, 012024 (2021)
24. Rogovyi, A., Korohodskyi, V., Khovanskyi, S., Hrechka, I., Medvediev, Y.: Optimal design of vortex chamber pump. J. Phys: Conf. Ser. **1741**, 012018 (2021)
25. Kosenko, N.: Polymorphism of aluminum oxide. IZV. universities. Chemistry and chemical technology, pp. 3–16 (2011)
26. Saravanan, P., Selvarajan, V., Rao, D.: Influence of process variables on the quality of detonation gun sprayed alumina coatings. Surf. Coat. Technol. **123**, 44–54 (2000)
27. Baldaev, L., Hamitsev, B., Baldaev, S., Prokofiev, M.: Influence of technological parameters of the process of detonation spraying of aluminum oxide on polymorphic transformations during the formation of the coating. Hard. Technol. Coat. **7**(151), 291–295 (2017)
28. Rogovyi, A., Korohodskyi, V., Medvediev, Y.: Influence of Bingham fluid viscosity on energy performances of a vortex chamber pump. Energy **218**, 119432 (2021)
29. Kundrák, J., Mitsyk, A.V., Fedorovich, V.A., Markopoulos, A.P., Grabchenko, A.I.: Modeling the energy action of vibration and centrifugal forces on the working medium and parts in a vibration machine oscillating reservoir with an impeller. Manufac. Technol. **21**(3), 364–372 (2021)
30. Cetegen, B., Semenov, S., Goberman, D.: Deposition of multi-layered alumina-titania coatings by detonation waves. Scripta Mater. **48**(10), 1483–1488 (2003)

Key Performance Indicators: A Perspective on Developing and Implementing Performance Indicators in a Textile Company Warehouse

Ana Francisca Costa[1](✉), Ana Pinto[1], Ana Catarina Mendes[1], Ana Rita Quinteiro[1], Carla Coelho[1], Carla Meira[1], Isac Pereira[1], Jorge Marinho[1], José Barbosa[1], Natacha Cunha[1], Rui Miranda[1], Nuno Drumond[2], and Rui M. Lima[1]

[1] Department of Production and Systems, Algoritmi Centre, School of Engineering, University of Minho, Campus de Azurém, 4800-058 Guimarães, Portugal
afscc16@gmail.com

[2] Continental – Indústria Têxtil do Ave, S.A., Rua António José da Costa, 200, 4760 – 608 Lousado, Portugal

Abstract. The performance of a company's warehouse influences its competitiveness. High-performance warehouse management allows the company to increase profit, reduce delivery time and provide better customer service. To have a perception of the warehouse management performance, it is necessary to identify the Key Performance Indicators (KPI) and be able to monitor them. Creating a performance monitoring system in an industrial organization requires a deep understanding of the organization, the need and culture, to dig the information and adapt the theoretical indicators to the needs of the organization. Thus, the main purpose of this article is to develop and implement a monitoring system for key performance indicators for an industrial organization in an operations area in which there are still no performance measurements/indicators, describing the main steps that allow the development of such a system. This system resulted from university-industry cooperation, which allowed to establish a set of KPIs and a Dashboard that led the company to easily start monitoring its performance.

Keywords: Key performance indicator · Warehouse performance measures · KPI dashboard · University-business cooperation

1 Introduction

Industrial organizations strive to increase their competitive advantage [1], which requires companies to improve their logistics performance [2], and, therefore, warehouse performance. The first step for improvement of performance is to measure it. A warehouse performance measurement is a method to measure activity performance, program or service which is provided by a warehouse [2].

These measures are meaningful if companies can "see" if their goals and objectives are properly accomplished, allowing them to compare achieved results with target objectives. Firms, activities and workers need indicators, as they fulfil the fundamental

J. Machado et al. (Eds.): icieng 2022, LNME, pp. 317–328, 2023.
https://doi.org/10.1007/978-3-031-09382-1_28

activities of measuring and in this way evaluating the status of the system, educating for a better service to the client because the measure indicates how is it possible to deliver value to customers and directing in a way that allows overcoming problems highlighted by the size of the gaps between the indicators and targets [3]. Even though literature exists that presents suggestions for KPIs for warehouse management, the process of implementing systems for monitoring those KPIs depends on contextual variables in each organization. Thus, describing the process of creating KPIs monitoring systems in different industrial organizations may help practitioners and other researchers. The main purpose of this article is to develop and implement a monitoring system for key performance indicators for an industrial organization in an operations area in which there are still no performance measurements/indicators, describing the main steps that allow the development of such a system. Moreover, this article identifies key performance indicators for a warehouse to help Logistics Operations Department to lead to better performance in this area. It should be noted that this article studies warehouse performance measurement in Continental-ITA, a warehouse of a textile company located in Lousado, Portugal and that the work was developed in cooperation with the Department of Production and Systems of the School of Engineering of the University of Minho.

2 Literature Review

Although the existing literature covers a wide variety of information on warehouse performance measurement, this review will focus on the following major themes: warehouse activities, warehouse performance measurement and meaningful indicators.

2.1 Warehouse Activities

According to Edward Frazelle, warehouse activity consists of five operations: receiving, putting away, storage, picking and shipping [4].

Receiving is the interface of a warehouse for incoming material flow. It involves the assignment of trucks to docks and the scheduling of unloading activities [5]. Put-away is the act of placing merchandise in storage. As consequence, storage is the physical containment of merchandise while it is waiting for demand or until it is released from storing. Order picking is the process of removing items from storage to meet a specific demand [4] and shipping is the activity that involves scheduling and assignment of trucks to deliver the orders, packing after picking and loading trucks [5].

2.2 Warehouse Performance Measurement

Performance measurement is used by organizations to ensure they are going in the right direction and achieving targets in terms of organizational goals and objectives [6].

David Parmenter [7] says that there are three types of performance measures:

1. Key result indicators (KRIs): tells how it was done.
2. Performance indicators (PIs): tells what to do.

3. Key Performance Indicators (KPIs): tells what to do to increase performance dramatically.

The same author defines KPIs as a set of measures that focus on those aspects of organizational performance that are the most critical for the current and future success of the organization. KPIs are rarely new to the organization because they are not always identified or active.

For Eric Peterson [8], key performance indicators are always rates, ratios, averages or percentages; they are never raw numbers because they do not provide context. Without context, 500 is just a number. Not good, not bad, therefore not informative. Raw numbers are necessary to provide context to the reports and to promote debates, but they are not key performance indicators. Thus, Key performance indicators are designed to summarize meaningfully compared data [8]. It is something that can be counted and compared and provides evidence of the degree to which a goal is being achieved in a specific period [12].

Some authors debate the differences between the term "Metric" and "KPI". Wayne Eckerson makes this distinction in a very peculiar way. For this author, the only difference between a metric and a KPI is that a KPI embodies a strategic objective and measures performance according to the goal [9], while a metric is nothing more than something that is measured [10]. Therefore, it is correct to say that the two terms are complementary, but should not be confused, since the "metric" is generic and the "KPI" is specific.

There is no agreement, among researchers in identifying which KPIs are suitable for warehouses [2]. The concept of performance measurement suggests that organizations use types of measures that fit their strategy, their organizational structure and the environmental uncertainty that they face [11].

Despite that, the following table shows a summary of 25 KPIs that are adequate for warehouses. This summary is presented by Kusrini, Novendri and Helia [2], and is based on the Frazelle model [4] (Table 1).

This summary can be used as a guide, but it is important to note that every company has its specific needs, which requires an adaptation to the company's specific context.

2.3 Meaningful Indicators

Good key performance indicators are well defined, well presented, create expectations and drive to actions [8]. As explained before, the golden rule for key performance indicators is that KPI must be based on objectives [12]. In a warehouse scenario, the objectives may be to reduce the stock, increase the number of units stored or increase the usage storage.

Another golden rule is the one that defends that KPIs must have owners: that means an individual one and not an entity such as a department [12]. For each KPI, there must also be a description of the KPI and its purpose, a label with a short description, an updater individual who is responsible for updating the KPI at the pre-defined periods, a precise calculation with a mathematical formula, a frequency for measuring and recording the KPI and data sources used for calculation [12].

The purpose of measuring and reporting performance is to enable an organisation to be managed effectively and be capable of taking decisions. Thus, key performance

Table 1. Warehouse key performance indicators [2].

Activity	Financial	Productivity	Utilization	Quality	Cycle time
Receiving	Receiving cost per line	Receipts per man-hour	%Dock door utilization	%Receipts processed accurately	Receipt processing time per receipts
Put away	Putaway cost per line	Putaway per man-hour	%Utilization of putaway labour and equipment	%Perfect putaway	Putaway cycle time (per putaway)
Storage	Storage space cost per item	Inventory per square foot	%Locations and cube occupied	% Locations without inventory discrepancies	Inventory days on hand
Order picking	Picking cost per order time	Order lines picked per man-hour	%Utilization of picking labour and equipment	%Perfect picking lines	Order picking cycle time (per order)
Shipping	Shipping cost per customer order	Orders prepared for shipment per man-hour	%Utilization of shipping docks	%Perfect shipments	Warehouse order cycle time

indicators should either drive to action or provide a warm, comforting feeling to the ones that are reading them [8]. It is important to know who to congratulate when a number improves by 5% or what to do if a number decline by 5%.

Finally, there should be an accurate report to present the KPI. Reading KPIs should be straightforward. Users should be able to look at a visual display and know instantly whether a process or project is on track [9]. Having reports that show graphics, the best and the worst scenarios and alerts, are winning reports. Many web analytics application vendors provide direct data access from Microsoft Excel which can simplify the report generation process and give a rich display [8].

3 Methodology

3.1 Problem Description

This project was developed in partnership with the company Continental-ITA (C-ITA), which is one of the Continental Group factories that operate in the textile area. It produces three major products: Tire Cord Fabrics, Tire Cord Cables and Knitted Fabrics.

Previous year, C-ITA raised 87 million euros due to its sales. The company is expecting sales growth this year. Due to this prevision, there is an urgent need to closely monitor all movements of the warehouse, for example, the use of storage space, the number of warehouse-ins and outs, the level of stock of each product and the quality service level of their internal logistics team. Despite having performance indicators, C-ITA has no performance indicators for the warehouse, has no objectives aligned with a way to measure them, has no way to compare results over time and has no visual display to show what is happening in the warehouse. All these constraints lead to the need to develop a monitoring system for key performance indicators. This need raised the opportunity to develop a research project aiming to answer the following questions: What are the warehouse key performance indicators (KPIs) most appropriate to this industrial organization? How could these KPIs be calculated? How could a monitoring system of warehouse KPIs be implemented in this type of organization?

3.2 Action Research

Action Research is an approach to research that seeks understanding and at the same time attempts to improve the situation under investigation. In action research, the researchers are not only observing; they are actively working in making the change happen, working cooperatively with the client, and continuously adjusting to new information and events [13]. It is a methodology that assumes that social phenomena are continually changing rather than static [14], and, so, this approach is the most adequate for this project, which ultimately aims to change the organization in such a way that it adopts a new system. In this way, the research project aims to iteratively develop a set of steps to support this change, helping other practitioners and researchers working in similar or comparable environments. The final solution for the monitoring system was qualitatively evaluated through open questions answered by the main users of the system.

4 Development of the Action-Research Project

For almost four months, there were weekly visits to Continental-ITA to gather with the responsible members of the Logistics Operations Department and contact with the company's daily life.

Each visit had different aims. Initially, it was necessary to understand the company's reality and its needs, so it could be possible to define objectives. So, in these early weeks, the aim was to get answers to the following questions: How often does the company want to measure indicators? Who will have the responsibility to measure it? What is the purpose of having indicators? Which activities need the most attention? How will data be acquired?

Regarding the first question, the company aimed to monitor the indicators every month and, as it is something new that needs to be tested, it was established that the operations manager was responsible for obtaining the measures.

In the early meetings, the company also shared some indicators they already had thought about their need. These indicators were: the amount of material received, movements made to fulfil the production, batches received in the finished goods inventory area, loaded trucks, maximum storage units, average stock per product, outbound warehouse stock, own goods, and stock on commission. In each indicator, it was necessary to relate the necessary workers, with the hours used and the activity performed, to realize, in the future, the ideal number of these variables to perform the activities efficiently.

Subsequently, it was necessary to collect the data and analyse them. In this case, the company had one requirement: all data must be obtained through its management system (SAP - Systems Applications and Products). Analysing data was, by far, the most challenging step. In academic problems, the data is clean and ready to use, but in companies, there is a need to look for the data and to filter information according to the needs and requirements. In this case, it was necessary to understand the reports, which meant it was required to be aware of the warehouse routine. It was important to be engaged with the company and to look at warehouse work to see the loads and unloads, the receiving and the shipments, the picking and the put-away in the lines and the daily usage of the storage space.

Observing activities was also helpful for grouping indicators into different categories. The criteria used to make the distinction was based on the different warehouse activities. With this distinction, a more efficient way was also created for the KPI owner to quickly understand the purpose of the measure.

Due to this observation, in the second cycle of the research project, it was decided that other indicators, such as ratio sales *vs.* finished goods or storage utilization were crucial for measuring warehouse performance. In turn, it was possible to understand that some indicators listed by the company were not adjusted to the requirements and the reality of the company. For example, there are no inventory records in the outbound warehouse because the system only provides a snapshot of the current warehouse state. Thus, instead of starting to measure this indicator, it is necessary to find a way to create records about this problem. Because of these new findings, it was necessary to restructure the information provided and put it into tangible objectives.

With the help of Logistics Operations Department members, it was possible to come to a common understanding and to develop eleven warehouse indicators, namely: the number of loaded batches, amount of unloaded batches, stock accuracy, FEFO overrule (First Expired, First Out - meaning goods are dispatched from the warehouse according to their expiration date), ratio sales *vs.* finished goods in stock, number of raw materials moved to the production area, average stock of each product, storage utilization and inventory per square meter.

Textile products are developed to meet certain parameters, e.g., conductivity, temperature resistance, etc., so, after a specific period, the fabrics start to lose their properties. If the material has its validation date expired, it is not possible to use it in the production/sell to the customer, which creates waste. In this sense, it is important to be aware of the validation date of the materials and act according to the FEFO.

After being acquainted with the data, it was time to develop the key performance indicator formulas. For that, it was used Microsoft Excel with three fundamental interfaces: data navigation, ways to calculate the indicators and a dashboard to display the performance indicators and enable monitoring of the warehouse performance.

5 Definition of the Warehouse KPIs

As said above, warehouse performance measurement is a novelty to Continental-ITA. An organizational novelty brings the need to deal with some resistance to the change. To minimize this struggle, it is important to make people feel part of the process and to involve them in it. It is important to talk with those who work in the warehouse, the middle management and the senior management. With everyone engaged in the process, there are more guarantees to successfully implement the change, because everyone knows why this measurement is necessary, are capable of reading and understanding the values and are truly connected to the defined objectives. Additionally, it is necessary to identify and make an interpretation of the existing data in the company, to integrate information and create useful indicators.

Once these steps are accomplished, it is important to have frequent meetings with the team and discuss the project. In the first meetings, a brainstorming strategy is important to identify the objectives and, with them, outline the key performance indicators to track. It is also useful to think that the company representatives will be the users of the system and so it is necessary to listen to their requirements and perceive their daily routine.

Besides that, it is important to understand that few indicators are always better than none, and, consequently, it is better to start slowly and have something that reflects the objectives defined, than want to have a large range of indicators and non-suitable ones.

As explained earlier in this article, a number without a context is just a number, so, to have an accurate reading of the performance, it is essential to define standards with which data will be compared to. Having standards is a key to tracking the performance and extracting meaningful analyses.

Finally, as important as having standards, is to have formulas to calculate the indicators. Table 2 shows the KPIs developed for the C-ITA's warehouse, which is described below.

Table 2. KPI's developed for the C-ITA's warehouse

Warehouse Activity	KPI	Formula
Receiving	Number of unloaded batches	$Q_{Unloaded}/(M * 8)$
Shipment	Number of loaded batches	$Q_{Loaded}/(M * 8)$
	The average stock of each product	$\frac{\text{Stk\_Avg}}{Tot\_Stk}$
Quality's Service	Stock accuracy	$\frac{Stk\_Dev}{Tot\_Stk} * 100$
	FEFO overrule	$\frac{Stk\_Disp}{FEFO\_events} * 100$
Internal Logistics team	Number of raw materials moved to the production area	$Q_{Moved}/(M * 8)$
	Storage utilization	$\frac{Stk\_Stg}{Stg\_Cap} * 100$
Storage	Ratio sales *vs.* finished goods in stock	$\frac{Q\_Sales}{\text{FG\_Stored}} * 100$
	Inventory per square meter[2]	$\frac{PU\_Stored}{Max\_PU}$

Regarding receiving and shipping, the company wants to control the number of loaded and unloaded batches, respectively, the quantity of raw material received (Q_Unloaded) and loaded (Q_Loaded). This indicator reflects the amount of material moved per person (M), per hour, considering an eight-hour shift.

The "Average stock of each product", shows the relationship between the average of stock (Stk_Avg) that is stored, for each type of product (Raw Materials, Chemicals and Finished Goods), and the total stock stored (Tot_Stk).

Concerning "Stock Accuracy", this is an indicator that reveals the percentage of items that are wrongly recorded in the database. To obtain this value, it is necessary to compare the number of deviations recorded (Stk_Dev), in kilograms, and the total stock stored (Tot_Stk).

The "FEFO Overrule" calculates the percentage of events in which FEFO's policy is not respected. To calculate this indicator, it is crucial to perceive the total number of dispatches that are made to production (Stk_Disp) and compare it with the number of times that the FEFO's rule is not respected (FEFO_Events).

Regarding the performance of the warehouse's team, it is important to determine the number of raw materials moved to the production area, to fulfil the needs of the production. This indicator is based on an eight-hour shift and reveals the number of goods sent to the production (Q_Moved), per person (M), per hour, considering an eight-hour shift.

Storage utilization reflects the percentage of storage space used (Stk_Stg), compared to the total usable space capacity of the warehouse (Stg_Cap). This indicator calculates the storage utilization of each type of product (raw material, chemicals and finished goods).

With "Ratio sales vs. finished goods in stock", it is possible to perceive the connection between the number of finished goods that the company has stored (FG_Stored), and the number of products sold (Q_Sales).

Finally, "Inventory per square meter" reflects the area used to store the inventory. This indicator allows the user to perceive the number of packing units stored (PU_Stored), compared to the total of packing units that are possible to store in that area (Max_PU). Once more, this indicator is designed to analyse each type of product (raw materials, chemicals and finished goods).

6 Visual Management of the Warehouse KPIs

According to Bell and Davidson [15], the visual is getting prevalent in our modes of communication and management. This means that visual strategies are powerful tools to lead to a faster understanding of potential issues. According to this fact, it is important to note that keeping KPIs visible to the organization is a way to ensure that they will be understood and fulfilled by the majority. In addition to visual management, it is also relevant to develop a way to intuitively track measures. If a company does not have ways to measure its performance, probably that is related to the time factor. Companies are always running against time, so to settle some changes, it is important to develop intuitive and time saver policies. That is why it is recommended to use interactive dashboards, with interactive graphics and meaningful colours (for example, red for what is not expected and green for what is a good practice). Using Microsoft Excel, companies can reach some really good results. At C-ITA, it was developed a Microsoft Excel spreadsheet solution with visual basic for applications (VBA) code to treat the data, that includes a dashboard, storage for monthly data, the formulas that calculate the indicators, and storage to import reports exported via the SAP system.

The dashboard represented in Fig. 1, with fictitious data for purpose of illustration, allows to monitor the KPIs monthly, showing the value from the current month, as well as its relationship with the value of the previous month. The user can understand if some data has increased or declined compared to the month before, by analysing the highlights in red and in green that appear in the dashboard.

Fig. 1. KPIs dashboard

The VBA code is used to save time for the user to easily navigate the data and, consequently, to guarantee that the monitoring is truly implemented. This code was developed to make it easier to use the system. This VBA code is specially developed for the company to filter its raw material, its finished goods, its different storage locations and to import data from the original SAP report to the spreadsheet that manipulates the data. So, having something intuitive, and a time-saver, is more likely to be implemented than having something that needs more steps to achieve the same goal.

7 Benefits for the Company

Project results are only meaningful if it is possible to look at them and perceive their usefulness. The question that emerged is "What are the benefits for the company engaged in the process?".

This question was researched both in a quantitative way, by the application of the model and in a qualitative way, using data collected through open questions answered by company representatives.

The team applied the performance indicators to one month data, which allowed, on one hand, to test the formulas developed and, on the other hand, to start creating a historic about these KPIs. Table 3 allowed the company to start monitoring the activities performed in the warehouse, verify the usefulness of the KPIs, and be aware of the tasks that needed more attention for potential improvements.

Table 3. KPIs calculated for one month

KPI	One month
Number of loaded batches (Kg)	2430,06
Number of unloaded batches (Kg)	2787,23
Picking productivity (%)	15%
Number of raw material moved to the production (Kg)	373,49
Ratio sales vs finished goods in stock (%)	110%
FEFO overrule (%occurrences)	27%
Stock sccuracy (%deviations)	3%
Stock RM (Kg)	587636
Stock Chemicals (Kg)	41335
Stock cable (Kg)	186676
Stock Ply Client type 1 (Kg)	1455669
Stock Ply Client type 2 (Kg)	0

In this sense, the company representatives were asked about what the main advantages of this type of approach were. According to one of the professionals:

> *"For C-ITA, as for any productive organization, the measurement of data for its analysis and correlation, to understand the performance standard, is fundamental so that leadership takes its places, communication flows and the operational and strategic parts are accomplished. Due to the competitive times, we are facing in the industry, the ability to predict, anticipate and, above all, know the company's performance, is the foundation for a sustainable value chain and for what we compromise to deliver, every day: create value. Thus, the restructuration and aggregation of the key performance indicators in a new format of monitorization create the opportunity to forecast events, through the analysis of a large amount of data and correlations." [Professional A]*

When asked about the perspective of the company regarding the applicability of the indicators developed, the company representatives showed appreciation for the solution developed, as can be seen by the following excerpts.

"Since it was a requirement in the initial proposal of the project, its potential was always high. Specifically focusing on the product delivered..., its applicability is immediate. This happens because, not only was delivered a new structure for the warehouse KPIs, based on what was suggested by C-ITA but it was also created a digital tool, connected with SAP so that the data collection and its monitorization is more efficient and user-friendly." [Professional A]

"The study developed is a first and major step for C-ITA towards a quantified and strategic methodology to measure performance in warehouse operations. We believe that with the KPIs described together with the corresponding tools to collect, manipulate and share the results, the company will be able to quickly start measuring performance and soon define targets based on those. It was another good example of how academics "join forces" with the industry to bring the best of the two worlds." [Professional B]

Based on these statements, it is possible to note that this approach is a "starting point" that can be explored to start measuring the performance of a specific area in this industrial organization. Although the project is still in the initial implementation phase, it is clear that, for this company, having tools that are engaged in the culture of the organization is a good strategy to guarantee the success of the project.

It is also relevant to note that different organizations have different needs and different goals. To this company, it was decided to measure the number of unloaded batches, the number of loaded batches, the stock accuracy, the FEFO overrule, the number of raw materials moved to the production area, the storage utilization, the average stock of each product, the ratio sales *vs.* finished goods in stock and the inventory per square meter.

In short, the research led to a prioritization of five important steps that can be considered when there is a need to develop and implement key performance indicators. These steps are: engage with the people and the process to be part of the solution; listen to the user and look for a solution that suits the requirements; design tools that are linked to the daily routine of the user, to ensure the efficient use of the work developed; adjust the work to the reality of the user and develop visual tools, in other to make things easy to understand and, consequently, to help on decision-making.

8 Conclusion

This article focused on the definition, development and implementation of a monitoring system for key performance indicators for an industrial company and deriving a set of steps that could be used in similar or comparable types of companies. This objective was accomplished, and the company verified the usefulness of the model and system developed.

It is also important to refer that this work had some limitations such as the fact that this research is based on the reality of one single company and responds to its specific needs. In the future, would be interesting to study, in detail, the impact of the KPIs in the company, develop a solution based on the Power BI, as well as make a detailed evaluation of the challenges of using the indicators in this and other companies.

Acknowledgements. This work has been supported by FCT – Fundação para a Ciência e Tecnologia within the R&D Units Project Scope UIDB/00319/2020.

References

1. Ilieş, L., Turdean, A.-M., Crişan, E.: Warehouse performance measurement – a case study. Reasearch Gate Net, pp. 1–5 (2009)
2. Kusrini, E., Novendri, F., Helia, V.N.: Determining key performance indicators for warehouse performance measurement – a case study in construction materials warehouse. MATEC Web of Conferences, pp. 1–4 (2018)
3. Franceschini, F., Galetto, M., Maisano, D.: Management by Measurement: Designing Key Indicators and Performance Measurement Systems. Springer Science & Business Media (2007)
4. Frazelle, E.H.: World-Class Warehousing and Material Handling, 2nd edn. McGraw-Hill, New York (2002)
5. Gu, J., Goetschalckx, M., McGinnis, L.F.: Research on warehouse operation: a comprehensive review. Eur. J. Oper. Res. **177**(1), 1–21 (2007)
6. Ishaq Bhatti, M., Awan, H.M., Razaq, Z.: The key performance indicators (KPIs) and their impact on overall organizational performance. Qual. Quant. **48**(6), 3127–3143 (2013). https://doi.org/10.1007/s11135-013-9945-y
7. Parmenter, D.: Key Performance Indicators: Developing, Implementing and Using Winning KPIs, 1st edn. Wiley, New Jersey (2007)
8. Peterson E.T.: The Big Book of Key Performance Indicators. 1st edn. Web Analytics Demystified Series (2006)
9. Eckerson W.W.: Performance Management Strategies: How to Create and Deploy Effective Metrics. The Data Warehouse Institute (2009)
10. Kerzner. H.: Project Management Metrics, KPIs, and Dashboards: A Guide to Measuring and Monitoring Project Performance. 1st edn. Wiley, New Jersey (2011)
11. Gosselin. M.: An empirical study of performance measurement in manufacturing firms. Int. J. Product. Perform. Manage **54**, 419–437 (2005)
12. Clive. K.: Developing Meaningful Key Performance Indicators/Strategy Resources/Intrafocus Ebooks-Downloads. https://www.intrafocus.com/services/developing-meaningful-kpis/. Accessed 28 Dec 2019
13. Coughlan P., Coghlan. D.: Action research for operations management. Int. J. Oper. Product. Manage. **22**(2), 220–240 (2002)
14. Easterby-Smith M., Thorpe R., Jackson P.R., Jaspersen. L.J.: Management & Business Research. 6th edn. SAGE (2018)
15. Bell E. and Davison. J.: Visual management studies: empirical and theoretical approaches. Int. J. Oper. Product. Manage. **15**, 355–375 (2013)

Composite Materials Based on GPPS/PE-HD Matrix Used in Automotive Industry

Lucia Knapčíková[1](✉), Justyna Trojanowska[2], Matúš Martiček[1], and Jozef Husár[1]

[1] Faculty of Manufacturing Technologies With a Seat in Prešov, Department of Industrial Engineering and Informatics, The Technical University of Košice, Bayerova 1, 080 01 Prešov, Slovak Republic
{lucia.knapcikova,jozef.husar}@tuke.sk, matus.marticek@student.tuke.sk

[2] Poznan University of Technology, ul. Piotrowo 2, 61-138 Poznań, Poland
justyna.trojanowska@put.poznan.pl

Abstract. Polymer composites have many advantages over metal and ceramic composites. The most important are high specific strength, relatively easy production and processing, corrosion resistance, better dynamic properties, good damping properties, high dimensional stability and the ability to produce complex parts at once or assemble them from a smaller number of elements (integral construction). High specific strength was the main reason reinforced plastics are suitable for applications that have to withstand impacts. The automotive industry is a complex industry that affects mechanical, electrical, chemical and, of course, plastics and rubber production. The requirements of the automotive industry have changed quite significantly recently. The most important factors influencing car design today are cost, final consumption and pollutant emissions. At the same time, consumption and emissions were not taken into account very significantly a few years ago. The presented paper aims to thermal analysis of composite materials based on polyethylene (PE) and polystyrene (PS) and Maleinanhydrid a Di-terc-butyl peroxide were used as additives.

Keywords: Composite materials · Thermal analysis · DSC · Automotive industry

1 Introduction

Plastics and polymer composites are becoming more and more popular in automobiles. They contain various functional materials that exhibit a wide range of desirable properties [1]. They are strong and light, resistant to chemicals and harsh environments, and are excellent thermal and electrical insulators. They can be transparent or opaque, soft, flexible, solid, or they can be of almost any shape and size. They are heat resistant and corrosion resistant and are recyclable [2]. Plastics are about seven times lighter than steel and nine times more diluted than copper and its alloys. Together with their cost-effectiveness, these advantages have made plastics the preferred material in a wide

J. Machado et al. (Eds.): icieng 2022, LNME, pp. 329–338, 2023.
https://doi.org/10.1007/978-3-031-09382-1_29

range of commercial applications [3]. The use of composite materials in the automotive industry is growing every year. This trend is expected to continue in the coming years due to increasing demand for new cars and an increase in the share of composites in car construction. In many applications, composite materials replace metallic materials, which contributes to significant weight reduction in new cars (up to 35%) [4]. Various new weight reduction technologies will help further reduce the consumption and emissions of future cars. By increasing the production volumes of composite materials, they will become even more competitive against metallic materials [2, 4]. By using composites, it is possible to reduce the resulting weight of components and thus of the entire car while maintaining or even improving the required characteristics. The vehicle's lower weight will thus be reflected mainly in the reduction of fuel consumption and air emissions, which is an important environmental aspect of sustainable development today [3, 4]. By reducing the vehicle's weight by 10%, its consumption will be reduced by up to about 7%, which of course, will already affect fuel consumption in real life [1].

2 Material and Methods

2.1 Input Materials and Additives

Polyethylene (PE) is a durable thermoplastic with a variable crystalline structure. It is one of the most commonly produced plastics globally (tens of millions of tons are produced in the world every year). Polyethylene is used in applications on films, tubes, plastic parts, laminates, etc., in several markets (packaging, automotive, electrical industry, etc.) [5, 6].

Polypropylene (PP) is a tough, rigid crystalline thermoplastic made from a propene (or propylene) monomer. It is a linear hydrocarbon resin. The chemical formula of polypropylene is (C_3 H_6) _n. Polypropylene is one of the cheapest polymers available today. PP belongs to the group of polyolefin polymers and is one of the three most commonly used polymers today. Polypropylene is used as plastic as well as a fibre in the automotive industry, industrial applications, consumer goods, etc. [6].

Polystyrene (PS) is a synthetic aromatic hydrocarbon polymer made from styrene monomer. Polystyrene can be solid or foamed. General-purpose polystyrene is clear, hard and rather brittle. It has a fairly low resistance to oxygen and water vapour and has a high glass transition temperature [4, 5].

Maleinanhydride it is a maleinanhydride, a colourless or white solid with a strong odor, and is a plasticizer. It is manufactured on a large scale for applications in coatings and polymers [5, 6].

Di-tert-butyl peroxide is primarily used as a polymerization initiator and crosslinking agent. The substance is used as a radical initiator to induce polymerization (free radical polymerization) for the production of plastics (polypropylene visbreaking, acrylic resins) and as a crosslinking agent for the production of resins (resin hardener) [5, 6].

2.2 Differential Scanning Calorimetry

Differential Scanning Calorimetry (DSC) is a thermal analysis technique that monitors how the thermal capacity changes with temperature. A sample of known weight is heated or cooled, and changes in its heat capacity are observed as changes in heat flow [4, 7]. This allows the detection of melts, glass transitions, phase changes and curing. Because of this flexibility, as most materials have some kind of transitions, DSC is used in many industries, including pharmaceuticals, polymers, food, paper, printing, manufacturing, agriculture, semiconductors, and electronics [6]. Based on the data obtained during the test, the TestExpert software creates a graph of the temperature dependence on the amount of heat transferred by the sample. DSC graphs give a clear visual idea of the composition of the tested samples. For this purpose, the twin-screw extruders Brabender, with a circular nozzle diameter of 8 mm was used. The produced samples (S1-S21) were manufactured in the following ratios: 10/90, 50/50 and 90/10, first without the additive, and then with the additive in the ratios: 10/90/5/1. The 5% of the total weight of the mixture of two polymers is the percentage of maleic anhydride and 1% of the total weight is the percentage of di-tert-butyl peroxide. Di-tert-butyl peroxide was used to activate the reaction between the polymers and the maleinanhydride (Table 1). The weight of the mixture of two polymers is always 300 g [8].

Table 1. Sample with selected types of fillers (additives) [6].

Sample	HDPE	PP	GPPS	HIPS	MA	DTBP
S1	100% (300 g)	–	–	–	5% (15 g)	1% (3 g)
S2	–	100% (300 g)	–	–	5% (15 g)	1% (3 g)
S3	–	–	100% (300 g)	–	5% (15 g)	1% (3 g)
S4	–	–	–	100% (300 g)	5% (15 g)	1% (3 g)
S5	10% (30 g)	90% (270 g)	–	–	5% (15 g)	1% (3 g)
S6	50% (150 g)	50% (150 g)	–	–	5% (15 g)	1% (3 g)
S7	90% (270 g)	10% (30 g)	–	–	5% (15 g)	1% (3 g)
S8	10% (30 g)	–	90% (270 g)	–	5% (15 g)	1% (3 g)
S9	50% (150 g)	–	50% (150 g)	–	5% (15 g)	1% (3 g)
S10	90% (270 g)	–	10% (30 g)	–	5% (15 g)	1% (3 g)
S11	10% (30 g)	–	–	90% (270 g)	5% (15 g)	1% (3 g)
S12	50% (150 g)	–	–	50% (150 g)	5% (15 g)	1% (3 g)
S13	90% (270 g)	–	–	10% (30 g)	5% (15 g)	1% (3 g)

(continued)

Table 1. (*continued*)

Sample	HDPE	PP	GPPS	HIPS	MA	DTBP
S14	–	10% (30 g)	90% (270 g)	–	5% (15 g)	1% (3 g)
S15	–	50% (150 g)	50% (150 g)	–	5% (15 g)	1% (3 g)
S16	–	90% (270 g)	10% (30 g)	–	5% (15 g)	1% (3 g)
S17	–	10% (30 g)	–	90% (270 g)	5% (15 g)	1% (3 g)
S18	–	50% (150 g)	–	50% (150 g)	5% (15 g)	1% (3 g)
S19	–	90% (270 g)	–	10% (30 g)	5% (15 g)	1% (3 g)
S20	–	–	10% (30 g)	90% (270 g)	5% (15 g)	1% (3 g)
S21	–	–	50% (150 g)	50% (150 g)	5% (15 g)	1% (3 g)
S22	–	–	90% (270 g)	10% (30 g)	5% (15 g)	1% (3 g)

Legend to the Table 1:
HDPE- High-Density Polyethylene
PP- Polypropylene
GPPS- General Purpose Polystyrene
HIPS-High Impact Polystyrene
MA- Maleinanhydride
DTBP- di-tert-butyl peroxide

With a set temperature too high in the extrusion, the polymer structure can be degraded, leading to degradation of the chemical and mechanical properties of the samples. On the other hand, at low temperatures in extruders, due to the melting point of one or two polymers, the material will not be able to melt properly and completely, which will also affect the ability to form new chains in the chemical structure of the blends [9, 10].

Another important factor influencing the final chemical structure of the samples is cooling. Large manufacturing plants that use extruders use "modular cooling die" that evenly cools the material throughout its area. In our case, the samples cooled on the conveyor belt and their structure could affect [11, 12]:

- changing air temperature,
- the temperature of the upper layer of the conveyor belt,
- humidity in the room.

3 Results and Discussion

As already mentioned, differential scanning calorimetry (DSC) analysis gives an idea of the structure of the material and its components. The DSC test is used for amorphous and crystalline polymers [6, 13]. Amorphous polymers in our cases include GPPS and HIPS, and all other polymers are as crystalline: HDPE and PP. The crystallization drop and melting peak only occur with polymers that can form crystals (crystalline polymers) [14]. Completely amorphous polymers will show no crystallization or melting. But polymers with both crystalline and amorphous domains will show all the transitions. If you look at the DSC graph (Fig. 1), it can see a big difference between the glass transition and the other two thermal transitions, crystallization and melting. As for the glass transition, there is no decline and no peak. It is because latent heat is not released or absorbed by the polymer during the glass transition. Both melting and crystallization involve the release or absorption of heat [15]. The only thing we see at the glass transition temperature is the change in the heat capacity of the polymer [13]. Figure 1 shows the difference between a crystalline and an amorphous polymer [16]. HDPE is a crystalline polymer, reaching a melting point (Tm), expressed by a characteristic peak at 133 °C. The GPPS material, in turn, is an amorphous polymer that has reached a glass transition (Tg) at 94 °C. A 50/50 blend of the two polymers show the black curve's result. The resulting material has a crystal structure, so melting at the peak temperature of 135 °C is assumed.

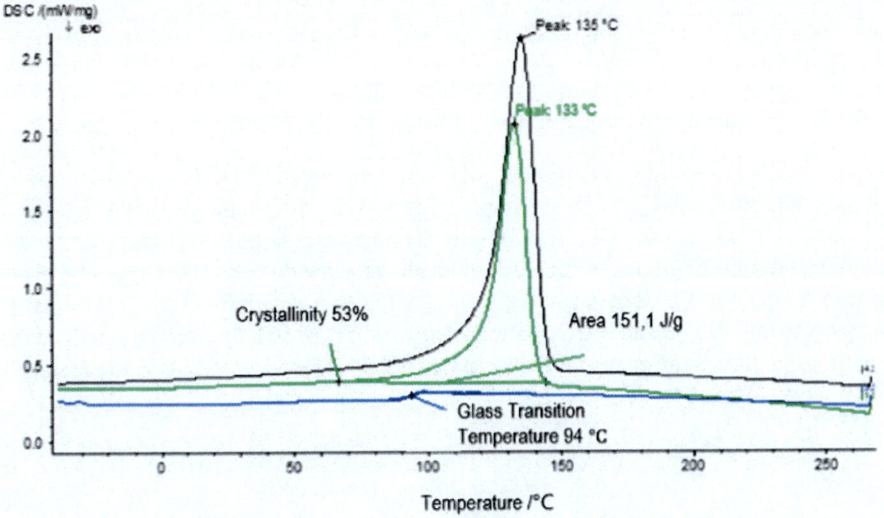

Fig. 1. DSC graph of sample S9: black curve - GPPS/PE-HD 50/50; green - HDPE; blue – GPPS [Authors own processing].

Upon addition to 5% Maleinanhydride, the DSC graph (Fig. 2) changes, resulting in a glass transition and melting of the material, which characterizes the presence in the structure of both S9 and amorphous and crystalline structures. It should also be noted that the glass transition temperature changes slightly (from 94 to 91 Å) and the melting peak does not change. By adding MA and the DTBP reaction catalyst, we combined the two structures in one sample.

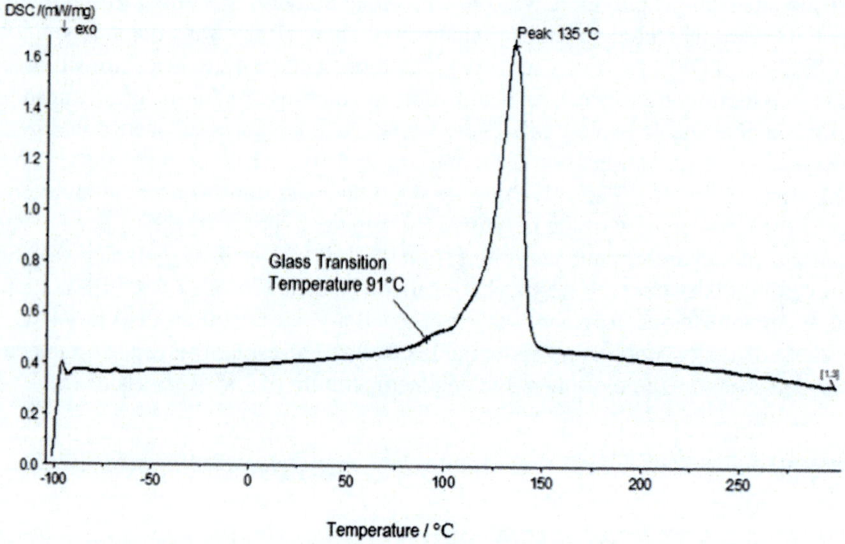

Fig. 2. Differential scanning calorimetry of sample No. S9 [Authors own processing].

If we compare the DSC graph of samples S21 and SP21, we can see that MA significantly affected the material structure. Figure 2 shows that both GPPS and HIPS have almost identical DSC graphs, which are amorphous polystyrene. Therefore, the glass transition temperature is practically the same for the two materials [12, 17]. However, if you look at Fig. 3, you can see that after vitrification [16, 18], the sample graphs with and without additives differ. It can be explained by the presence of MA in the material structure (Fig. 4).

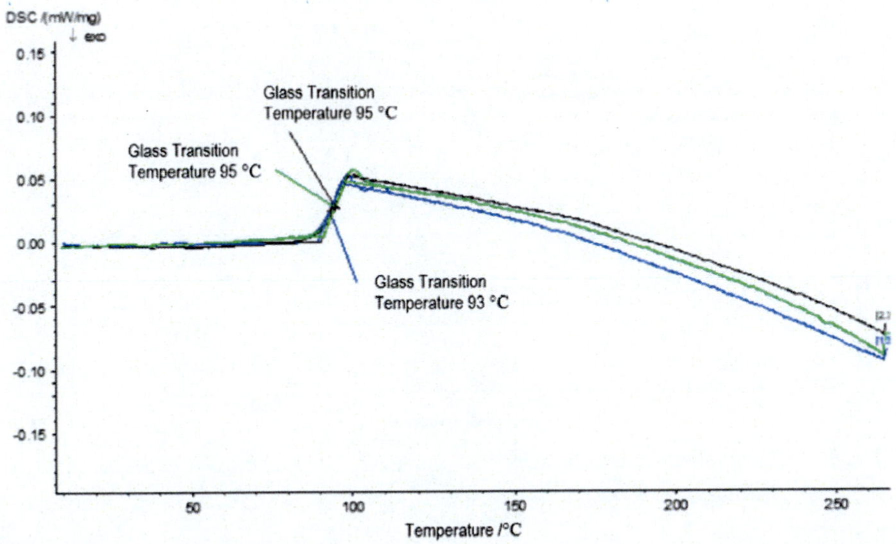

Fig. 3. DSC graph of sample S21: black curve - GPPS / HIPS 50/50; green - HIPS; blue – GPPS [Authors own processing].

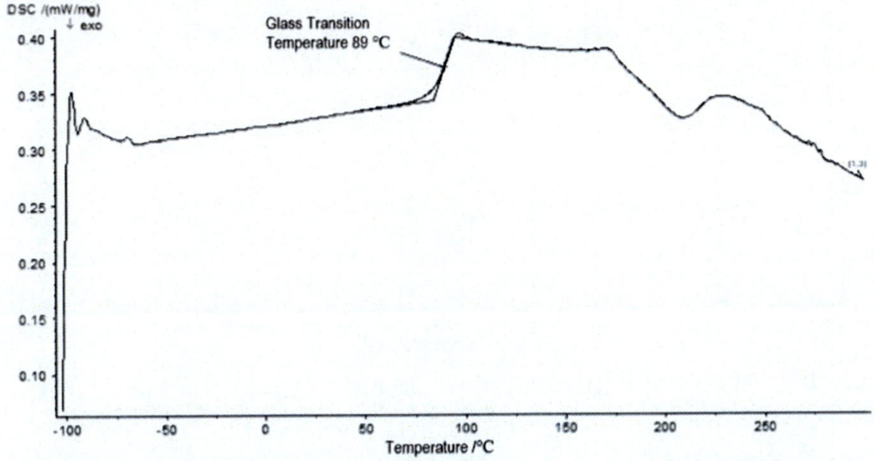

Fig. 4. Differential scanning calorimetry of sample No. S21 [Authors own processing].

DSC descriptions of samples S3 (Fig. 5) and S4 (Fig. 6) are also given. It can be also noticed from these graphs that after the glass transition phase of the material, which is inherent in an amorphous polymer [17, 19], significant fluctuations in the value of heat flow in the material begin, leading to the formation of "peaks" and "depressions".

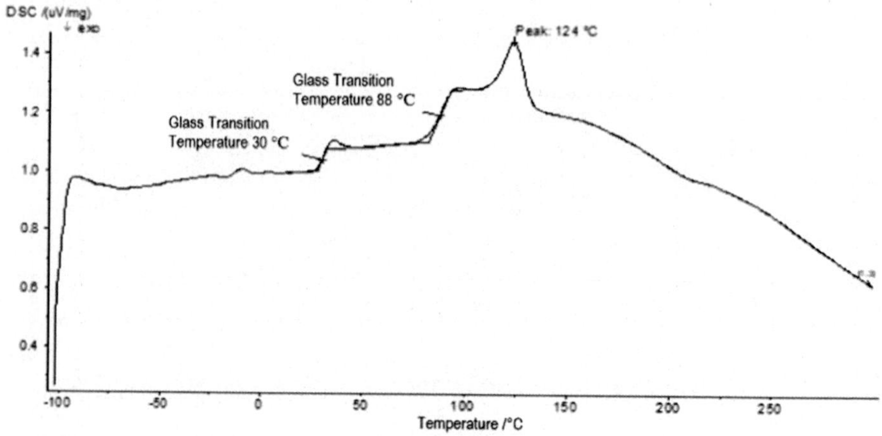

Fig. 5. Differential scanning calorimetry of sample No. S3 [Authors own processing].

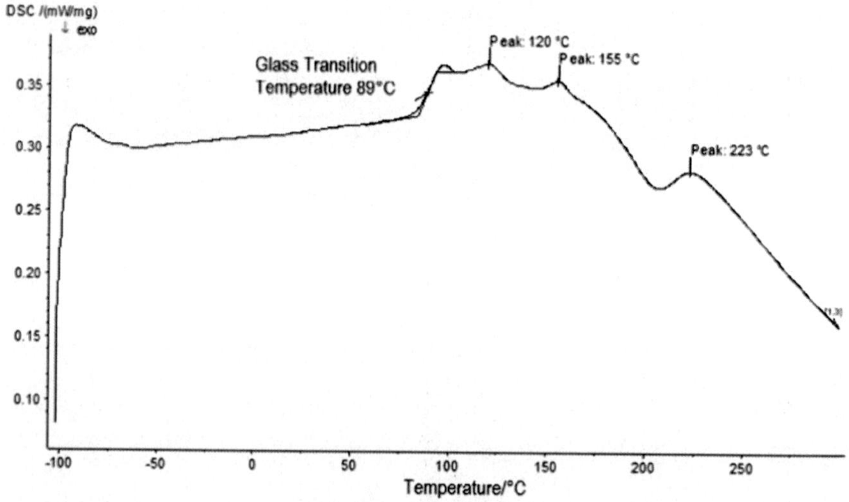

Fig. 6. Differential scanning calorimetry of the sample No. S4 [Authors own processing].

4 Conclusion

The research used selected types of polymers, their composition and properties, and the properties of MA and DTBP fillers. Differential scanning calorimetry was used for analyzing the chemical-physical properties of the materials. The test results of samples of polymers and polymer mixtures without fillers were also compared with samples with filler. DSC analysis in some cases shows that a mixture of two structures in one sample can be obtained by adding MA amorphous and crystalline. Research in the future may be extended where other types of polymers (ABS acrylonitrile-butadiene-styrene, PVC polyvinyl chloride, etc.) or fillers can be used. It is also possible to change the ratio of polymers in the mixture (20/80, 30/70, etc.), to change the percentage of MA filler in the mixture. It is also possible to change the method of sample production (using an injection moulding machine) and the cooling process and to determine the effect of new factors on the final result of the resulting material. The obtained results can be applied in future work by designing components that are not mechanically stressed and where a large change in the structure of the material is not expected (from a physicochemical point of view), like interior parts, covering parts in the automobile.

Acknowledgement. This paper is part of a project that has received funding from the European Union´s Horizon 2020 research and innovation programme under grant agreement No.723274.

LESS THAN WAGON LOAD

http://www.lessthanwagonload.eu

References

1. John, S., Duane, P.: Modern Styrenic Polymers: Polystyrene and Styrenic Copolymers. Wiley, West Succex (2003). ISBN: 978-0-471-49752-3. [2020-7-3]. https://onlinelibrary.wiley.com/doi/book/10.1002/0470867213
2. Pacaiova, H., Sinay, J., Turisova, R., et al.: Measuring the qualitative factors on copper wire surface. Measurement **109**, 359–365 (2017)
3. Maul, J. et al.: Ullmann's Encyclopedia of Industrial Chemistry: Polystyrene and Styrene Copolymers. Wiley-VCH, Weinheim (2007). ISBN: 978-3-527-31602-1. [2020-9-3]. Available on https://onlinelibrary.wiley.com/doi/abs/https://doi.org/10.1002/14356007.a21_615.pub2
4. Poklemba, R. et al.: Design of bed machine for machine tool based on polymer concrete mixtures. TEM J. **9**(1), 25–29 (2020)
5. Hunt, T.P.: Encyclopedia of Separation Science. Polymer additives: supercritical fluid chromatography. ICI Technology Middlesbrough, UK (2003). ISBN: 978-0-12-226770-3. [cit 2020-5-4]. https://doi.org/10.1016/B0-12-226770-2/03351-2
6. Smokov, I.: Examination of selected properties of thermoplastic materials and their application to the automotive industry. Bachelor thesis, FVT TUKE (2020)
7. Lohbeck, K. et al.: Ullmann's Encyclopedia of Industrial Chemistry: Maleic and Fumaric Acids. Wiley-VCH, Weinheim (2007). ISBN: 978-3-527-31602-1. [2020-9-3]. https://onlinelibrary.wiley.com/doi/abs/https://doi.org/10.1002/14356007.a16_053
8. Knapčíková, L. et al.: Experimental study of thermoplastics material reinforced by various types of high-strength fibres. MM Sci. J. **2016**, 1043–1046. ISSN 1803-1269 (2016)
9. Ostroha, R., Yukhymenko, M., Lytvynenko, A., Bocko, J., Pavlenko, I.: Granulation process of the organic suspension: fluidized bed temperature influence on the kinetics of the granule formation. In: Ivanov, V., et al. (eds.) DSMIE 2018. LNME, pp. 463–471. Springer, Cham (2019). https://doi.org/10.1007/978-3-319-93587-4_48

10. Mesaros, P., et al.: The impact of information and communication technology on cost reducing in the execution phase of construction projects. TEM J. **9**(1), 78–87 (2020)
11. Kanáliková, A., Behúnová, A., Mandičák, T., Mesároš, P.: New technologies for sustaining development in infrastructure, logistics and construction industry. Acta Logistica **6**(4), 171–177 (2019)
12. Harper, C.A.: Handbook of Plastics Technologies, pp. 15–35. The McGraw- Hill Companies, United States of America. ISBN 0-07-146068-3 (2006)
13. Hosford, W.F.: Mechanical Behavior of Materials. USA, Cambridge University Press, New York (2010)
14. Vasiliev, V.V., Morozov, E.V.: Mechanics and Analysis of Composite Materials. Elsevier, London (2001)
15. Straka, M., Rosova, A., et al.: Principles of computer simulation designfor the needs of improvement of the raw materials combined transport system. Acta Montanistica Slovaca. **23**, 163–174 (2018)
16. Fiebig, S., Sellschopp, J., Manz, H., Vietor, T., Axmann, K., Schumacher, A.: Future challenges for topology optimization for the usage in automotive lightweight design technologies. In: Proceedings of the 11th world congress on structural and multidisciplinary optimization, 7–12 June 2015, vol. 142, pp. 1–8. Sydney, Australia (2015)
17. Sinay, J., Balazikova, M., Dulebova, M., et al.: Measurement of low-frequency noise during CNC machining and its assessment. Measurement **119**, 190–195 (2018)
18. Straka, M., Khouri, S., et al.: Utilization of computer simulation for waste separation design as a logistics system. Int. J. Simul. Model. **17**(4), 83–596 (2018)
19. Barry, C.M.F., Orroth, S.A.: Processing of thermoplastics, In: Harper, C.A. (ed.): Modern Plastics Handbook. USA, ISBN 0-07-026714-6 (2000)

On the Suitability of Instrumented Orthogonal Cutting Tests Towards Cutting Load Modelling in Turning Operations

T. Silva[1(✉)], V. Sousa[1], F. Silva[1], and A. M. P. Jesus[1,2]

[1] INEGI, University of Porto, R. Dr. Roberto Frias 400, 4200-465 Porto, Portugal
tesilva@inegi.up.pt

[2] Faculty of Engineering, University of Porto, R. Dr. Roberto Frias 400, 4200-465 Porto, Portugal

Abstract. Cutting load monitoring is of utmost importance in the machinability assessment in metal cutting operations. It allows for the analysis of important aspects such as the cutting stability or tool geometry, thus enabling a constant improvement of machining operations. In order to facilitate the benchmark of different cutting scenarios orthogonal cutting is often used, given its simplicity when compared with industrial machining operations of more intricate dynamics. In this study, the specific cutting pressure estimates obtained from orthogonal cutting experiments are compared with the ones obtained from turning operations for 3 selected alloys, in order to evaluate the suitability this parameter towards mechanistic modelling.

Keywords: Metal cutting · Modelling · Specific cutting pressure · Orthogonal cutting · Turning

1 Introduction

Monitoring and prediction of cutting loads that are developed during machining operations is of utmost importance, particularly when optimizing a certain machining/cutting process. These forces provide valuable information regarding the cutting stability of a given process, difficulty in cutting a material, the cutting tools' sustained wear and, consequently, tool-life [1]. Moreover, their prediction constitutes the foundation of more advanced and capable modelling. Still, cutting load monitoring can be quite difficult, especially for industrial machining operations, given their more intricate dynamics, thus hindering the acquisition of accurate and reliable cutting force values and the process optimization. Modelling of turning operations is currently most conducted through numerical simulation (i.e. Finite Element Method), which despite constituting a valuable tool can be a time-consuming method for cutting load estimation [2]. Orthogonal cutting is commonly seen as an experimental setup that helps portraying and

J. Machado et al. (Eds.): icieng 2022, LNME, pp. 339–350, 2023.
https://doi.org/10.1007/978-3-031-09382-1_30

simplifying the complex dynamics of machining operations. It can be employed, for instance, to facilitate the benchmarking of different operative conditions on similar tool-workpiece material pairs, on more intricate machining operations. In sum, orthogonal cutting has been used to characterize the machinability of materials [3], analysis of cutting mechanics and even on the prediction and modelling of cutting forces that are developed during machining of many materials [4,5], using various tool geometries and coatings [6], and for a wide variety of cutting conditions. Moreover, orthogonal cutting has been heavily employed for the prediction of cutting forces generated during machining of carbon fibre reinforced plastics, offering some insight on the generated surface quality and even on the cutting mechanics [7].

In this work, load measurements have been conducted under orthogonal cutting conditions. One of the primary goals is to infer on the machinability evaluation through the estimation and analysis of specific cutting pressure, K_c for distinct operative parameters. It can be defined as the material's resistance to cutting and it can be estimated through Eq. 1, where F_c is the principal cutting force and A_c the uncut chip cross-section. This area can be calculated as the product of chip width (b) and uncut chip thickness (t_0) for the orthogonal cutting and between depth of cut (a_p) and feed rate (f) in turning operations. The specific cutting pressure can be directly related to the specific cutting power, providing valuable and relevant information when studying/optimizing cutting processes (energy minimization), not only for turning or orthogonal cutting, but also to model other cutting processes such as milling [8], focusing for instance on machining parameters optimization [9], tool geometry [10], lubricating conditions [11] and even machine energy consumption [12].

$$K_c = \frac{F_c}{A_c} \tag{1}$$

For the sake of completeness and material diversity, a NiCr tool steel (AISI P20), MgSi aluminium alloy (AW 6082) and a leaded brass alloy, (CW 614N) were selected. As regards the materials' state, the AISI P20 tool steel was used in the hardened and tempered (delivery) condition and the aluminium alloy in the extruded state. Moreover, the same materials were submitted to instrumented longitudinal turning operations, enabling a comparison of the obtained data with the previously estimated machinability indicators. Chip morphology is compared between cutting processes in order to infer on chip formation similarities. In addition, surface roughness is measured in the machined samples (for turning operations) and the results are compared with the theoretical estimation of Ra.

2 Experimental Details

2.1 Workpiece Materials

In order to focus this study on metallic materials with well-known distinct values of mechanical strength, the group of alloys described in Table 1 was selected.

Table 1. Mechanical properties of the considered metallic alloys.

Material	Density [g/cm^3]	Elastic modulus [GPa]	Yield strength [MPa]	Tensile strength [MPa]	Hardness [Brinell]
AISI P20 [13]	7.80	205	900	1020	290–330
AW 6082 [14]	2.70	70	260	320	95
CW 614N [15]	8.40	97	220	430	>90

Despite the existence of some data regarding the specific cutting pressure of material groups (such as carbon steel or structural aluminium alloys), very little data is found when referring to specific alloys. By describing the exact materials, this study contributes to a more detailed modelling of cutting forces for those alloys, providing their specific cutting pressures. For that, samples of the same batch were used to prepare specimens for both orthogonal cutting ($30 \times 24 \times 2$ mm plates) and turning operations (ø25×120 mm rods).

2.2 Orthogonal Cutting

Orthogonal cutting tests were carried out on mechanical shaping machine. Its very high stiffness and robustness are important features that aim at dealing with high cutting loads, whilst enabling similar speed conditions to industrial machining operations, as previously shown in other studies [3]. Machine instrumentation allowed for load measurement in the three relevant directions, which was accomplished using a piezoelectric dynamometer (Kistler 9257B) and corresponding signal amplification (Kistler 5070A). Data conversion, storage and analysis was enabled using Advantech 4711A acquisition system. A scheme of the mechanical shaping machine dynamics is presented in Fig. 1a, as well as an illustration of the instrumentation chain. Cutting speed of the ram can be selected through convenient gear selection. Despite the non-constant speed on each stroke, the material specimens were placed at central stroke position, where speed is most stable. A cutting speed of approximately 35 m/min was selected. Material specimens were prepared into thin-plates of each considered material. Since the thickness of the plate will correspond to the cutting width parameter (used to calculate the specific cutting pressure), it was ground to 2 mm. In addition, a cutting length of 24 mm was defined, which corresponds to the specimens' length.

A clamping vise was built for specimen fixation and tightened to the dynamometer, which in turn was bolted to the shaping machine table, as shown in Fig. 1b. All components were aligned with the ram movement. The shaping machine tool slide allows for the definition of the uncut chip thickness, t_0, through its vertical displacement. Given that its scale resolution (0.1 mm) is not appropriate for a thorough control of the t_0, a Mitutoyo dial indicator with a 0.005 mm resolution was clamped to the tool slide for improved t_0 definition. The dial indicator probe had the vise surface has a reference, which was ground for better accuracy. Despite the possibility of defining approximate nominal values of t_0, its real values were monitored using a verification methodology based on the amount of

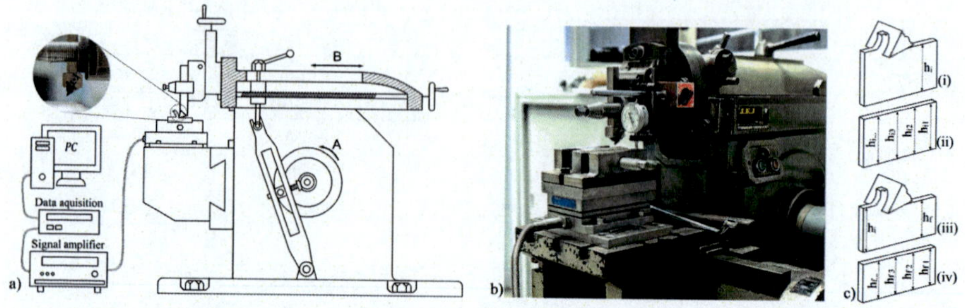

Fig. 1. Scheme of mechanical shaping machine dynamics: the bull gear (A) rotation, powered by an electrical motor, translates into a linear and cyclical movement of the ram (B), back and forward (a); orthogonal cutting experimental setup showing load cell, vise and dial indicator probe on tool slide (b); illustration of t_0 measurement procedure (c).

removed material from the workpiece after each cut. This methodology, illustrated in Fig. 1c, starts with the positioning and levelling of the specimen on the clamping vise, followed by an initial cut to ensure parallelism between the specimen top surface and the ram path. After that, the specimen is unclamped and placed on a profile projector for its height measurement h_i. Successive measurements (in 1 mm intervals) are performed (ii) in order to be able to calculate the effective t_0 along the cutting length. A Mitutoyo toolmaker's microscope with a 2x magnification lens was used for measuring the linear height distances along the specimens' cutting length. Due to being zones of transient cutting, due to chip formation evolution and its cutting end, the specimen extremities were excluded from the measurement stage. Once measured, the specimen is re-clamped on the vise, using a gauge block to guarantee its horizontal alignment and the experimental cutting test takes place with the advance of the cutting tool and the formation of a chip that is collected. Lastly, a final measurement is led for the same successive 1 mm intervals, enabling the calculation of the effective t_0 by subtracting each corresponding initial height to the final height values (h_i-h_f).

With respect to cutting tools, specially-designed carbide inserts (AlTiN coated through physical vapour deposition) with three available cutting edges of 4 mm in length with a rake angle (α_r) of 5°, were used in the orthogonal cutting tests, manufactured by Palbit S.A. The inserts were mounted on a tool holder with a V pocket with their inverse geometry (ensuring rigidity), which in turn was fixed to the shaping machine ram (see Fig. 1a).

2.3 Turning Tests

The turning experimental tests may translate a more accurate measure of a real-life machining operation, thus providing a good measure of the materials' machinability. On the other hand, the controlled conditions and simplified kinematics of orthogonal cutting (relatively to real machining operations) promote a better understanding of the underlying physics of cutting and facilitate the comparison with numerical models. It is, however, necessary to understand how compatible these cutting operations are and if the same machinability tendencies are identified between different cutting operations. For that, the materials machinability was assessed by means of longitudinal turning tests carried out on a universal lathe EFI DU20 with 5.9 kW power in similar conditions to the previously described orthogonal cutting tests (e.g. dry cutting operations, workpiece-tool relative positioning, flat tool rake face).

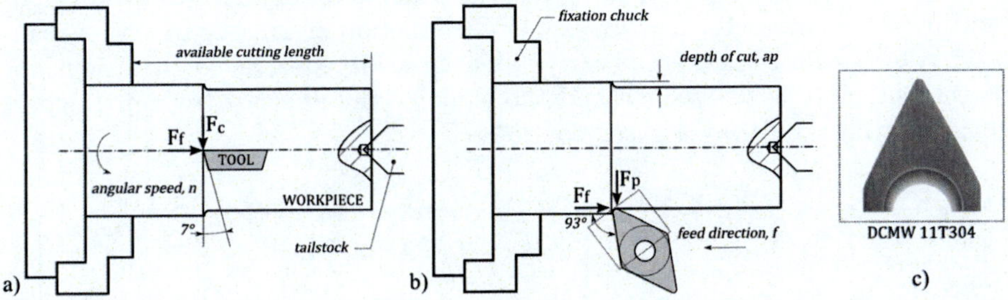

Fig. 2. Scheme of the turning setup: (a) front view and (b) top view of tool-workpiece relative position; rake surface of the used tool (c).

With regards to load measurement and acquisition, and identical setup to the one in orthogonal cutting tests was employed. The load cell was mounted on the lathe's tool carriage and levelled in order to ensure the alignment of the cutting tool with the centre of rotation of the fixation chuck, as illustrated in the scheme of Fig. 2, where it is possible to infer on cutting force (F_c), feed force (F_f) and passive force (F_p) directions in the turning operations. An ISO DCMW 11T304 (Ti(C,N) CVD coated) cutting insert with a flat top, 0.4 mm nose radius was mounted on a tool holder providing a −3° side cutting edge angle, 0° cutting edge inclination and 7° back relief angle. Tail stock was used to minimize chatter due to vibration. The complete combination of two rotational speeds (1800 and 2500 rpm) and three feeds (0.05, 0.1 and 0.2 mm/rev) were explored for the same fixed depth of cut (0.5 mm). Since the tests were performed under constant rotational speeds, the cutting speeds decrease along the successive cutting passes due to the diameter reduction. The 2500 rpm resulted in approximate cutting speeds of 200 m/min; the 1800 rpm resulted in approximate cutting speeds of 100 m/min. Initial and final diameters of the samples were considered for average cutting speed calculation. These cutting speeds are significantly higher than in orthogonal cutting. Still, it will be interesting to understand how representative the cutting forces from the orthogonal cutting can be for turning, under higher cutting speeds. Surface roughness (R_a) was measured using a Mitutoyo SJ-210 stylus contact tester, which allow to infer on the surface integrity of the employed operative conditions.

3 Results and Discussion

3.1 Cutting Load

Orthogonal cutting conditions were successfully employed using the developed experimental setup, for a large range of uncut chip thickness. Both transient beginning of chip formation (for cutting length <5 mm) and steady-state metal cutting conditions can be clearly identified from both cutting force (F_c) and feed force (F_f) evolution, in function of instantaneous t_0, as illustrated in Fig. 3a, where an example of the obtained data from each orthogonal cutting test is presented. Passive force (F_p) is not represented due to being null (as orthogonal conditions were imposed). The average F_c in function of average t_0 is shown in Fig. 3b, where a clear linear trend can be identified. Moreover, a close relationship between the materials' strength and the slope of the linear regression (higher strength for higher slope) is observed.

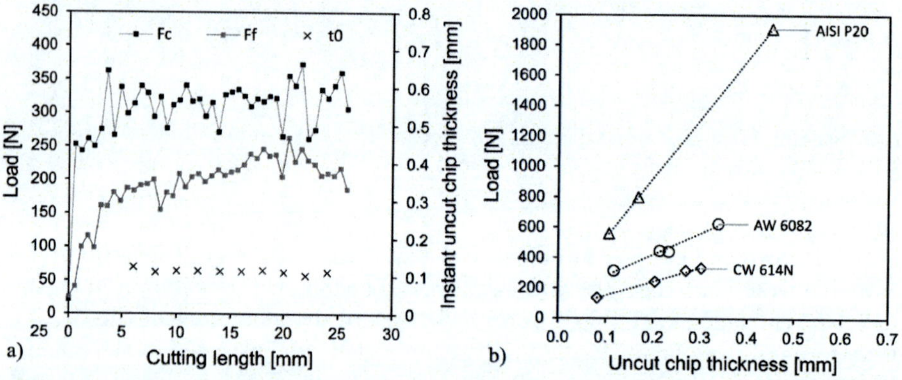

Fig. 3. Experimental cutting loads for orthogonal cutting of AW 6082 showing instant $t0$ in function of cutting length (a); principal cutting force (F_c) value in function of workpiece material and t_0 (b).

Figure 4a shows an example of the cutting load signal registered by the instrumentation system in function of cutting time, for turning operation. Stable (steady-state) cutting conditions were achieved, which allowed for the estimation of average cutting force in function of operational parameters and workpiece material, as shown in Fig. 4b. Standard deviation is quite constant and in the range of 4 to 6 N for all average F_c values. A clear influence of the feed rate on the principal cutting force is discerned (higher load for higher feed rate). On the other hand, doubling cutting speed accounts for less than 10% variation on the principal cutting load.

3.2 Specific Cutting Pressure

Figures 5a and 5b present the specific cutting pressure results obtained from both orthogonal cutting and turning experimental setups, respectively. A correspon-

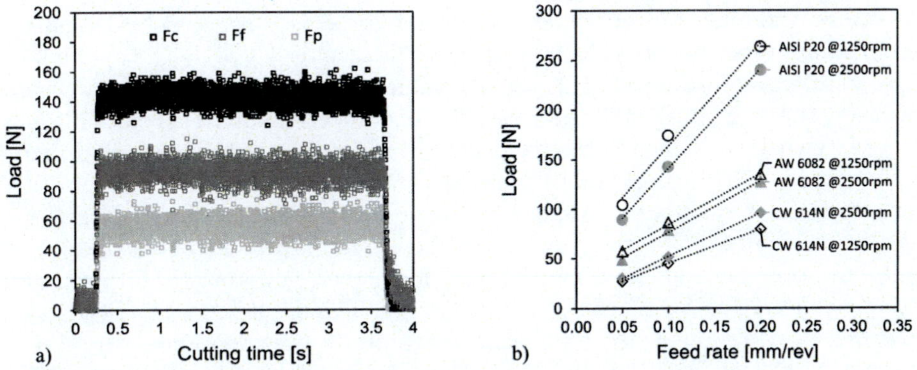

Fig. 4. Cutting load evolution observed in longitudinal turning of AISI P20 steel with $f = 0.10\,\mathrm{mm/rev}$, $a_p = 0.5\,\mathrm{mm}$ and $n = 2500$ rpm (a); principal cutting force (F_c) value in function of workpiece material, rotational speed, n, and feed rate, f (b).

dence of K_c values with the material strength of each alloys can be discerned. A standard deviation in the range of 60 to 250 MPa for all individual average K_c values was calculated. Also interesting to note, is that the influence of cutting speed seems to be more significant in the AISI P20 steel, which may indicate a higher sensitivity to deformation rate strengthening/thermal softening balance. The significantly distinct values of K_c depending on the operation are a consequence of the well-known size effect in metal cutting. This means that for increasing t_0 or f, the specific cutting pressure will be smaller, tending to a plateau. Two main reasons contribute to such effect. The first is related with the existence of material defects, such as grain boundaries and their less likely occurrence for smaller sizes of removed material [16]. The second is related to the intrinsic plastic deformation of the cutting process: for smaller uncut chip thick-

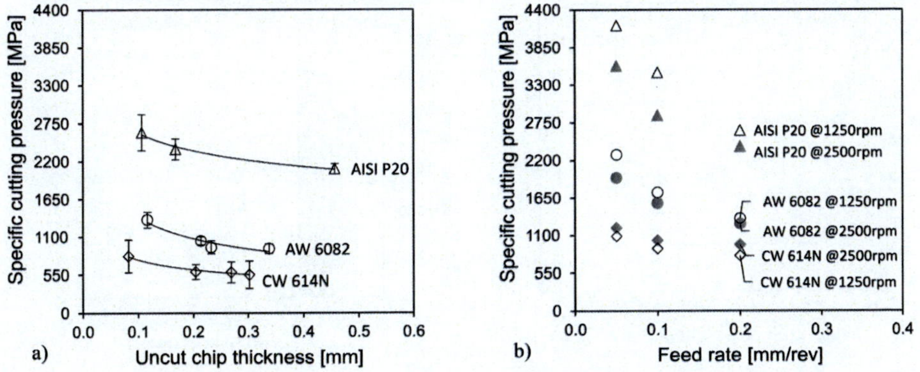

Fig. 5. Specific cutting pressure curves obtained in function of t_0 for orthogonal cutting (a) and in function of f for turning (b) experimental setups.

ness values, the effective rake angle becomes increasingly negative, enhancing the ploughing effect at tool-chip interaction [17].

In addition, according to [18], the K_c value, for increasing uncut material dimension (i.e. t_0, f) should theoretically tend to the material's mechanical strength, typically obtained through tensile testing. In order to be able to compare results from both orthogonal and turning experiments, K_c is plotted in function of uncut chip cross-section area, as shown in Figs. 6a, 6b and 6c. Despite the smaller chip areas in longitudinal turning operations ($<0.1\,\mathrm{mm}^2$) there is a noticeable continuity and coherence of the curves plateau for higher chip areas in orthogonal cutting configuration for all considered materials. Though it is well known that tool geometry can significantly alter the K_c magnitude, the adoption of very simple geometries, such as cutting inserts without rake surface discontinuities (flat rake faces), is commonly employed to obtain reference values of specific cutting pressure. Moreover, in this study, the identical tool-workpiece relative positioning additionally contributes to the compatibility in-between cutting operations, thus minimizing tool geometry effects in the intended comparison of Fig. 6.

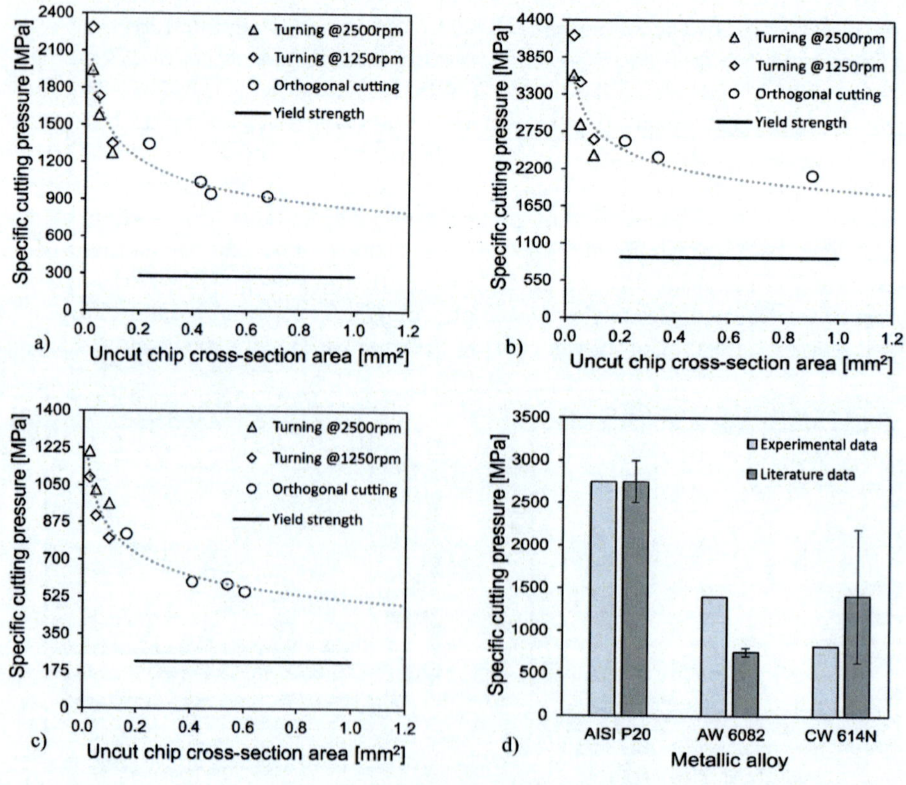

Fig. 6. Comparison of specific cutting pressure results with yield strength of AW 6082 (a), AISI P20 (b) and CW 614N (c); comparison between literature values and measured values of K_c, for uncut chip cross-section area reference of $0.1\,\mathrm{mm}^2$ (d).

Despite its decreasing tendency with uncut chip cross-section area increase, the K_c results in this research work are consistently higher (averagely 30%) than the materials' yield strength. It is generally accepted that the formation of new surfaces (and, thus, the required energy for such phenomenon) as well as friction on the secondary and tertiary shear zones account for the higher values of K_c, when compared to the materials mechanical strength obtained from tensile tests, where both friction and new surfaces formation effects are relatively negligible. Even though the obtained experimental values are very close to literature data [19–21] for the steel, a significant mismatch is noticed for the two other tested materials (refer to Fig. 6d). Such fact highlights the need for a more accurate depiction of K_c (for particular alloys) due to the lack of data accuracy.

The inter-operation compatibility is further enhanced by the similarities in chip formation of each material. Figure 7 shows the chip morphology of each material in both orthogonal (Fig. 7a) and longitudinal turning (Figs. 7b and 7c). In the latter configuration chip representative examples are show for both tested cutting speeds. While for aluminium and steel alloys continuous chip formation is observed (regardless of the cutting operation), segmented/loose chips are obtained for the brass alloy. Moreover, a tendency for less chip curvature can be noticed for increasing chip cross section in both orthogonal cutting and turning operations of aluminium and steel. With regards to cutting speed, a chip morphology transition (from curled to ribbon) is identified for the highest tested chip cross section of the selected aluminium and steel materials.

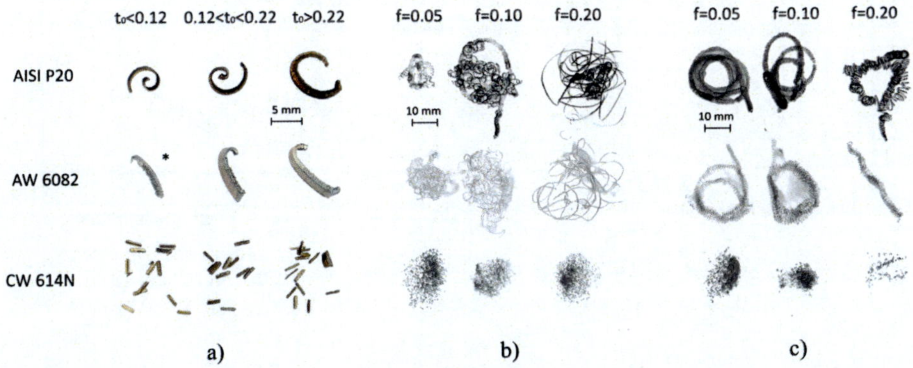

Fig. 7. Chip morphology of each considered material in both orthogonal cutting conditions (a) and longitudinal turning at approximately 200 m/min (b) and 100 m/min (c); *chip is not complete.

3.3 Roughness Measurement

In order to evaluate surface integrity of the machined samples and understand the appropriateness of the selected operative conditions, roughness measurements were performed (along the longitudinal direction of turning samples). Those measurements were also compared with theoretical estimates of Ra and were derived from traditional and widely used equations (refer to Eq. 2), well described in literature, accounting solely for geometric factors, namely tool insert nose radius (r_e) and feed rate (f) [19,22].

$$R_a = \frac{f^2}{32r_e} \tag{2}$$

Roughness values were similar among machined materials and the effect of cutting speed does not seem to have significant influence. It is, though, observed that R_a values for the brass alloy are slightly higher, which may be related with the distinct (segmented) chip formation mechanism, when compared with the continuous chips of the steel and aluminium alloys. As regards the theoretical predictions, and given that size effect phenomena and material performance are not accounted for in the estimates, surface roughness is underestimated for finishing cutting conditions (i.e. 0.1 mm/rev) and overestimated for higher values of the feed per revolution, as illustrated by Fig. 8.

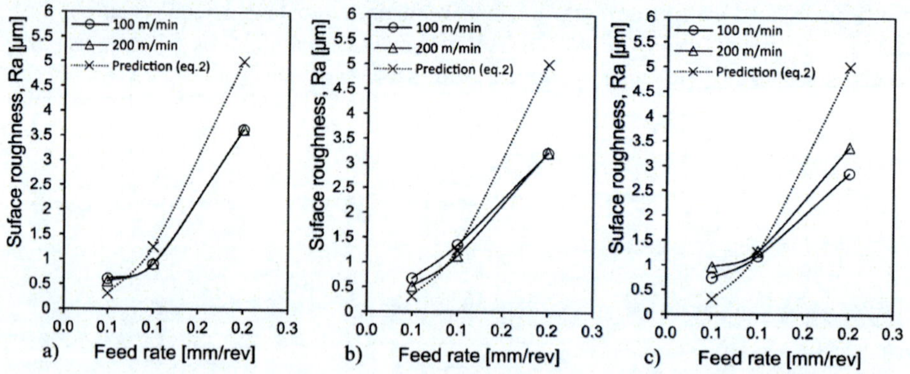

Fig. 8. Comparison between measured R_a surface roughness and theoretical prediction based on tool radius and feed rate, for AISI P20 (a), AW 6082 (b) and CW 614N (c).

4 Conclusions

The developed experimental setup successfully enabled orthogonal cutting conditions for a wide range of uncut chip thickness. The following conclusions can be drawn:

– Continuity among the estimated specific cutting pressure by means of orthogonal cutting and turning operations was observed, which allows for the validation of a straightforward experimental setup towards cutting analysis, promoting optimization procedures.

– Despite its higher influence on the tested steel, cutting speed does not present significant expression on the calculated specific cutting pressure;
– Specific cutting pressure shows consistently higher values (of approximately 30%) to the material's mechanical strength obtained from tensile testing;
– Despite its simplistic (solely geometric) prediction, the estimated surface roughness values (Ra) is only consistent with the actual measure surface roughness values for intermediate feed rates, given the unaccounted size effect of metal cutting (under and overestimated for lower and higher feed levels).

Acknowledgements. The MAMTool (PTDC/EME-EME/31895/2017) project is acknowledged, funded by Programa Operacional Competitividade e Internacionalização, and Programa Operacional Regional de Lisboa funded by FEDER and National Funds (FCT).

References

1. Sousa, V.F.C., et al.: Cutting forces assessment in CNC machining processes: a critical review. Sensors **20**(16), 4536 (2020)
2. Tzotzis, A., et al.: 3D-FEM approach of AISI-52100 hard turning: modelling of cutting forces and cutting condition optimization. Machines **10**(2), 74 (2022)
3. Silva, T.E.F., Rosa, P.A.R., Reis, A.R., de Jesus, A.M.P.: Machinability of the 18Ni300 additively manufactured maraging steel based on orthogonal cutting tests. In: Machado, J., Soares, F., Trojanowska, J., Ottaviano, E. (eds.) icieng 2021. LNME, pp. 1–13. Springer, Cham (2022). https://doi.org/10.1007/978-3-030-79165-0_1
4. Boujelbene, M.: Investigation and modeling of the tangential cutting force of the Titanium alloy Ti-6Al-4V in the orthogonal turning process. Procedia Manuf. **20**, 571–577 (2018)
5. Sela, A., et al.: A mechanistic model to predict cutting force on orthogonal machining of Aluminum 7475–T7351 considering the edge radius. Procedia CIRP **82**, 32–36 (2019)
6. Chen, Y., Wang, J., An, Q.: Mechanisms and predictive force models for machining with rake face textured cutting tools under orthogonal cutting conditions. Int. J. Mech. Sci. **195**, 106246 (2021)
7. An, Q., et al.: Study on the cutting mechanics characteristics of high-strength UD-CFRP laminates based on orthogonal cutting method. Compos. Struct. **131**, 374–383 (2015)
8. Wang, L., et al.: Modeling and analysis of specific cutting energy of whirling milling process based on cutting parameters. Procedia CIRP **80**, 56–61 (2019)
9. Wang, B., et al.: Proper selection of cutting parameters and cutting tool angle to lower the specific cutting energy during high speed machining of 7050–T7451 aluminum alloy. J. Clean. Prod. **129**, 292–304 (2016)
10. Boyea, J., et al.: Studies of end mill flute geometry enhancements on specific cutting energy and chip formation. Procedia CIRP **101**, 286–289 (2021)
11. Ali, M.A.M., Azmi, A.I., Khalil, A.N.M.: Specific cutting energy of Inconel 718 under dry, chilled-air and minimal quantity nanolubricants. Procedia CIRP **77**, 429–432 (2018)
12. Parida, A.K., Maity, K.: Study of machinability in heat-assisted machining of nickel-base alloy. Measurement **170**, 108682 (2021)

13. Uddeholm. Uddeholm Impax Supreme, 7 Set (2011). https://www.ramada.pt/ficheiros_upload/pb_impax_supreme_english160604080325.pdf
14. PolyLanema. Liga de Aluminio AW 6082, 2 Fev (2022). https://www.polylanema.pt/pt/aluminios-tecnicos/placas/aw-6082/
15. Ramada. Latao T - Ligas Nao Ferrosas Para Moldes e Mecanica Geral, 11 Dez (2017). https://www.ramada.pt/pt/produtos/acos/ligas-nao-ferrosas-para-moldes-e-meca-a2nica-geral/lata-83o-t.html
16. Shaw, M.C.: The size effect in metal cutting. Sadhana **28**(5), 875–896 (2003)
17. Arsecularatne, J.A.: On tool-chip interface stress distributions, ploughing force and size effect in machining. Int. J. Mach. Tools Manuf. **37**(7), 885–899 (1997)
18. Taniguchi, N.: The art of nanotechnology [Lecture] (1993)
19. Groover, M.P.: Groover's principles of modern manufacturing: materials, processes, and systems. Wiley Global Education (2016)
20. Mitsubishi Materials Corporation. Formulae for cutting power. http://www.mitsubishicarbide.net/contents/mhg/enuk/html/product/technical_information/information/formula4.html
21. Sumitool. Technical guiding reference: The basis of turning, turning edition. https://www.sumitool.com/en/downloads/cutting-tools/general-catalog/assets/pdf/GC_N2_en.pdf
22. Toenshoff, H.K., Denkena, B.: Basics of cutting and abrasive processes (2013)

Author Index

A
Agrawal, Rajeev, 116
Alionte, Andreea Dana, 240
Antosz, Katarzyna, 68
Apostolescu, Tudor Catalin, 91
Azevedo, José, 80

B
Balashov, Artem, 279
Barbosa, José, 317
Barbosa, Marta L. S., 58
Bartolomeu, Tiago, 80
Bastos, João A., 161
Benchea, Marcelin, 47
Brito, Marlene F., 161
Bujoreanu, Carmen, 47
Burduk, Anna, 279

C
Canacheu, Cristian, 240
Cartal, Laurentiu Adrian, 91
Carvalho, Joana, 136
Carvalho, Vítor, 228
Ciobanu, Alexandru Costin, 240
Coelho, Carla, 317
Cordeiro, Rafael, 1
Costa, Ana Francisca, 317
Costa, Nelson, 216
Costa, Rúben D. F. S., 58
Cunha, Natacha, 317

D
Dawod, Nazem, 91
de Sousa, Filipe Soares, 293
Dias, Gil, 228
Dontu, Grigore Octavian, 91
Drumond, Nuno, 317

F
Falkenhain, Paul, 1
Faria, Ricardo, 103
Ferreira, Ana C., 103, 266
Ferreira, Fernanda A., 250
Ferreira, José Soeiro, 293
Ferreira, Luís P., 58
Ferrera, Conrado, 209
Forte, Luís D., 103

G
Gapinski, B., 173
Gomes, Beatriz Especial, 250
Grămescu, Bogdan, 91
Guerreiro, Rita, 161

H
Horky, Roman, 200
Hren, Iryna, 200
Husár, Jozef, 329

I
Ianuş, Gelu, 47

J. Machado et al. (Eds.): icieng 2022, LNME, pp. 351–353, 2023.
https://doi.org/10.1007/978-3-031-09382-1

J
Jain, Jinesh Kumar, 116
Jasiulewicz-Kaczmarek, Małgorzata, 68
Jaskevič, Martin, 259
Jesus, A. M. P., 339

K
Kharlamov, Yuriy, 305
Knapčíková, Lucia, 329
Krol, Oleg, 305
Kumar, Manoj, 116
Kusmierczak, Sylvia, 200

L
Leão, Celina P., 266
Lima, Maria Margarida, 293
Lima, Rui, 209
Lima, Rui M., 317

M
Machado, José, 24
Margolis, Iara, 216
Marinho, Jorge, 317
Martiček, Matúš, 329
Martins, Tiago, 228
Mathia, T. G., 173
Meira, Carla, 317
Meißner, Sven, 1
Mendes, Ana Catarina, 317
Mendes, Joaquim, 36
Miranda, Rui, 317
Moraru, Edgar, 91
Murzac, Ion, 240
Musiał, Kamil, 279

N
Naprstková, Nataša, 200
Nițu, Nicolae Adrian, 47
Novotný, Jan, 259
Nunes, Ana Catarina, 293

O
Oliveira, Cristina, 293
Oliveira, Marisa, 250
Öztürk, Elif Göksu, 293

P
Pasternak, Viktoriya, 24
Pata, Arminda, 13
Pereira, Armanda, 173, 228
Pereira, Isac, 317
Pereira, Maria Teresa, 250
Pinal, Diego, 216
Pinto, Ana, 317
Pinto, Bebiana, 58
Pinto, Paulo, 187
Porkuian, Olga, 305
Providência, Bernardo, 216
Puga, Hélder, 209

Q
Quinteiro, Ana Rita, 317

R
Ribeiro, João, 209
Rizescu, Ciprian Ion, 148
Rizescu, Dana, 148
Rocha, Pedro Filipe, 293
Rodrigues, Ana Maria, 293
Romanchenko, Oleksiy, 305
Rysińska-Wojtasik, Dominika, 279

S
Sá, José C., 58
Santos, André S., 161
Scholz, Sebastian, 1
Seabra, Eurico, 36
Silva, Agostinho, 13
Silva, Diogo, 187
Silva, F., 339
Silva, Francisco J. G., 58
Silva, João, 187
Silva, Patrícia, 36
Silva, Rute, 216
Silva, T., 339
Singhal, Tejendra Singh, 116
Soares, Filomena, 228
Soares, Margarida, 136
Sokolov, Volodymyr, 305
Sousa, Bruno, 161
Sousa, V., 339
Souza, Andrews, 209
Souza, Maria Sabrina, 209
Svirzhevskyi, Kostiantyn, 24
Swojak, N., 173
Szelewski, M., 173

T
Teixeira, José, 80, 187
Teixeira, Senhorinha, 80, 187
Trojanowska, Justyna, 329

U
Ungureanu, Liviu Marian, 240

V
Văleanu, Dana Mirela, 240
Valério, Nádia, 136
Varela, Leonilde R., 161
Vilarinho, Maria Cândida, 136
Vysloužil, Tomáš, 259

W
Wieczorowski, M., 173

Z
Zabolotnyi, Oleg, 24
Zadorozhnikova, Irina, 24